BIOACTIVE COMPOUNDS FROM PLANT ORIGIN

Extraction, Applications, and Potential Health Benefits

Innovations in Plant Science for Better Health: From Soil to Fork

BIOACTIVE COMPOUNDS FROM PLANT ORIGIN

Extraction, Applications, and Potential Health Benefits

Edited by

Hafiz Ansar Rasul Suleria, PhD

Colin Barrow, PhD

AAP | APPLE ACADEMIC PRESS

Apple Academic Press Inc.
3333 Mistwell Crescent
Oakville, ON L6L 0A2
Canada USA

Apple Academic Press Inc.
1265 Goldenrod Circle NE
Palm Bay, Florida 32905
USA

Exclusive worldwide distribution by CRC Press, a member of Taylor & Francis Group

International Standard Book Number-13: 978-1-77188-786-1 (Hardcover)
International Standard Book Number-13: 978-0-42902-928-8 (eBook)

Library and Archives Canada Cataloguing in Publication

Title: Bioactive compounds from plant origin : extraction, applications, and potential health benefits / edited by Hafiz Ansar Rasul Suleria, PhD, Colin Barrow, PhD.

Names: Suleria, Hafiz, editor. | Barrow, Colin J., editor.

Series: Innovations in plant science for better health.

Description: Series statement: Innovations in plant science for better health | Includes bibliographical references and index.

Identifiers: Canadiana (print) 20190130784 | Canadiana (ebook) 20190130830 | ISBN 9781771887861 (hardcover) | ISBN 9780429029288 (ebook)

Subjects: LCSH: Plant bioactive compounds—Separation. | LCSH: Phytochemicals. | LCSH: Medicinal plants

Classification: LCC QK898.B54 B56 2019 | DDC 572/.2—dc23

CIP data on file with US Library of Congress

Apple Academic Press also publishes its books in a variety of electronic formats. Some content that appears in print may not be available in electronic format. For information about Apple Academic Press products, visit our website at **www.appleacademicpress.com** and the CRC Press website at **www.crcpress.com**

INNOVATIONS IN PLANT SCIENCE FOR BETTER HEALTH: FROM SOIL TO FORK BOOK SERIES

Series Editor-in-Chief:

Dr. Hafiz Ansar Rasul Suleria
Honorary Fellow at the Diamantina Institute,
Faculty of Medicine, The University of Queensland (UQ), Australia
email: hafiz.suleria@uqconnect.edu.au

The objective of this new book series is to offer academia, engineers, technologists, and users from different disciplines information to gain knowledge on the breadth and depth of this multifaceted field. The volumes will explore the fields of phytochemistry, along with its potential and extraction techniques. The volumes will discuss the therapeutic perspectives of biochemical compounds in plants and animal and marine sources in an interdisciplinary manner because the field requires knowledge of many areas, including agricultural, food, and chemical engineering; manufacturing technology along with applications from diverse fields like chemistry; herbal drug technology; microbiology; animal husbandry; and food science; etc. There is an urgent need to explore and investigate the innovations, current shortcomings, and future challenges in this growing area of research.

We welcome chapters on the following specialty areas (but not limited to):

- Food and function
- Nutritional composition of different foods materials
- Functional and nutraceutical perspectives of foods
- Extraction of bioactive molecules
- Phytomedicinal properties of different plants
- Traditional plants used as functional foods
- Phytopharmacology of plants used in metabolic disorders

- Importance of spices and medicinal and functional foods
- Natural products chemistry
- Food processing waste/byproducts: management and utilization
- Herbals as potential bioavailability enhancer and herbal cosmetics
- Phytopharmaceuticals for the delivery of bioactives
- Alternative and complementary medicines
- Ethnopharmacology and ethnomedicine
- Marine phytochemistry
- Marine microbial chemistry
- Other related includes nuts, seed spices, wild flora, etc.

About the Book Series Editor-in-Chief

Dr. Hafiz Suleria is an eminent young researcher in the field of food science and nutrition. Currently, he is an Honorary Fellow at the Diamantina Institute, Faculty of Medicine, The University of Queensland (UQ), Australia. Before joining the UQ, he worked as a lecturer in the Department of Food Sciences, Government College University Faisalabad, Pakistan. He also worked as a Research Associate in a PAK-US Joint Project funded by the Higher Education Commission, Pakistan, and Department of State, USA, with the collaboration of the University of Massachusetts, USA, and the National Institute of Food Science and Technology, University of Agriculture, Faisalabad, Pakistan.

Dr. Suleria's major research focus is on food science and nutrition, particularly in screening of bioactive molecules from different plant, marine, and animal sources, using various cutting-edge techniques, such as isolation, purification, and characterization. He also did research work on functional foods, nutraceuticals, and alternative medicine. He has published more than 60 peer-reviewed scientific papers in different reputed/impacted journals. He is also in collaboration with more than five universities where he is working as a co-supervisor/special member for PhD and postgraduate students and also involved in joint publications, projects, and grants.

Expertise: Food Science and Nutrition, Functional Foods, Bioactive Molecule Research, Marine Bioactives, Nutraceuticals, Medicinal Plants, Natural Product Chemistry, and Drug Discovery

OTHER BOOKS ON PLANT SCIENCE FOR BETTER HEALTH BY APPLE ACADEMIC PRESS, INC

Bioactive Compounds of Medicinal Plants: Properties and Potential for Human Health
Editors: Megh R. Goyal, PhD and Ademola O. Ayeleso, PhD

Plant- and Marine-Based Phytochemicals for Human Health: Attributes, Potential, and Use
Editors: Megh R. Goyal, PhD and Durgesh Nandini Chauhan, MPharm

Human Health Benefits of Plant Bioactive Compounds: Potentials and Prospects
Editors: Megh R. Goyal, PhD and Hafiz Ansar Rasul Suleria, PhD

Human Health Benefits of Plant Bioactive Compounds
Editors: Megh R. Goyal, PhD and Hafiz Ansar Rasul Suleria, PhD

Plant Secondary Metabolites for Human Health: Extraction of Bioactive Compounds
Editors: Megh R. Goyal, PhD, P. P. Joy, PhD, and Hafiz Ansar Rasul Suleria, PhD

Bioactive Compounds from Plant Origin: Extraction, Applications, and Potential Health Benefits
Editors: Hafiz Ansar Rasul Suleria, PhD and Colin Barrow, PhD

Phytochemicals from Medicinal Plants: Scope, Applications, and Potential Health Claims
Editors: Hafiz Ansar Rasul Suleria, PhD, Megh R. Goyal, PhD, and Masood Sadiq Butt, PhD

The Therapeutic Properties of Medicinal Plants: Health-Rejuvenating Bioactive Compounds of Native Flora
Editors: Megh R. Goyal, PhD, PE, Hafiz Ansar Rasul Suleria, PhD, Ademola Olabode Ayeleso, PhD, T. Jesse Joel, PhD, and Sujogya Kumar Panda, PhD

The Role of Phytoconstitutents in Health Care: Biocompounds in Medicinal Plants
Editors: Megh R. Goyal, PhD, Hafiz Ansar Rasul Suleria, PhD, and Ramasamy Harikrishnan, PhD

Assessment of Medicinal Plants for Human Health: Phytochemistry, Disease Management, and Novel Applications
Editors: Megh R. Goyal, PhD and Durgesh Nandini Chauhan, MPharm

CONTENTS

ABOUT THE EDITOR

Hafiz Ansar Rasul Suleria, PhD, is the Alfred Deakin Research Fellow at Deakin University, Geelong, Victoria, Australia. Recently, he has completed a postdoctoral fellowship at the Department of Food, Nutrition, Dietetic and Health at Kansas State University, Manhattan, Kansas, USA. He is an Honorary Fellow in the Diamantina Institute, Faculty of Medicine, The University of Queensland, Australia. Dr. Suleria has been awarded an International Postgraduate Research Scholarship (IPRS) and an Australian Postgraduate Award (APA) for his PhD research at the University of Queensland (UQ) School of Medicine, the Translational Research Institute (TRI), in collaboration with the Commonwealth and Scientific and Industrial Research Organization (CSIRO, Australia).

Before joining the UQ, he worked as a Lecturer in the Department of Food Sciences at the Government College University, Faisalabad, Pakistan. He also worked as a Research Associate in the PAK-US Joint Project funded by the Higher Education Commission, Pakistan, and the Department of State, USA, with the collaboration of the University of Massachusetts, USA, and the National Institute of Food Science and Technology, University of Agriculture, Faisalabad, Pakistan.

His major research focus is on food nutrition, particularly in the screening of bioactive molecules—isolation, purification, and characterization using various cutting-edge techniques from different plants, marine, and animal sources, *in vitro*, *in vivo* bioactivities, cell culture and animal modeling. He has worked on functional foods and nutraceuticals, food and function, and alternative medicine.

Dr. Suleria has published more than 60 peer-reviewed scientific papers in different reputed/impact journals. He is also in collaboration with more than ten universities where he is working as a co-supervisor/special member for PhD and postgraduate students and also involved in joint publications, projects, and grants.

Readers may contact him at: hafiz.suleria@uqconnect.edu.au

ABOUT CO-EDITOR

Colin Barrow, PhD, is the Alfred Deakin Professor and Chair of Biotechnology at Deakin University, Geelong, Victoria, Australia, where he is Director of the Centre for Chemistry and Biotechnology. He has previously published two books, including one entitled *Marine Nutraceuticals and Functional Foods*. Dr. Barrow has an H-index of 43, with more than 300 peer-reviewed publications and over 8000 total citations.

Dr. Barrow has held senior roles in both industry and academia. Prior to joining Deakin University, he spent eight years leading the research and development at Ocean Nutrition Canada (ONC) as Executive Vice President of Research and Development. ONC was the world leader in research and commercialization of large scale omega-3 processing and functional food ingredient development and the largest marine biotechnology company in Canada.

He was previously on the Expert Advisory Committee for the Canadian Natural Health Product Directorate (NHPD), is a founding member of the International Society for Nutraceuticals and Functional Foods (ISNFF), and was a member of Board of Directors for the International Society for the Study of Fatty Acids and Lipids (ISSFAL). Professor Barrow was awarded the Nova Scotia Biotechnology and Life Sciences Industry Association (BIONOVA) award for Research Excellence in 2007. He was appointed Guest Professor at Yunnan Minzu University (2015 to 2018), at Qingdao University (2014 onwards), and at the Oil Crops Research Institute Chinese Academy of Agricultural Sciences (2017 to 2022).

His research focuses on food biotechnology and the application of nanomaterials and enzymes for industrial purposes, including structure-function of lipases and graphene for biological applications, as well as a broad spectrum of natural products chemistry, biological chemistry, and food chemistry. Professor Barrow has a PhD in chemistry from the University of Canterbury in New Zealand and an MBA from Penn State University in the USA.

Readers may contact him at: colin.barrow@deakin.edu.au

CONTRIBUTORS

Huma Bader Ul Ain, PhD
Research Associate, Institute of Home and Food Sciences, Government College University, Faisalabad, Pakistan. E-mail: humahums@yahoo.com

Kashif Akram, PhD
Assistant Professor, Department of Food Sciences, Cholistan University of Veterinary and Animal Sciences Bahawalpur, Pakistan. E-mail: kashifakramft@gmail.com

Muhammad Sajid Arshad, PhD
Assistant Professor, Institute of Home & Food Sciences, Government College University, Faisalabad, Pakistan. E-mail: sajid_ft@yahoo.com

Muhammad Umair Arshad, PhD
Director, Institute of Home and Food Sciences, Government College University, Faisalabad, Pakistan. E-mail: umairfood1@gmail.com

Kanza Aziz Awan, PhD
National Institute of Food Science and Technology, Faculty of Food, Nutrition & Home Sciences, University of Agriculture, Faisalabad, Pakistan. E-mail: kanza_awan@live.com

Colin Barrow, PhD
Alfred Deakin Professor, Faculty of Science, Engineering and Built Environment, Centre for Chemistry and Biotechnology, School of Life and Environmental Sciences, Deakin University, 75 Pigdons Road, Geelong, VIC 3216, Australia. E-mail: colin.barrow@deakin.edu.au

Masood Sadiq Butt, PhD
Professor, National Institute of Food Science and Technology, Faculty of Food, Nutrition & Home Sciences, University of Agriculture, Faisalabad, Pakistan. E-mail: drmsbutt@yahoo.com

Farid Chemat, PhD
University of Avignon, UMR 408 – INRA, GREEN Extraction Team, F-84000 Avignon, France. E-mail: farid.chemat@univ-avignon.fr

Afka Deen, BSc
Division of Nutritional Biochemistry, National Institute of Fundamental Studies, Hantana Road, Kandy 20000, Sri Lanka. E-mail: fathimaafka@yahoo.com

Ahmad Din, PhD
Ayub Agricultural Research Institute, Faisalabad, Pakistan. E-mail: ahmaddinft@gmail.com

Muhammad Umar Farid
Lecturer, Department of Animal Sciences, CVAS-Jhang campus, University of Veterinary & Animal Sciences, Lahore, Pakistan. E-mail: umar.farid@uvas.edu.pk

Umar Farooq, PhD
Associate Professor, Department of Food Science & Technology,
MNS-University of Agriculture Multan, Pakistan. E-mail: umar.farooq@mnsuam.edu.pk

Iahtisham-Ul-Haq, PhD
Head of Department/Assistant Professor, Department of Diet and Nutritional Sciences,
Faculty of Health and Allied Sciences, Imperial College of Business Studies, Lahore, Pakistan.
E-mail: iahtisham@hotmail.com

Zafar Hayat, PhD
Professor, Department of Animal Sciences, CVAS-Jhang campus,
University of Veterinary & Animal Sciences, Lahore, Pakistan.
E-mail: zafar.hayat@uvas.edu.pk

Ali Imran, PhD
Assistant Professor, Institute of Home & Food Sciences, Government College University,
Faisalabad, Pakistan. E-mail: dr.aliimran@gcuf.edu.pk

Muhammad Imran, PhD
Associate Professor, Faculty of Allied Health Sciences,
University Institute of Diet & Nutritional Sciences, The University of Lahore, Pakistan.
E-mail: mic_1661@yahoo.com

Muhammad Imran, PhD
Institute of Home and Food Sciences, Faculty of Life Sciences, Government College University,
Faisalabad 38000, Pakistan. E-mail: imran@gcuf.edu.pk

Nida Kanwal, PhD
Faculty of Food, Nutrition and Home Sciences, National Institute of Food Science and Technology,
University of Agriculture, Faisalabad 38040, Pakistan. E-mail: nidakanwal26@gmail.com

Anees Ahmed Khalil, PhD
Assistant Professor, Faculty of Allied Health Sciences,
University Institute of Diet & Nutritional Sciences, The University of Lahore,
Pakistan. E-mail: aneesahmedkhalil@gmail.com

Muhammad Kamran Khan, PhD
Faculty of Life Sciences, Institute of Home and Food Sciences, Government College University,
Faisalabad 38000, Pakistan. E-mail: mk.khan@gcuf.edu.pk

Anam Latif, PhD
Faculty of Food, Nutrition and Home Sciences, National Institute of Food Science and Technology,
University of Agriculture, Faisalabad 38040, Pakistan. E-mail: anamlatif101@gmail.com

Ruvini Liyanage, PhD
Division of Nutritional Biochemistry, National Institute of Fundamental Studies,
Hantana Road, Kandy 20000, Sri Lanka. E-mail: ruvini@ifs.ac.lk

Athar Mahmud, PhD
Professor, Department of Poultry Production, University of Veterinary & Animal Sciences,
Lahore, Pakistan. E-mail: atharmahmud@uvas.edu.pk

Hussan Munir, MSc (Hons.)
Al-Shaheer Foods, Karachi, Pakistan. E-mail: hussan.munir@ascfoods.com

Farhan Saeed, PhD
Assistant Professor, Institute of Home and Food Sciences, Government College University, Faisalabad, Pakistan. E-mail: f.saeed@gcuf.edu.pk

Saira Sattar, MS
Faculty of Life Sciences, Institute of Home and Food Sciences, Government College University, Faisalabad 38000, Pakistan. E-mail: zarlish1627@gmail.com

Afshan Shafi
Lecturer, Department of Food Science & Technology, MNS-University of Agriculture Multan, Pakistan. E-mail: afshaeman@gmail.com

Aamir Shehzad
Assistant Professor, Faculty of Food, Nutrition and Home Sciences,
National Institute of Food Science and Technology, University of Agriculture,
Faisalabad 38040, Pakistan. E-mail: draamir@uaf.edu.pk

Hafiz Ansar Rasul Suleria, PhD
Alfred Deakin Postdoctoral Research Fellow, School of Life and Environmental Sciences,
Faculty of Science, Engineering and Built Environment, Centre for Chemistry and Biotechnology,
Deakin University, 75 Pigdons Road, Geelong, VIC 3216, Australia.
E-mail: hafiz.suleria@uqconnect.edu.au

Tabussam Tufail, PhD
Visiting Lecturer, Research Associate, Institute of Home and Food Sciences,
Government College University, Faisalabad, Pakistan.
E-mail: tabussamtufail@gcuf.edu.pk

Rizliya Visvanathan, MPhil
Division of Nutritional Biochemistry, National Institute of Fundamental Studies,
Hantana Road, Kandy 20000, Sri Lanka. E-mail: frizlya@gmail.com

Asna Zahid, PhD
Human Nutrition and Dietetics, Faculty of Food, Nutrition and Home Sciences,
National Institute of Food Science and Technology, University of Agriculture,
Faisalabad 38040, Pakistan. E-mail: asna.zahid52@gmail.com

Aurang Zeb, PhD
Professor, Faculty of Allied Health Sciences, University Institute of Diet & Nutritional Sciences,
The University of Lahore, Lahore, Pakistan. E-mail: drzebkhattak@gmail.com

ABBREVIATIONS

3,4-DHS	3,4-dihydroxy-trans-stilbene
4,40-DHS	4-hydroxy-trans-stilbene
ABA	abscisic acid
ABTS	2,2-azinobis-3-ethylbinzothiazoline-6-sulfonic acid
ACE	angiotensin converting enzyme
ACF	aberrant crypt foci
AChE	acetylcholinesterase
ACR	acrylamide
ADP	adenosine diphosphosphate
AGE	aged garlic extract
AGPs	antibiotic growth promoters
AIDS	acquired immune deficiency syndrome
ALA	alpha lipoic acid
ALT	alanine aminotransferase
AMK	AMP-activated protein kinase
AMP	adenosine monophosphate
AMR	antimicrobial resistance
APC	allophycocyanin
AST	aspartate aminotransferase
ATP	adenosine triphosphate
B-PE	B-phycoerythrin
BCBD	β-carotene bleaching test
BDNF	brain-derived neurotrophic factor
BHA	butylated hydroxyl anisole
BHT	butylated hydroxyl toluene
BOD	biodegradable
cAMP	cyclic adenosine monophosphate
CAT	catalase
CCl4	carbon tetrachloride
CD4	cluster of differentiation-4
CHD	coronary heart diseases
CI	confidence interval
CNS	central nervous system

COX-2	cyclooxygenase-2
CP	cyclophosphamide
CSE	conventional solvent extraction
CVD	cardiovascular disease
DADS	diallyl disulfide
DAS	diallyl sulfide
DATS	diallyl trisulfide
DEAE	diethylaminoethyl
DHA	docosahexaenoic acid
DMH	1,2-dimethylhydrazine
DMPD	N-N dimethyl-P-phenylenediamine
DNA	deoxyribonucleic acid
DPPH	1,1-diphenyl-2-picrylhydrazyl
DSSC	dye-sensitized solar cell
DW	dry weight
E102	tartrazine
E110	sunset yellow FCF
E127	erythrosine
E129	allura red
EAAE	enzyme-assisted aqueous extraction
EACC	Ehrlich ascites carcinoma cell
EACP	enzyme-assisted cold pressing
EAE	enzyme-assisted extraction
ECM	extracellular matrix
EDR	endothelium-dependent relaxation
EE	ether extract
EFSA	European Food Safety Authority
eNOS	endothelial nitric oxide synthase
EPA	eicosapentaenoic acid
EPA	Environmental Protection Agency
ER	endoplasmic reticulum
ERK	extracellular signal-regulated kinases
ESVAC	European Surveillance of Veterinary Antimicrobial Consumption
EU	European Union
FAS	fatty acid synthase enzyme
FCAT	Freund's complete adjuvant test
FDA	Food and Drug Administration

FRAP	ferric reducing antioxidant power
GABA	gamma-aminobutyric acid
GADD	growth arrest and DNA damage
GC-MS	gas chromatography mass spectrometry
GIT	gastro intestinal tract
GLUT	glucose transporter
GPx	glutathione peroxidase
GR	glutathione reductase
GRAS	generally recognized as safe
GRP	glucose-regulated protein
GRx	glutathion reductase
GSH	glutathione
GSH-Px	glutathione peroxidase
GSP	grape seeds proanthocyanidins
H1703	lung squamous cell carcinoma
HCC	hepatocellular carcinoma
HCT116	human colon cancer cell
HDFa	human dermal fibroblasts adult
HDL	high-density lipoproteins
HepG2	human hepatocarcinoma cell
HFD	high fructose diet
HHP	high hydrostatic pressure
HHPE	high hydrostatic pressure extraction
HIF-1α	hypoxia inducible factor-1α
HIV	human immunodeficiency virus
HMGCR	3-hydroxy-3-methylglutaryl-coenzyme A reductase
HO-1	heme oxygenasse-1
HPLC	high-performance liquid chromatography
HRS	hydroxyl radical scavenging
HSV-1	herpes simplex virus-1
Iκβ	inhibitor of κβ
IC50	inhibitory concentration 50
IDDM	insulin-dependent diabetes mellitus
IDF	International Diabetes Federation
IL	interleukin
IL-1β	interleukin-1 beta
IL-6	interleukin-6
IPC	International Poultry Council

IVD	intervertebral disc
JNK	c-Jun N-terminal kinases
K562	human chronic myeloid leukemia cell
KK-Ay	type 2 diabetic model
LDH	lactate dehydrogenase
LDL	low-density lipoproteins
LLE	liquid–liquid extraction
LLL	combination of linoleic, linoleic, and linoleic
LO	lipooxygenase activity
LPS	lipopolysaccharide
MAE	microwave-assisted extraction
MAP	microwave-assisted processing
MAPK	mitogen-activated protein kinase
MCF-7	human mammary cancer cell
MDA	malondialdehyde
MDA-MB-231	breast tumor cell lines from pleural effusions
MDM-LDL	malondialdehyde-modified-LDL
MDR-1	multidrug resistant protein-1
MEP	mevalonic acid pathway
MF	mounting frame
MFRM	mango fruit reject meal
MHG	microwave hydrodiffusion and gravity
MIC	minimum inhibitory concentration
MMC	mitomycin C
MMP	mitochondrial membrane potential
MMP 9	matrix metalloproteinase 9
mRNA	messenger ribonucleic acid
mTOR	mammalian target of rapamycin
NF-κB	nuclear factor kappa-light-chain-enhancer of activated B cells
NIDDM	non-insulin-dependent diabetes mellitus
NMDA	m-methyl-D-aspartate
NMR	nuclear magnetic resonance
NO	nitric oxide
NOS	nitric oxide synthase
NPQ	non-photosynthetic quenching
NQO-1	NADPH: quinone oxidoreductase 1
NSCLC	nonsmall cell lung cancer cell

NSS	neurological severity score
ODC	ornithine decarboxylase
OH	ohmic heating
OLL	combination of oleic, linoleic, and linoleic
ORAC	oxygen radical absorbance capacity
ORCs	olfactory receptor cells
OVA	ovalbumin
p-38	mitogen-activated protein kinases
PABC	pro-oxidant–antioxidant balance
PAL	phenylalanine ammonialyase
PEF	pulsed electric field
Phospho-AKT	protein kinase B (PKB) or serine/threonine-specific protein kinase
PLE	pressurized liquid extraction
PLL	combination of palmitic, linoleic, and linoleic
POL	combination of palmitic, oleic, and linoleic
POP	persistent organic pollutants
PP	phenylpropanoid
PPARγ	peroxisome proliferator-activated receptors
PTZ	pentylenetetrazol
PUFA	polyunsaturated fatty acids
R-PE	R-phycoerythrin
RAAS	renin angiotensin aldosterone system
RE	retinol equivalents
ROS	reactive oxygen species
RSV	respiratory syncytial virus
SAC	S-allyl cysteine
SAMC	S-allylmercaptocysteine
SF	supercritical fluids
SFA	saturated fatty acids
SFE	supercritical fluid extraction
SI	Stimulation Index
SIRT	sirtuin (silent mating-type information regulation 2 homolog)
SNPs	silver nanoparticles
SOD	superoxide dismutase
STZ	streptozotocin
TAC	total antioxidant capacity

TAG	triacyl glycerol
TBARS	thiobarbituraic acid-reactive species
TBI	traumatic brain injury
TC	total cholesterol
TEAC	trolox equivalent antioxidant capacity
TG	triglycerides
TGF	tumor growth factor
TLR	toll-like receptor
TNF	tumor necrosis factor
TPC	total phenolic content
TrKB	tropomyosin receptor Kinase B
TUNEL	terminal deoxynucleotidyl transferase dUTP nick end labeling
U937	human histiocytic lymphoma cell line
UAE	ultrasound-assisted extraction
UCP	uncoupling protein
UFA	unsaturated fatty acids
UNESCO	United Nations Educational, Scientific and Cultural Organization
UPE	ultrahigh pressure extraction
USDA	United States Department of Agriculture
UV	ultraviolet
UVB	ultraviolet B
VDC	voltage-dependent Ca^{2+} channel
VLDL	very low-density lipoproteins
WAT	white adipose tissues
WHO	World Health Organization
XO	xanthine oxidase

PREFACE

We introduce this book volume under the book series Innovations in Plant Science for Better Health: From Soil to Fork, published by Apple Academic Press. This book mainly covers the current scenario of the research and case studies and the importance of phytochemicals in therapeutics, under two main parts: Part I: Extraction of Bioactive Compounds and Their Applications, and Part II: Bioactive Compounds and Health Claims.

Part I describes the advances in the extraction of bioactive compounds from various sources. Advanced extraction techniques such as enzyme-assisted, microwave-assisted, ultrasound-assisted, pressurized liquid extraction and supercritical extraction techniques are described in detail. Natural products and their bioactive compounds are being increasingly utilized in preventive and therapeutic medication. Bioactive compounds have been utilized for the production of pharmaceutical supplements and more recently as food additives to increase the functionality of foods.

Part II covers the role of different bioactive compounds and their health-promoting potential for lifestyle diseases. The incorporation of any functional foods, nutraceuticals, and bioactives in the daily diet is a beneficial endeavor to help prevent the progression of chronic disorders. This section explains the botany, physical characteristics, uniqueness, uses, distribution, importance, phytochemistry, traditional importance, nutritional importance, bioactivities, and future trends of different functional foods. Functional foods, beyond providing basic nutrition, may offer a potentially positive effect on health and cure various disease conditions such as metabolic disorders, cancer, and chronic inflammatory reactions.

This book volume sheds light on the potential of plants for human health from different technological aspects and contributes to the ocean of knowledge on food science and nutrition. We hope that this compendium will be useful for students and researchers of academia as well as for persons working with the food, nutraceuticals, and herbal industries.

The contributions by the cooperating authors to this book volume have been most valuable in the compilation. Their names are mentioned in each chapter and in the list of contributors. We appreciate you all for having patience with our editorial skills. This book would not have been written without the valuable cooperation of these investigators—many of whom are renowned scientists—who have worked in the field of food science, biochemistry, and nutrition throughout their professional careers. I am glad to introduce my mentor and a coeditor, Prof. Colin Barrow, who brings his expertise and innovative ideas on bioactive compound, drug discovery, and separation sciences in this book.

The goal of this book volume is to guide the world science community on how bioactive compounds can alleviate us from various conditions and diseases.

We will like to thank editorial and production staff, and Ashish Kumar, Publisher and President at Apple Academic Press, Inc., for making every effort to publish this book when all are concerned with health issues.

We request that readers offer us constructive suggestions that may help to improve future works.

I thank Dr. Megh R. Goyal for his leadership and for inviting me to join his team. He is a world-renowned scientist and engineer with expertise in agricultural and biological engineering. Truly he is giver and a model for budding scientists. I am on board to learn.

— Hafiz Suleria

PART I
Extraction of Bioactive Compounds and Their Applications

CHAPTER 1

EXTRACTION OF BIOACTIVE MOLECULES: CONVENTIONAL VERSUS NOVEL METHODS

MUHAMMAD KAMRAN KHAN[1,*], FARID CHEMAT[2], SAIRA SATTAR[1], and MUHAMMAD IMRAN[1]

[1]*Faculty of Life Sciences, Government College University, Institute of Home and Food Sciences, Faisalabad 38000, Pakistan*

[2]*University of Avignon, UMR 408 – INRA, GREEN Extraction Team, F-84000 Avignon, France*

[*]*Corresponding author. E-mail: mk.khan@gcuf.edu.pk*

ABSTRACT

Bioactive compounds are widely distributed in plant sources and are the most abundant secondary metabolites of the plants. Based on chemical characterization, they include: glycosides, flavonoids, tannins, terpenoids, lignans, alkaloids, peptides, and others. The extraction of bioactive molecules has drawn increasing attention due to their antioxidant and therapeutic potential through their interactions with vascular endothelial cells to prevent cardiovascular disorders and cancer. Although the development in chromatographic and spectrometric analytical techniques has significant contribution in the detection of bioactive components, the success depends on the method of extraction as two-third part of analytical work is required to get these components. In conventional extraction methods, several combinations of temperature, solvent, agitation speed, and extraction time have been optimized to get maximum yields, but these conventional proposals are creating burden on environment due to high temperatures for a long time and secondly, affect the heat-sensitive bioactive components. Alternatively, application of novel techniques like ultrasound, microwave, high pressure, pulsed electric field, supercritical fluid, and others is more welcoming due to less environmental burden in the form of lessened usage of organic solvents, low working temperatures, short duration, and improved quality and yield with high selectivity to compounds of interest. The efficiencies of both either conventional or novel extraction methods are typically based on the input parameters, complexity of plant source, structural conformation of bioactive components, and skill to up-scale them. This chapter aims to discuss basic mechanisms involved in the extraction of bioactive molecules from plant material through conventional as well as novel extraction techniques.

1.1 INTRODUCTION

Bioactive compounds are secondary metabolites in plants that are prompting toxicological and pharmacological effects in both animals and man. The production of these secondary metabolites within the plants took place as a result of primary metabolic and biosynthetic pathways for compounds related with growth and development of the plant, therefore, they are considered as by-products of plant cells metabolism.[7] These bioactive compounds are well known due to their significant functionality in plants, for instance, (1) during photosynthesis, flavonoids act as free radicals scavengers, (2) terpenoids can appeal pollinators or seed dispersers, (3) alkaloids repels insects or herbivore animals, and (4) likewise, other secondary metabolites indulge in many different functions within the plants.[90]

Majority of the plants produce bioactive compounds including food and feed plants, but the higher concentration of bioactive molecules are present in medicinal or poisonous plants.[83,98] Various studies confirm that these secondary compounds have a protective action on human health and are the key elements of a healthy and balanced diet.[20,143] They have beneficial effect on prevention of cardiovascular diseases, inflammation, glucose intolerance, and obesity.[12,162] According to World Health Organization, the approximate share of world's population that relies on their primary health care from plant-derived natural medicines is 65–80%.[158,48] For instance, flavonoid is most important phytochemical capable of controlling the incidence of diabetes, cancer, and cardiovascular diseases.[71] Catechin, another bioactive molecule present in green tea, is effective against obesity.[91] Bioactive compounds also possess antioxidant activity which decreases the deleterious effects of substances with high oxidative potential. This antioxidant property is principally designated to their redox potential which lets them to work as reducing agents.[83] Thus being an important constituent of human diet, the extraction of biochemical molecules from natural plant sources is of keen interest for the researchers.

1.2 CATEGORIES OF BIOACTIVE MOLECULES

Classification of plants bioactive molecules depends on different criteria. They could be presented clinically, toxicologically, pharmacologically, or botanically; but it is complicated as even chemically related compounds can possess different clinical outcomes or even less genetically related species could produce dissimilar bioactive molecules.[7] Therefore, it is preferable to classify them according to the chemical classes and biochemical pathways. According to Croteau et al.[37] and Taiz et al.,[141] bioactive molecules from plant sources are categorized into: (1) terpenes and terpenoids (roughly 25,000 compounds) produced by mevalonic acid and non-mevalonate (MEP) pathways, (2) alkaloids (roughly 12,000 compounds) produced through shikimic acid pathway, and (3) phenolic compounds (roughly 8000 compounds) produced by malonic acid and shikimic acid pathways.

1.2.1 TERPENES

Among the natural products, the terpenoids occupied the largest group, also used in number of industrial sectors as fragrances, spices, and flavors also used in cosmetics and perfume-making industries. Huge types of terpenoids have been identified, that is, 25,000; these compounds have major role in defense against biotic and abiotic factors and also act as source of attraction for the insects of pollination.[136] Terpenoids have been studied for their role in medicine and biotechnology. Terpenes are hydrocarbon-based compounds which contain a structure derived from isoprene which give rise to another structure which may also be divided into isopentane units.[119]

There are two major pathways to produce terpenes, that is, conventional acetate-mevalonic acid and non-mevalonic acid pathways. The conventional acetate-mevalonic acid pathway is operated in the cytosol and mitochondria of plant cells; there, various compounds are synthesized, namely, sterols, sesquiterpenes, and ubiquinones.[85] While the non-mevalonic acid pathway takes place in plastids of plant cells and prepare compounds like hemi-, mono-, sesqui-, and diterpenes, additionally carotenoids and phytol tail of chlorophyll.[22]

Odors and flavors of terpenoids are strong. These compounds have wide range of their action which is being used for herbal remedies. Among these variations, particular examples are antibacterial, antineoplastic, antiviral effects along with metabolic stimulation which are very important. Toxicity of these compounds is associated with their concentrated form as volatile oils. Terpenoids mainly belongs to family *Lamiaceae* (thyme family) but may be present in other similar families.[7,15]

In fragrance and perfume industries, monoterpenoids are major constituent of various essential oils while other acyclic compounds are geraniol, linalool, and myrcene. Camphor, menthol, limonene, and pinene are cyclic structures. Due to higher boiling point, diterpenes do not possess the characteristics of essential oils and are considered as component of plant resins.[50] Terpenoids including squiterpenes, compounds with three isoprene units are mostly present in aliphatic bi- and tricyclic forms. Farnesol, a aliphatic bicyclic form, is an important intermediate in terpenoid synthesis. Arteether, extracted from *Artemisia annua*, is a sesquiterpene lactone derived from artemisinin and nowadays, used as an antimalarial drug. Triterpenes (C30) are mainly composed of six isoprene structures and are formed from squalene during biosynthesis. Such compounds have higher melting points, are colorless, occur in solid form, and mainly found in resins, cork, and cutin. Steroids, saponins, and cardiac glycosides are produced from triterpenoids which are pharmacologically active. From seeds of *Azadirachta indica* a powerful insect ant-feedent is produced. Among other tri-terpenoids, cucurbitacins and limonins are effective insect steroid hormone antagonists.[102]

Plant steroids which are hydroxylated at C3 position are classified as sterols. Steroids are altered forms of triterpenes and in animals act as necessary hormones, for example, estrogens such as progesterone and androgens such as testosterone, coenzymes, and provitamins. Diosgenin is the important source of much progesterone which is derived semisynthetically. On the other hand, amaryllidaceae, dioscoreaceae and liliaceae, and dicot families (Solanaceace and Scrophulariaceace. Saponins) are important source of saponins (C27). These compounds are composed of mainly two parts, namely, aglycone (genin or triterpene) and glycone (sugar). Important preparations based upon saponins are licorice (*Glycerrhiza glabra*), primula root (*Primula*), sarsaparilla root (*Sarsaparilla*), ginseng (*Panax ginseng*) and ivy leaves (*Hedera*). Glycyrrhizins are the salt form of ammonium and calcium with glycyrrhizic acid, and on sucrose scale, they are 50–100 times sweeter.[152]

1.2.2 ALKALOIDS

The alkaloids have potent activity and bitter taste. They are heterocyclic compounds containing nitrogen. In more than 150 families, 12,000 types of these compounds are present in plants. *Papaveraceae, Apocynaceae, Ranunculaceae, Fabaceae, Rubiaceae, Solanaceae*, and *Rutaceae* are important families of alkaloids, while less common lower plants and fungi (ergot alkaloids) also contain these compounds.[77] Alkaloids are present in isomeric forms as salts of organic acids like malic, oxalic, lactic, citric, tannic, tartaric, and other acids in plants. On the other hand, few weak basic alkaloids (such as nicotine) present freely in plant systems. Some members of alkaloids are also found in glycosidic form with galactose, glucose, and rhamnose such as solanine. They also occur in the form of amides (piperine), and as esters (cocaine and atropine) of organic acids.[82,101]

Plants contain alkaloids in their various parts such as large amounts of these compounds are present in seeds (nux vomica, *Areca*), stem bark (cinchona and pomegranate), and roots (aconite and belladonna). Alkaloids are abundant in dicots as compared to monocots.[126] Alkaloids are used as narcotics, stimulants, poisons, and pharmaceuticals due to their potent activity. Some of the most common examples of alkaloids which are being used are the anticancer agent—vinblastine, the muscle relaxant—(+)tubocurarine, analgesics—codeine and morphine, the antiarrhythmic agent—ajmalicine, the gout suppressant—colchicine, the sedative—scopolamine, and the antibiotic—sanguinarine. Caffeine in coffee and tea along with nicotine in all preparations such as chewing, smoking, etc. are extensively used on daily basis.[100]

Different clinical properties are found in different alkaloid groups. For instance, tropane alkaloids are abundantly found in *Solanaceae* family, for example, in *Atropa belladonna, Datura* spp., and *Hyoscyamus niger*. The compounds of these alkaloid groups contain anticholinergic effect to lessen the smooth muscle spasms, pain, and hypersecretion; therefore, these compounds are extremely medicinally important. *Asteraceae* (daisy family), particularly *Boraginaceae* (borage family) and *Senecio* spp. (Ragworts) are the good source of pyrrolizidine alkaloids. After bioactivation, they exert adverse effects on human health. Isoquinoline alkaloids are present in *Berberidaceae* (barberry family) and *Papaveraceae* (poppy family). Such compounds have wide range of biochemical effects in humans by controlling different malady conditions (cancer cells, bacteria, and pain) along with improvement in bone marrow leucocytes and myocardial contractility.[109] *Coffea arabica* (coffee) and *Theobroma cacao* (cacao) are the main sources of methylxanthine alkaloids which show an important impact on neurological systems of humans and animals. Similarly another group of alkaloids, pseudoalkaloids, which are chemically close to alkaloid, affect the central nervous system. These compounds are synthesized by species in *Apiaceae* (carrot family), for example, *Cicuta virosa* (cowbane) and *Conium maculatum* (hemlock).

1.2.3 POLYPHENOLS

Polyphenols are widely distributed in nature. They are the secondary compounds of plant kingdom. Almost 8000 types of phenolic compounds are identified and classified into various subgroups based on the number of phenol rings present and the structural elements which bind such rings to one another. These classes include phenolic acids (hydroxycinnamic acids and hydroxybenzoic acids), flavonoids (flavonols, flavanols, flavanones, flavones, proanthocyanidins, and isoflavones), tannins, stilbenes, and lignans. These classes of polyphenols are present in plants and in various foods of plant origin.[98,99] Simple phenolic compounds have at least one OH-group bounded to an aromatic ring, such as catechol while majority of compounds contain C6C1 carbon skeleton having carbonyl group bounded to the aromatic ring.[60] Mostly, phenolic compounds are prepared through shikimate pathway, but sometimes a few phenolic compounds, for example, orcinols and quinones, are synthesized by the polyketide pathway. Phenolics synthesized from either pathway shared common structure such as flavonoids, stilbenes, pyrones, and xanthones.[91] Majority of phenolic compounds are present in leaves, woody parts of plants such as barks, stems, flowering tissues, etc.[74] Phenolics add taste, color, and nutritional properties to the fruit.[30]

Flavonoids compounds are composed of two phenolic rings joined through a pyranring and proanthocyanidins, the polymers of flavonoid units both of which occurred in glycosidic forms. Any compound containing phenol group acts as an antioxidant. Other actions include reducing inflammation and carcinogenicity. Isoflavones are also known as phytoestrogens. A long range of pigments is present in plants, for example, flavonoids and proanthocyanidins. *Fabaeceae* (bean family) are the main source of the isoflavones.[36]

Tannins exist in two types: condensed and hydrolyzable, depending on their structural complexity. Condensed tannins are large oligomers of flavonoid units, whereas hydrolyzable tannins are composed of glycosidic center (commonly glucose) with several catechin/phenolic acid derivatives. Solubility of tannins decreased with the increase in size of the molecule. Tannins could be antinutritional as they can bind with proteins and minerals while bigger tannins are served as astringents in various diseases (diarrhea, transudate, and skin bleeding). These compounds are present in wide range in plant kingdom. *Fagaceae* (beech family) and *Polygonaceae* (knotweed family) are few examples of plants containing tannins.[14]

Lignans contain different functional groups and consist of two phenylpropanoid units to form an 18-carbon skeleton. These compounds are present within the cell membrane and perform specification functions as they contain lipophilic properties.[52] Lignans are present in different concentrations in different plant species but higher amounts are discovered in oilseeds. Phytoestrogenic, cathartic, or antineoplastic effects are associated with lignans.[65]

1.2.4 OTHER BIOACTIVE COMPOUNDS

1.2.4.1 PROTEINS AND PEPTIDES

Proteins perform an extremely important role in food and feed. Proteins components are absorbed into the blood from intestinal tissues and provide building blocks of the body protein. Besides these, many proteins also act as bioactive molecules.[108] These bioactive proteins are unable to hydrolyzed in GIT, rather than absorbed in blood and exert their particular function in the body. Such proteins are produced by *Euphorbiaceae* (spurge family) and *Ricinus communis* (castor bean). For example, ricin tends to prevent the synthesis of proteins and produce gradual effects in animals and humans. These proteins exist in minor quantity in seeds of several species of *Fabaceae* (bean family). Symptoms related to colic and other metabolic disorders may produce if seeds are not heat treated to inactivate lectin.[79]

1.2.4.2 GLYCOSIDES

Glycosides may originate from various types of secondary metabolites which are bound with a monosaccharide, oligo-saccharide, or uronic acid. Therefore, it contains two groups, first one is glycine (saccharide or uronic acid part) while remaining part is known as aglycon. Cyanogenic glycosides, cardiac glycosides, anthraquinone glycosides, saponins, and glucosinolates are the some main groups of glycosides. Flavonoids are also found as glycosides. After intake of glycosides, it hydrolyzes in the colonic part, while the more hydrophobic glycosides (aglycone) might be absorbed.[160]

Steroidal structure is present in aglycones of cardiac glycosides. They inhibit the Na^+/K^+-ATPase-pumps operated in the cell membranes. Aglycones of cyanogenic glycosides are derived from the amino acids.[112] Hypothyroidism may result as these compounds may interfere with utilization of iodine. Sulfur containing amino acids are present in amino acid-derived aglycones which have pungent smell. In various cells, these compounds exert a complex effect on cytochrome (P450 isoforms) and therefore it decreases the hepatic bioactivation of environmental procarcinogens. Majority of saponins (soap-forming compound) are present as glycosides. Emulsifying properties are associated with saponin glycosides which are comparatively big molecules having hydrophilic and hydrophobic aglycone parts.

1.3 IMPORTANCE OF EXTRACTION

Extraction is the primary stage in medicinal plant research with the significant impact on the final outcome. Bioactive compounds contain the pool of molecules having broad diversity of functionalities and structures that present an important role in the production of food additives, functional foods, and nutraceuticals. The distribution of bioactive compounds in nature vary according to their concentration, some of them are present at low level, whereas some compounds, such as polyphenols, can be found in higher concentration. Therefore, to obtain these compound in adequate level, huge harvesting is required which is quite complicated and unbeneficial related to cost.[72] Although the development in chromatographic and spectrometric analytical techniques has significant contribution in the detection of bioactive components, still success depends on the method of extraction as two-third part of analytical work is required to get these components.[11] The innate obstacles in producing and screening required bioactive compounds have led to the advancement of the novel extraction technologies.

Currently, many researchers and industrialists are involve in finding out various methods to explore the potential of bioactive compounds from natural sources for the prevention and treatment of various human diseases and to meet other needs. The efficiency of these compounds to interact with different biological molecules including DNA and proteins for the production of preferred outcome allows them to be fully utilized in designing therapeutic agents derived from natural products.[6] Hence for this purpose, the extraction of bioactive molecules from plant sources along with the estimation of their quantitative and qualitative properties is important for exploration of new biomolecules to be used by agrochemical and pharmaceutical industry.[68] According to UNESCO in most developing countries, 80% of the world's population relies on the use of herbal products on regular basis to keep good health.[27] The thousands of chemical compounds present in these plants are used for different infectious diseases. These phytochemicals possess beneficial biological activity such as antioxidant, antimicrobial, anticancer, analgesic, antidiarrheal, and wound healing..[45]

Nowadays, the food industry is focusing to manufacture and develop different functional food products. The growing interest of consumer for healthy food makes this new class of food product successful in the market. In general, these functional foods include various different types and proportions of bioactive compounds.[15] Therefore, functional food can be defined as a new product in which bioactive compounds from different natural sources are incorporated to formulate food with a specific function.[41]

Plaza et al.[118] discuss three main factors to be fulfilled by a functional food. First of all, the effect of functional food should be different as compared to normal diet. Second, there should not be any side effect of developed functional product, and third, it should be beneficial in reducing risk of developing pathological condition and should also help to improve physiological function. Therefore, the desired biomolecules that possess biological activity such as antiviral, antioxidant, antihypertensive, antidiabetic, and so on are extracted from plant sources to be used in formulating different functional products as not all bioactive compounds cover all these aspects.[118] These functional foods help in decrease in cholesterol levels, maintaining remission of Crohn's disease, alleviation of lactose intolerance, inhibition of cancer cell proliferation in vitro and in vivo, and faster relief from diarrhea.[101,62,111] The functional foods that are presently accessible in the market are bakery products, drinks, meat products, cereals, eggs, and spreads.[137] Among all these products, beverages are the most convenient because of their comparative easy handling, processing, and formulation with more complex processed foods.[41,62]

The conventional techniques for the extraction of bioactive compounds used frequently are liquid–liquid or solid–liquid extraction, Soxhlet extraction, maceration, and hydrodistillation, and the advanced methods include subcritical and supercritical extractions, pressurized liquid extraction, ultrasound- and microwave-assisted extractions (MAEs), pulsed electric field extraction, and ohmic extraction. The novel extraction methods are basically used to enhance the release of compounds from the plant matrix. Therefore in the next few years, these technologies could provide an eco-novel approach to boost the production of particular compounds for use as constituents in the manufacturing of functional foods or as nutraceuticals.

1.4 CONVENTIONAL EXTRACTION TECHNIQUES

Plant extraction is a pragmatic exercise in light of the fact that distinctive solvents are used at different conditions, for example, temperature and time of extraction.[68] The time, temperature, solvent, pressure, and the matrix properties of plant part are the most well-known variables influencing extraction methods.[63] The extracting methods, utilized to separate bioactive mixes from plant sources, mostly depend on extracting power of various solvents being used and also on the use of heat and mixing. The most common conventional techniques in order to extract bioactive compounds are Soxhlet extraction, liquid–liquid extraction, evaporation, maceration, and hydrodistillation. In conventional extraction methods, several combinations of temperature, solvent, agitation speed, and extraction time have been optimized to get maximum yields, but these conventional proposals are creating burden on environment due to high temperatures for a long time and also affect the heat-sensitive bioactive components. Some bioactive compounds extracted through these conventional techniques are given in Table 1.1.

1.4.1 SOXHLET EXTRACTION

Soxhlet extraction is a most commonly used conventional technique to extract different compounds from the plant materials. It is an ordinary case of a comprehensive solid–fluid extraction. This procedure relies on the exchange of the target compound(s) from the sample (solid) to appropriate organic solvent(s). It is guaranteed that the extraction solvent (fluid) remains reliably in contact with the sample during this thermal extraction process.[25]

It was specifically designed to extract lipids but currently, it is also used to extract other components from the plant parts. Numerous valuable bioactive compounds were extracted by using the Soxhlet extraction technique from different plant sources.[161] To develop alternative extraction techniques Soxhlet extraction is used as model. Soxhlet extraction technique is effectively used if desired compound has good solubility in solvent and impurities are insoluble in that solvent.[35]

Soxhlet extraction technique is effectively used to extract components from solid sample like sediment, soil, and indoor dust samples. Other semisolid samples including sewage sludge, blood, milk, and fat can also be efficiently used to extracts components by using Soxhlet technique.[42] Soxhlet extraction apparatus consists of a thimble (sample holder), distillation flask, siphon, and condenser as it is evident from Figure 1.1.

Its operation is quite simple; generally, little amount of dry sample is kept in the inner side of the thimble (Fig. 1.1) which is then placed in a distillation flask containing desired solvent. After achieving specific temperature, the solution aspired through siphon. This siphon is used to mix back of the solution in the distillation flask which brings extracted component into the solvent. During operation, desired component remains in distillation flask in the form of solute while solvent moves to the bed of the plant material. This operation is repeated several times until the extraction of desired component is achieved.[56,157] The main advantage of this technique is use of single batch of solvent which is recycled again and

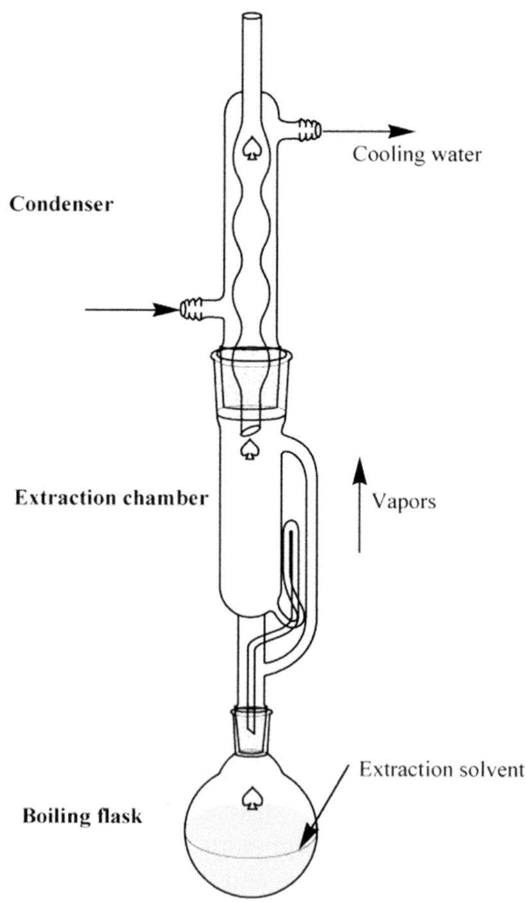

FIGURE 1.1 Soxhlet extractor.

again during the process. Many disadvantages are also associated with this technique, such as, heat-sensitive compounds cannot be extracted by using Soxhlet apparatus because prolonged heating time may destroy the heat-labile compounds[139] This technique is also not suitable for industrial applications as it have long extraction time and utilizes high amount of hazardous solvents along with other disadvantages.[16]

1.4.2 LIQUID–LIQUID EXTRACTION

For the efficiently extract the bioactive compounds, sample can be pretreated. Liquid–liquid extraction is a useful pre-treatment technique which is most commonly used now a day. It can increase the selectivity by separating analyte from matrix of the sample or by concentrating the desired analyte from high sample volume. This technique also associated with number of disadvantages including the manual working of this mass transfer operation which is laborious and time consuming along with higher demand of chemicals that can adversely affect to the operator, also expensive and cause environmental pollution.[107] Different strategies were used to minimize the risk of abovementioned disadvantages by reducing solution consumption, reduction of operator invention and expose along with increasing the sample rating etc.[31]

A modified liquid–liquid extraction technique, for example, flow-based LLE has been successfully used in different industries like food analysis, pharmaceutical, and clinical, among others which reduces the environmental pollution.[64] A conventional liquid–liquid extraction technique consists of three important components, that is, a phase segmenter, a phase separator, and an extraction coil (Fig. 1.2). Liquid sample is introduced either in a flow process or in a specifically defined volume, into an aqueous stream (which acts as reagent and carrier stream). After introduction of the liquid sample, homogenization process is carried out which results in the formation of reaction zone, directed toward the segmenter part of the liquid–liquid extraction. In this part, the two streams of aqueous and organic immiscible phases remain in contact and a single flow of alternate reproducible zones of both phases is produced. Consequently, in the extraction coil, mass transfer between the two phases multiple interfaces created by the segmentation process is taking place. Finally, in the phase separator, the little–liquid and organic phase parts are continuously divided into individual streams, one of that contains the analyte directed toward the detector for detection.[23]

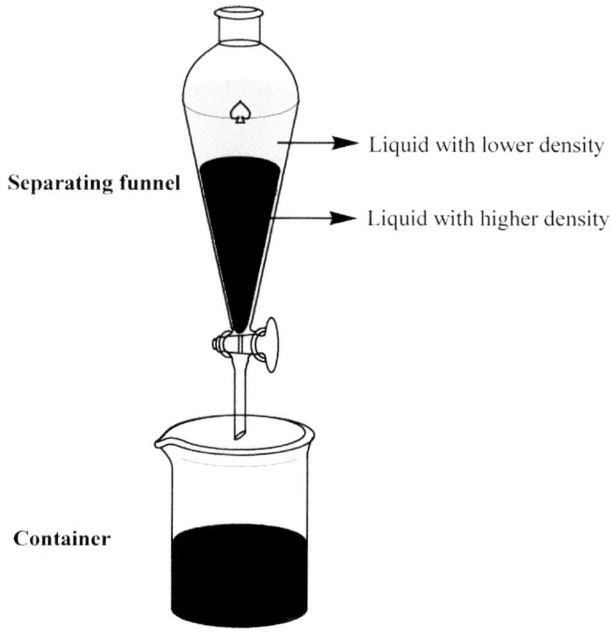

FIGURE 1.2 Liquid–liquid extractor.

1.4.3 MACERATION

Homemade tonics are prepared by using the maceration technique. It is very famous and inexpensive technique to isolate the oil or other desired components. Its operation is quite simple. Whole sample or coarse powder was kept in the solvent containing stoppered, which remain consistent with solvent for specific time period. During this period, frequent agitation is given until soluble matter dissolved as evident from the Figure 1.3. Heat-sensitive compounds can efficiently extracted by using this technique for example thermolabile drugs.[110] Several steps are associated with maceration process at small scale. Grinding of desired sample is first step which is carried out to increase the surface area. In second step suitable solvent is added. In third step of maceration, the marc which is the solid by-product of this extraction course is forced to recover large quantity of desired solutions. In forth step, impurities are separated from the obtained strained and the press out liquid by using filtration process. Extraction process help the extraction of bioactive molecules in two ways: (1) by increasing diffusion and (2) by removing concentrated solution from the solid surface for fetching new solvent to get high yield.

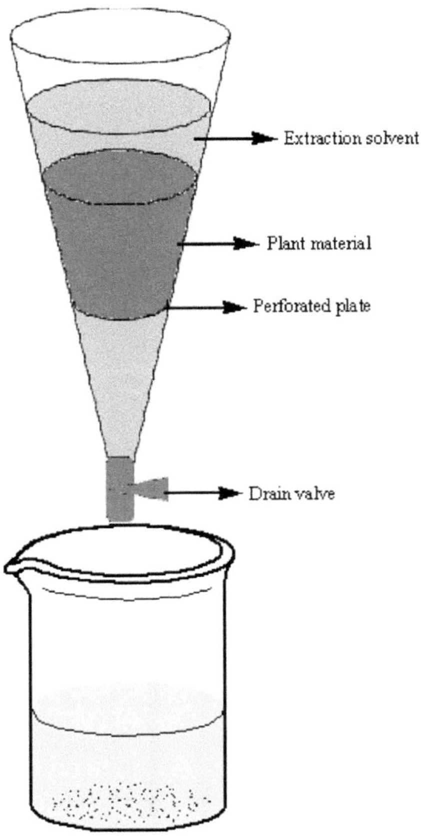

FIGURE 1.3 Apparatus for maceration.

1.4.4 HYDRODISTILLATION

Hydrodistillation is a conventional technique for extraction of basic oils and bioactive molecules from plants. It does not include organic solvents in it and usually takes place before plant material gets dehydrated. There are three kinds of known hydrodistillation methods: steam refining, steam and water refining, and direct water refining.[146] In hydrodistillation, to begin with, the plant material is placed in a batch system compartment; second, adequate amount of water is included and then brought to boil.[125] The setup includes condenser and a decanter to gather the condensate and to isolate bioactive molecules from solvent (Fig. 1.4). On the other hand, direct steam is infused into the plant sample. Steam and hot water go about as the principle powerful factors to free bioactive compounds of plant tissue. The vapor mixture of water and oil condenses by indirect cooling of water. The mixture streams from condenser to a container where immiscible part (oil and bioactive mixes) isolates naturally from the water.[135] Three fundamental physicochemical procedures are involved in hydrodistillation: (1) hydrodiffusion, (2) hydrolysis, and (3) disintegration by heat. In this technique, some volatile components may be lost due to high extraction temperature and this drawback restrains its utilization for the extraction of heat-sensitive compounds.

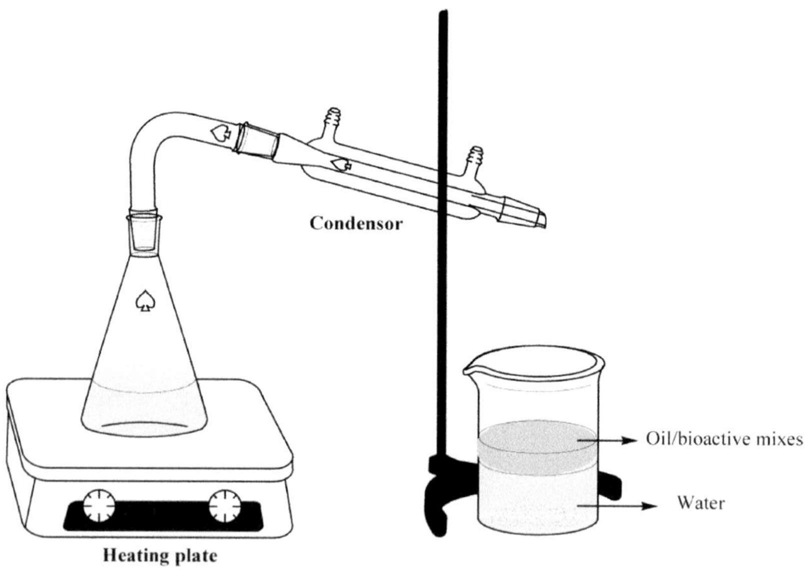

FIGURE 1.4 Apparatus for hydrodistillation.

TABLE 1.1 Biomolecules Extraction by Conventional Extraction Methods.

Conventional method	Bioactive compound	Optimal conditions	Remarks	References
Soxhlet extraction	Squalene, n- hexadecanoic acid, tetramethyl-2-hexadecen-1-1 and octadecatrienoic acid from agarwood leaves	Solvents: distilled water, hexane, isopropanol and ethanol for 6 h at 60°C	Extraction using hexane as solvent gave highest percentage of oil yield	[87]
	Tocopherol, phytol, β-sitosterol, γ-sitosterol from aromatic plant	Solvent: 200 mL of pentane for 4 h at 40°C	Oleoresins were obtained from 10–40 g ground plants	[61]
Liquid–liquid extraction	Flavanols, kaempferol, quercetin, apigenin, luteolin and myricetin from different vegetables	Solvent: 40 mL 80% methanol, 10 mL 6 M HCl and 80 mg ascorbic acid for 2 h at 90°C	Effective in releasing the aglycone and increasing recovery rate of phenolics from 76.1% to 98.6%	[66]
	Polyphenols and carotenoids of blackberry, naranjilla, and tree tomato	Solvent: 60 mL of 70% aqueous acetone containing 2% formic acid, twice for 15 min at 40°C	Major phenolics were extracted in higher amount from fruits	[106]
Maceration	Phenolic acids and flavonoids of Kinnow peel	Solvent: ethanol, methanol, acetone, and ethyl acetate, macerated for 20 h at 40°C	Extraction with 80% ethanol resulted in the highest yield (18.46%)	[128]
	Phenolic compounds and flavonoids from *Eryngium creticum*	Solvent: ethanol, macerated for 48 h with agitation of 360 rpm	About 410.93 mg total yield of bioactive compounds was extracted	[163]

TABLE 1.1 *(Continued)*

Conventional method	Bioactive compound	Optimal conditions	Remarks	References
Hydrodistillation	Volatile oil of rosemary	Solvent: water for 10 min	Almost 80% oil was extracted including 31.9% 1,8-cineol, 19.7% camphor, 12.8% α-terpineol, 12.2% borneol	[18]
	Essential oil from peel of *C. microcarpa*	Solvent: water for 8 h	94% limonene, β-myrcene (1.8%), linalool (0.4%), and α-terpineol (0.3%) were extracted	[103]

1.5 NOVEL EXTRACTION TECHNIQUES

Conventional extraction has become a least concerned method in food industry due to its various drawbacks like high cost, longer extraction time, thermal decomposition of thermolabile compounds, low extraction selectivity, requirement of high purity solvent, and its evaporation at high amount[56]; therefore, some novel and promising nonconventional extraction techniques are introduced to overcome these limitations. The development of these nonconventional methods came into being during the last 50 years. They acquire less time, give better quality and yield of extract and considered as more environmental friendly due to use of organic and synthetic chemicals in lesser amount. The most frequently used novel methods to extract the bioactive compounds from plant sources are ultrasound,[58,148] ohmic heating,[84] microwave heating,[76] pulsed electric field,[144] extrusion,[97] supercritical fluids,[156,59] and digestion using enzymes.[57] These novel extraction methods include safe solvents auxiliaries, less hazardous chemical synthesis, use of renewable feedstock, design for energy efficiency, lessen derivative formation, design to reduce degradation, and timely analysis for decreased level of pollution and inherently safer chemistry to prevent accident.

These novel nonconventional extraction techniques are further divided into thermal and non- thermal extraction. Thermal extraction methods includes: (1) MAEs and (2) ohmic heating whereas non-thermal extraction has vast methodologies: (1) ultrasound-assisted extraction (UAE), (2) pulse electric field-assisted extraction, (3) supercritical fluid extraction, (4) high-pressure-assisted extraction, and others.

1.5.1 THERMAL EXTRACTION METHODS

1.5.1.1 MICROWAVE-ASSISTED EXTRACTION

Microwaves have been utilized since World War II following the advancement of radar innovation, and later the main application of microwaves concerned household ovens. The utilization of microwave vitality as a warming source in expository research facilities began in the late 1970s and was applied to corrosive digestions.[2] The improvement of MAEs was first reported by Ganzler et al.[54,53]

The MAE is additionally considered as a novel technique for extricating solvent items into a liquid from an extensive variety of materials utilizing microwave vitality.[115] Microwaves are electromagnetic fields in the frequency range from 300 MHz to 300 GHz. They are composed of two oscillating fields that are opposite, for example, magnetic field and electric field. The principle of microwave heating depends on its immediate effects on polar materials.[89] Electromagnetic vitality is changed over to heat after ionic conduction and dipole rotation mechanism.[70] Heat is produced during ionic conduction mechanism as a result of the resistance of medium to flow ion. Alternatively, ions keep their path along field signs which change repeatedly. This regular change of direction brings about collapse amongst molecules and thus creates heat.

Patil and Shettigar[116] revealed a novel, microwave-assisted solvent extraction advancement known as microwave-assisted processing (MAP). The high-valued compounds are incorporated by microwave-assisted processing from natural sources like phytonutrients, nutraceuticals, and functional foods and from biomass like pharmaceutical actives. There are three successive steps of MAEs which are supposed to be fulfilled for advance extraction as described by Alupului et al.[9]: first, under expanded pressure and temperature, solutes should detached from binding sites of plant matrix; second, dispersion of solvent crosswise over plant matrix; third, discharge of solutes into solvent from plant matrix. A number of focal points of MAE have been depicted by Cravotto et al.,[34] for example, rapid increase in temperature to get bioactive compounds; increased extract yield, smart equipment size, and decreased thermal gradients. MAE can extricate bioactive compounds more quickly and a superior recuperation is expected than conventional extraction techniques. It is a specific method to remove natural and organometallic compounds that are more integral. MAE is additionally perceived as a green innovation since it lessens the utilization of natural solvent.[9] Very recently, a novel technique called microwave hydrodiffusion and gravity (MHG) has been introduced in the market of extraction by one of the contributor of this book, Prof. Farid Chemat (Fig. 1.5). The efficiency of bioactive extraction of microwave-assisted processing is listed in Table 1.2.

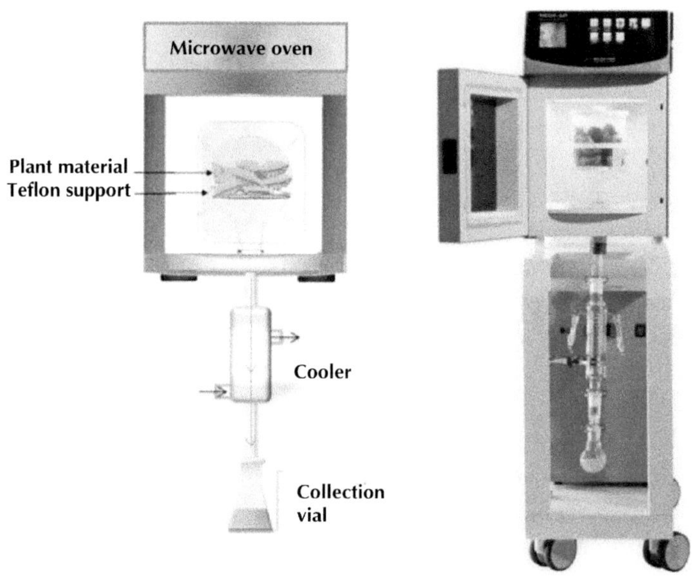

FIGURE 1.5 (See color insert.) Microwave hydrodiffusion and gravity (MHG) extraction system.

1.5.1.2 OHMIC HEAT-ASSISTED EXTRACTION

Ohmic heating, also termed as electroconductive heating, uses the innate electrical confrontation of plant material to create heat[153] (Fig. 1.6). Most plant materials contain ionic components, for example, acids and salts, permitting the conduction of electrical current.[114] This phenomenon can be used to create heat inside the item, changing the electrical vitality into thermal vitality, and hence heating plant materials at outstandingly quick rates without the requirement for a heating medium. This procedure avoids extreme thermal impairment to heat-sensitive constituents, for example, pigments and vitamins.[26,134] They have appeared to be mellow handling advancements protecting sensory, nutritional, structural, and functional properties of items superior to traditional methodologies.[80,151] Specifically, the extraction procedure has been utilized to expand the productivity of solute dispersion all through the membrane (electro-osmosis impact), bringing about a superior quality product.[122,44,17] Additionally, these are environmental-friendly technique, be it by enhancing the general energy effectiveness of the procedure or by diminishing the utilization of nonsustainable resources, decreasing ecological impression, while lessening handling costs and enhancing the added value of the product.[117] OH is generally known for its ability to give quick, homogeneous and exact heating wherever direct application of electrical energy to the food guarantees an exceptionally productive energy exchange. Thus, OH is as of now being effectively implemented in food handling industry.[133,131] More examinations have discovered that OHM has been appeared to build the extraction yields of rice grain oil and bioactive substances from rice wheat,[84,95] polyphenols from red grape pomace,[46] so on, as listed in Table 1.2.

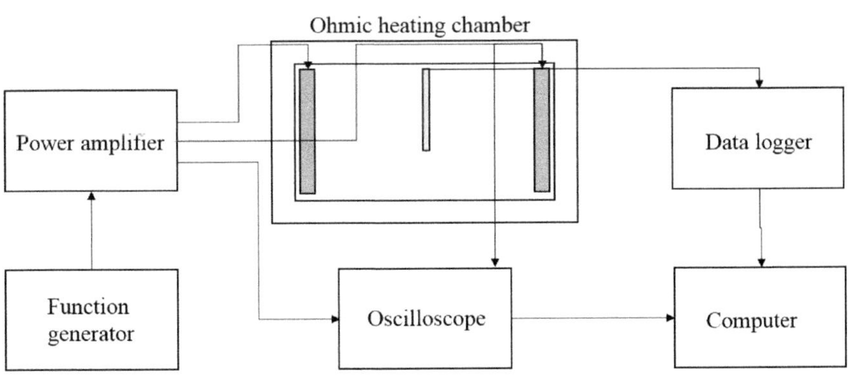

FIGURE 1.6 Schematic diagram of high-frequency ohmic system.

TABLE 1.2 Biomolecules Extraction by Novel Thermal Extraction Methods.

Thermal method	Bioactive compound	Optimal conditions	Remarks	References
Microwave-assisted extraction	Oleanolic acid and Ursolic acid from *Ocimum sanctum*	Solvent: ethanol and methanol, power = 272 W time = 3 min, solid to solvent ratio = 1:30	Highest extraction of OA and UA were observed with ethanol (89.64% and 86.76%	[142]
	Polyphenols from olive tree oil	Solvent free, power = 250–350 W, time = 2–3 min, amount of sample = 5–10 g	Both TPC and oleuropein were extracted at 250 W, 2 min, with 5 g sample	[130]
	Chlorogenic acid, caffeine and TPC from green coffee beans	Solvent: water, power = 800 W, time = 5 min, temperature = 50°C	Highest yield of chlorogenic and caffeine with MAE (31–62% and 22–40%, respectively) with radical scavenging activity of 75%	[145]
	Flavonoids from young barley leaves	Solvent: water, time= 11.12 min power = 1.27 W g^{-1}, liquid–solid ratio = 34.02 mL g^{-1}	MAE extraction was more efficient than conventional, with maximum flavonoids yield of 80.78 ± 0.52%	[55]
Ohmic heat-assisted extraction	Anthocyanins from black rice bran	MC = 30% and 40%, electric field strength = 50–200 V cm^{-1}	Both low and high concentration bioactive compounds were extracted. The highest level of bioactive compounds were observed at OHM with 40% MC (E = 50, 100 , and 150 V cm^{-1}) and 30% MC (E = 100, 150, and 200 V cm^{-1})	[96]

TABLE 1.2 *(Continued)*

Thermal method	Bioactive compound	Optimal conditions	Remarks	References
	Polyphenols from red grape pomace	Electric field strength = 100–800 V/cm, ethanol in water e/w = 00–50%	Cell membrane denatured to maximum extent. The highest extraction yields were obtained at 400 V/cm followed by a diffusion step for 60 min at 50°C and with a solvent composed of 30 % of e/w	[46]
	β-carotene and lycopene from gacaril oil	Electrical field strengths = 5.6 and 11.2 V/cm, garlic powder to n-hexane solvent = 1:7 (7 h), 1:6 (6 h) and 1:5 (5 h)	Highest extraction efficiency (81.40%) in ohmic treatment as compare to conventional	[1]
	Anthocyanins & total phenolic content from colored potato	Electric field strength = 0–30 V/cm, temperature = 30–90°C, Time = 0–10 min, water-based solvent	High recovery yield with less energy consumption, reduced treatment time, and no utilization of organic solvents	[117]

MAE, microwave-assisted extraction; MC, moisture content; OA, oleanolic acid; OHM, ohmic heating, TPC, total phenolic content; UA, ursolic acid.

1.5.2 NONTHERMAL EXTRACTION METHODS

1.5.2.1 ULTRASOUND-ASSISTED EXTRACTION

The improvement of ultrasound innovation is not new but the fact is that in recent times, the developments in the utilization of power ultrasound have seen to be accomplished.[138] In this sense, particular consideration has been given to its utilization in the recuperation of bioactive compounds from various plants sources.

UAE of polyphenols is a nonconventional procedure that includes blending the specimen with organic solvent in a beaker and setting it in an ultrasonic bath with preset time and temperature (Fig. 1.7). Its better extraction effectiveness is identified with the phenomenon of acoustic cavitation. At the point when the ultrasound power is adequate, the sound waves traveling in rarefaction and compression cycles can develop microbubbles in the fluid. Once framed, bubbles will assimilate the vitality from the sound waves and develop during the extension cycles and recompress during the pressure cycle. Further, bubbles may begin another rarefaction cycle or fall driving stun influxes of extraordinary states of weight and temperature (about 1000 atmosphere and around 4000 K of temperature).[47,138,88] Subsequently, the bursting of cavitation bubbles produces microjets of liquid which can hit the surface of the plant matrix stimulating extraction of bioactive compounds from the specimen to the solvent medium.[38]

UAE has been proposed as an extroverted alternative to conventional extraction, furnishing higher recuperation of focused compounds with less utilization of solvent as well as quicker analysis of bioactivity properties. Mostly, the UAE procedure term is under 60 min; however, the extraction yield is 6–35% higher than that acquired utilizing conventional extraction methods with longer extraction time of at least 12 h.[147,69] These days, UAE is broadly utilized for the extraction of valuable compounds. For instance, it has been utilized for the extraction of oil[3], proteins,[124] polysaccharides-protein complex,[29] sugars,[75] etc. However, the extraction of antioxidants, for example, polyphenols has been uniquely tended to enhancing their recuperation based on their yield and antioxidant potential through experimental design. Table 1.3 shows a portion of the cases of bioactive particles separated by UAE.

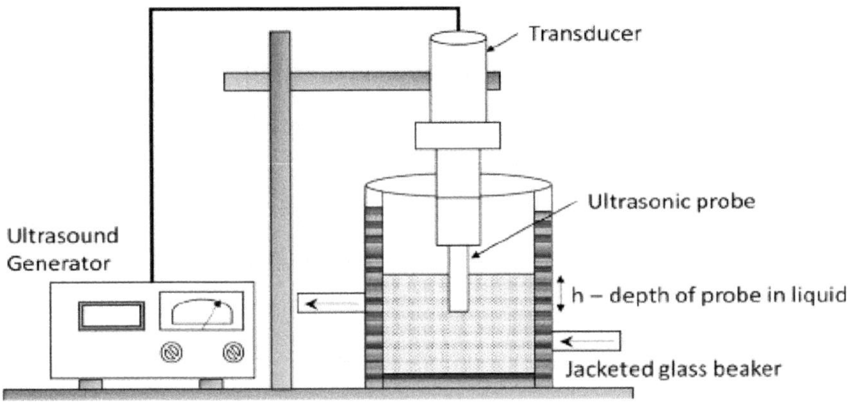

FIGURE 1.7 Schematic diagram of ultrasound-assisted extraction using a probe system.

1.5.2.2 PULSE ELECTRIC FIELD-ASSISTED EXTRACTION

The pulsed electric field (PEF) technique was perceived as valuable for enhancing the drying, dispersion, and pressing in the previous decade.[150,10,149] PEF is a rising innovation in recent years that has increased interest in the food industry for enhancing mass transfer operations.[123,81,44] The procedure depends on the utilization of external electric fields that instigate the electroporation of eukaryotic cell layers, improving the dispersion of solutes (Fig. 1.8). This permeabilization of cell membranes can be accomplished at moderate electric fields. PEF treatment includes the utilization of brief length (from μs to ms) electric field pulses of direct force (0.5–10 kV/cm) to plant tissues set between two anodes, causing the permeabilization not just of the cell membrane[19] but also of cell vacuoles[49] where a few metabolites are contained. PEF can destroy the layer structure of plant material thus expands mass exchange during extraction for improving extraction and diminishes extraction time. PEF has been connected to enhance arrival of intracellular components from plant cell by increasing permeability of cell membrane.[144] PEF application at a direct electric field (500 and 1000 V/cm; for 10^{-4}–10^{-2} s) is seemed to harm cell layer of plant tissue with little temperature increment.[49,86] Because of this reason, PEF can limit the debasement of heat-sensitive compounds.[4] PEF is additionally appropriate on plant materials as a pretreatment procedure preceding customary extraction to bring down extraction exertion.[94]

In particular, the upgradation in the extraction yield of phenolic compounds, anthocyanins, flavonoids, sugars, and proteins from food processing and agricultural by-products of various food commodities has been accounted for when PEF-assisted extraction with solvents has been utilized; some of them are mentioned in Table 1.3.

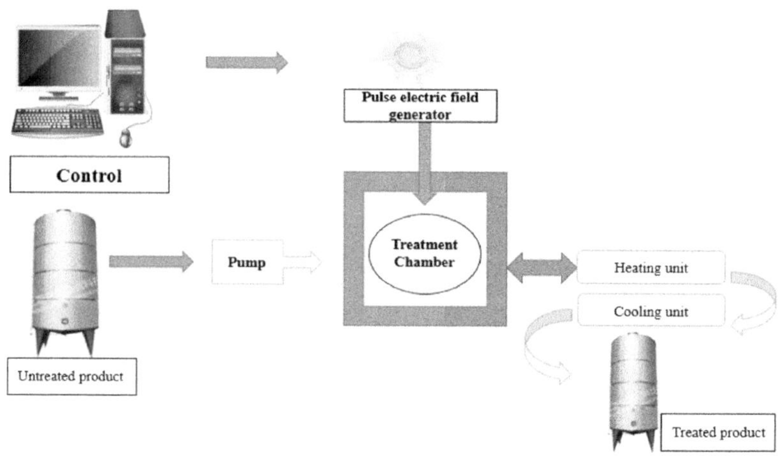

FIGURE 1.8 (See color insert.) Schematic diagram of pulsed electric field-assisted extraction.

1.5.2.3 SUPERCRITICAL FLUID EXTRACTION

Supercritical state is achieved when a substance is subjected to pressure and temperature beyond its critical point. Critical point is characterized as the characteristic pressure (Pc) and temperature (Tc) above which particular gas and fluid stages do not exist.[67] This is the most mechanically propelled extraction framework.[116] Supercritical fluid extraction (SFE) includes utilization of gases, typically CO_2, and compacting them into a thick fluid. This fluid is then pumped through a barrel containing the material to be separated. From that point, the concentrate-loaded fluid is drawn into a partition chamber where the concentrate is isolated from the gas and the gas is recuperated for reutilization (Fig. 1.9). Solvent properties of CO_2 can be controlled and balanced by shifting the pressure and temperature that one works at. The upsides of SFE are the adaptability it offers in pinpointing the constituents you need to separate from a given material and the way that your finished result has for all intents and purposes—no solvent deposits left in it. Supercritical carbon dioxide (SC–CO_2) is an appealing contrasting option to natural solvents since it is economical, nonexplosive, nontoxic, and has the capacity to solubilize lipophilic substances, and can be effortlessly expelled from the last items.[129,156,155] The drawback is that this innovation is very costly. There are numerous different gases and fluids that are exceedingly productive as extraction solvents when put under pressure.[116]

Nowadays, SFE is tremendously utilized as a part of numerous modern applications including coffee decaffeination, unsaturated fat refining and the extraction of fundamental oils and flavors from characteristic sources with potential use in functional foods and nutraceuticals.[39] This technique is a vital option over regular conventional extraction strategies utilizing organic solvents for extricating biologically dynamic compounds.[155] In any case, to build up an effective SFE, a few components should be taken into mind including SFs, cosolvents, crude materials, and extraction conditions for the extraction of a specific compound of interest in order to augment the extraction.[24] A few investigations have depicted the extraction of various common bioactive compounds utilizing supercritical liquid in Table 1.3.

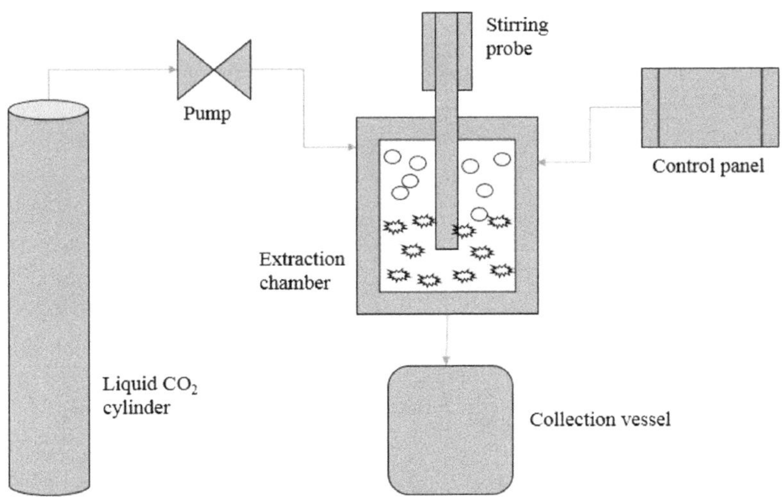

FIGURE 1.9 Scheme of a supercritical fluid extractor plant.

1.5.2.4 HIGH-PRESSURE-ASSISTED EXTRACTION

High-pressure processing is a food processing strategy that has indicated extraordinary possibilities in the food industry. Like heat treatment, high pressure can also be useful in denaturation of proteins, inactivation of microorganisms, and increasing the shelf life of processed items.[21,93] The utilization of high-pressure treatment in the extraction of bioactive compounds from plant material is a new strategy, which has been known as ultrahigh pressure extraction (UPE)[73] or high hydrostatic pressure extraction (HHPE).[159] The studies demonstrated that high-pressure system could abbreviate processing time, get higher extraction yields, utilization of lower power, have fewer impurities in the extraction fluid, and have no negative impact on the movement and conformation of bioactive segments.[33] Most importantly, this extraction method could be worked at ambient temperature with no thermal procedure, aside from the temperature rise occur because of the pressure.[32,132] UPE works at high pressure (normally 100–600 MPa) and low temperatures (generally up to 60°C) to extricate rapidly with low volumes of natural solvents and gives recuperations like other different techniques.[165] Parameters that essentially influence these recuperations are temperature, pressure, solvent, the number of cycles, extraction time, etc.[28] Every parameter can be advanced independently or by utilizing an experimental design (Fig. 1.10).

The pressure difference between inside and outside of the plant material system is extensive under HPE conditions. This pressure difference can pervade the solvent to move quickly through the broken films into plant material and enhance the mass transfer rate of solute or the rate of disintegration, which prompts reduced extraction time utilizing HPE, contrasted with ordinary extraction forms.[164] Besides, HPE can inactivate degrading enzymes, which may clarify the greater extraction yield and antioxidant activity as compared to other conventional extraction methods.[5] High pressure likewise can diminish the pH of the solvent during extraction and this decrease may improve the extraction of bioactive molecules because these are steadier at low pH.[93,78,33,13]

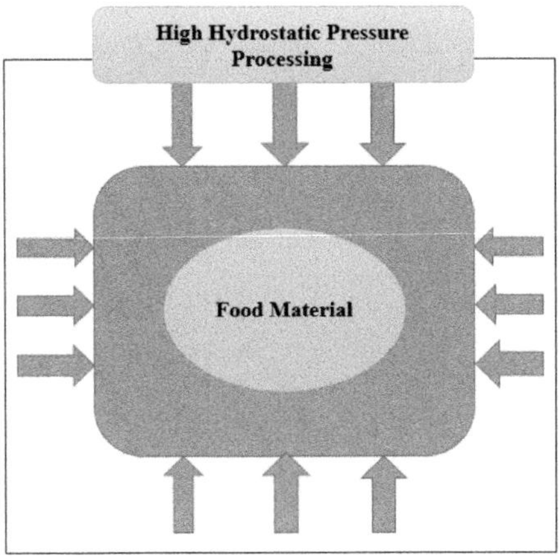

FIGURE 1.10 (See color insert.) Schematic diagram showing mechanism of high-pressure system.

TABLE 1.3 Biomolecules Extraction by Novel Nonthermal Extraction Methods.

Thermal method	Bioactive compound	Optimal conditions	Remarks	References
Ultrasound-assisted extraction	β-carotene from *Spirulina platensis* alga	Solvent: n-heptane, temperature = 30°C, electrical acoustic intensity = 167 W/cm^2	Ultrasonication showed 12 times increase in extraction yield of β-carotene (47.10%)	[43]
	Total phenols, antioxidant capacity, chlorogenic acid, caffeic acid, catechin, and epicatechin from carrot pomace	Solvent: ethanol 13–97%, time = 3–37 min, extraction temperature = 10–60°C	UAE decreases the extraction time and increases the extraction yield	[69]
	Fatty acids, b-sitosterol, a-tocopherol, squalene, total phenolics and carotene from palm pressed fiber	Solvent: ethanol, time= 2 h, temperature = 20 ± 2°C, electrical acoustic intensity = 36–204 W.cm^{-2}	Bioactive compounds possessing antioxidant activity were identified and quantified	[40]

TABLE 1.3 *(Continued)*

Thermal method	Bioactive compound	Optimal conditions	Remarks	References
Pulsed electric field-assisted extraction	Anthocyanins from purple-fleshed potato	Solvent: water 48% and ethanol 96%, time = 60–480 min, temperatures = 10–40°C, 5–35 pulses of 3 µs, 35 kV voltage	Extraction yield of anthocyanin increases by PEF at lower temperature with water as solvent	[121]
	Anthocyanins and flavonoids from plum and grape peels	Solvent: water, temperature = 70°C, 6 µs pulse width, 25 kV voltage	Chamber of larger diameter allows high number of pulses which helps in increasing the recovery of anthocyanins, flavonoids, and phenols from both fruits	[104]
	Anthocyanin from red cabbage	Solvent: water, 2.5 kV/cm electric field strength; 15 µs pulse	About 16–889 µg/mL anthocyanins were extracted, which are 2.15 times enhanced by PEF in water	[51]
Supercritical fluid extraction	Oleic acid, sterols and tocopherols from *Moringa oleifera*	Tempperature = 30°C, time = 300 min, pressure = 350 W, solvent/solid ratio = 1329.77 g CO_2/g	About 72.26–74.72% oleic acid (a health promoting fatty acid) is extracted	[127]
	Tocopherol from *Chenopodium quinoa*	Temperature = 130°C, time = 55–180 min, pressure = 185 W, solvent/solid ratio = 8.02–67.5 g CO_2/g	With comparison to hexane extraction, vitamin E yield increases four times by SFE	[120]
	Linolenic acid from *Gynostemma pentaphyllum*	Temperature = 43°C, time = 160 min, pressure = 320 W, solvent/solid ratio = 1483.11 g CO_2/g	About 95.69% of unsaturated fatty acids content was produced as compare to conventional methods	[154]

TABLE 1.3 *(Continued)*

Thermal method	Bioactive compound	Optimal conditions	Remarks	References
High pressure-assisted extraction	Total phenolic, flavonoid, tannin and Antioxidant activity from fig	Solvent = ethanol 15–48%, time = 18 and 29 min, pressure = 600 MPa	High pressure led increase of 8–11% of total phenolics, flavonoids and tannins content and 8–13% of antioxidant activity when compared to extracts performed at 0.1 MPa	[8]
	Catechins from green tea	Solvent: ethanol 50%, time = 15 min, temperature = 20°C, pressure = 600 MPa	Extraction yield was greater in shorter time as compare to conventional reflux extraction method	[73]
	Lycopene from tomato waste	Solvent: ethyl lactate, time = 10 min, pressure 700 MPa	HP-assisted extraction led to higher yields (from 2% to 64%	[140]

PEF, pulsed electric field; SFE, supercritical fluid extraction; UAE, ultrasound-assisted extraction.

1.6 SUMMARY

The efficiencies of either conventional or novel extraction techniques typically rely upon the basic input parameters: understanding the idea of plant lattice; chemistry of bioactive segments, and capability to up-scale them. Besides, choice of reasonable extraction process and advancement of different parameters are significant for upscaling purposes, that is, from bench scale to pilot plant level. Utilization of green extraction systems, for example, UAE,[105,92] MAE,[54] and SFE[113] has been quickly and constantly expanding all-inclusive for phytochemical handling of therapeutic plants as these strategies are quick when contrasted with conventional techniques. Likewise, these methods are ecologically amicable as far as energy and solvent utilization are concerned. Yield is likewise practically identical to traditional extraction and in some cases, it is much higher. Green extraction systems other than enhancing the yield and quality would have the capacity to save time and energy.

KEYWORDS

- **bioactive molecules**
- **terpenes**
- **alkaloids**
- **polyphenols**
- **Soxhlet**
- **liquid–liquid extraction**
- **maceration**
- **hydrodistillation**
- **ultrasound**
- **microwave**
- **supercritical fluid extraction**
- **pulsed electric field**

REFERENCES

1. Aamir, M.; Jittanit, W. Ohmic Heating Treatment for Gac Aril Oil Extraction: Effects on Extraction Efficiency, Physical Properties and Some Bioactive Compounds. *Innov. Food Sci. Emerging Technol.* **2017,** *41,* 224–234.
2. Abu-Samra, A.; Morris, J. S.; Koirtyohann, S. Wet Ashing of Some Biological Samples in a Microwave Oven. *Anal. Chem.* **1975,** *47* (8), 1475–1477.
3. Adam, F.; Abert-Vian, M.; Peltier, G.; Chemat, F. "Solvent-free" Ultrasound-Assisted Extraction of Lipids from Fresh Microalgae Cells: A Green, Clean and Scalable Process. *Biores. Technol.* **2012,** *114,* 457–465.
4. Ade-Omowaye, B.; Angersbach, A.; Taiwo, K.; Knorr, D. Use of Pulsed Electric Field Pre-treatment to Improve Dehydration Characteristics of Plant Based Foods. *Trends Food Sci. Technol.* **2001,** *12* (8), 285–295.
5. Ahmed, J.; Ramaswamy, H. S. Changes in Colour During High Pressure Processing of Fruits and Vegetables. *Stewart Postharvest Rev.* **2006,** *5* (9), 1–8.
6. Ajikumar, P. K.; Tyo, K.; Carlsen, S.; Mucha, O.; Phon, T. H.; Stephanopoulos, G. Terpenoids: Opportunities for Biosynthesis of Natural Product Drugs Using Engineered Microorganisms. *Mol. Pharmaceut.* **2008,** *5* (2), 167–190.

7. Aksel, E.; Jones, J. L. Advances in Lead-Free Piezoelectric Materials for Sensors and Actuators. *Sensors* **2010,** *10* (3), 1935–1954.
8. Alexandre, E. M.; Araújo, P.; Duarte, M. F.; de Freitas, V.; Pintado, M.; Saraiva, J. A. High-Pressure Assisted Extraction of Bioactive Compounds from Industrial Fermented Fig by-Product. *J. Food Sci. Technol.* **2017,** *54* (8), 2519–2531.
9. Alupului, A.; Calinescu, I.; Lavric, V. Microwave Extraction of Active Principles from Medicinal Plants. *UPB Sci. Bull. Series B,* **2012,** *74* (2), 1454–2331.
10. Angersbach, A.; Heinz, V.; Knorr, D. Effects of Pulsed Electric Fields on Cell Membranes in Real Food Systems. *Innovat. Food Sci. Emerg. Technol.* **2000,** *1* (2), 135–149.
11. Azmir, J.; Zaidul, I.; Rahman, M.; Sharif, K.; Mohamed, A.; Sahena, F.; Jahurul, M.; Ghafoor, K.; Norulaini, N.; Omar, A. Techniques for Extraction of Bioactive Compounds from Plant Materials: A Review. *J. Food Eng.* **2013,** *117* (4), 426–436.
12. Baretić, M. Targets for Medical Therapy in Obesity. *Digest. Dis.* **2012,** *30* (2), 168–172.
13. Basak, S.; Ramaswamy, H. Effect of High Pressure Processing on the Texture of Selected Fruits and Vegetables. *J. Text. Stud.* **1998,** *29* (5), 587–601.
14. Baxter, H.; Harborne, J. B.; Moss, G. P. *Phytochemical Dictionary: A Handbook of Bioactive Compounds From Plants*. CRC Press: Padstow, UK, 1998.
15. Bernal, W.; Auzinger, G.; Dhawan, A.; Wendon, J. Acute Liver Failure. *Lancet* **2010,** *376* (9736), 190–201.
16. Bimakr, M.; Rahman, R. A.; Taip, F. S.; Ganjloo, A.; Salleh, L. M.; Selamat, J.; Hamid, A.; Zaidul, I. Comparison of Different extraction Methods for the Extraction of Major Bioactive Flavonoid Compounds from Spearmint (*Mentha spicata* l.) Leaves. *Food Bioprod. Process.* **2011,** *89* (1), 67–72.
17. Boussetta, T.; Raad, H.; Lettéron, P.; Gougerot-Pocidalo, M. -A.; Marie, J. -C.; Driss, F.; El-Benna, J. Punicic Acid a Conjugated Linolenic Acid Inhibits TNFα-Induced Neutrophil Hyperactivation and Protects from Experimental Colon Inflammation in Rats. *PloS One* **2009,** *4* (7), e6458.
18. Boutekedjiret, C.; Bentahar, F.; Belabbes, R.; Bessiere, J. Extraction of Rosemary Essential Oil by Steam Distillation and Hydrodistillation. *Flav. Fragr. J.* **2003,** *18* (6), 481–484.
19. Bouzrara, H.; Vorobiev, E. Solid–Liquid Expression of Cellular Materials Enhanced by Pulsed Electric Field. *Chem. Eng. Process. Proc. Intensif.* **2003,** *42* (4), 249–257.
20. Brandt, K.; Christensen, L. P.; Hansen-Møller, J.; Hansen, S. L.; Haraldsdottir, J.; Jespersen, L.; Purup, S.; Kharazmi, A.; Barkholt, V.; Frøkiær, H. Health Promoting Compounds in Vegetables and Fruits: A Systematic Approach for Identifying Plant Components with Impact on Human Health. *Trends Food Sci. Technol.* **2004,** *15* (7), 384–393.
21. Buzrul, S.; Alpas, H. 18 Treatment of Foods Using High Hydrostatic Pressure. *Progr. Food Preserv.* **2012,** 375, 375–388.
22. Callisto, M.; Gonçalves Jr, J. F.; Graça, M. A. Leaf Litter as a Possible Food Source for Chironomids (diptera) in Brazilian and Portuguese Headwater Streams. *Revista Brasileira Zoologia,* **2007,** *24* (2), 442–448.
23. Cantwell, F. F.; Losier, M. Liquid—Liquid Extraction. *Compr. Anal. Chem.* **2002,** *37,* 297–340.

24. Cape, S. P.; Villa, J. A.; Huang, E. T.; Yang, T. -H.; Carpenter, J. F.; Sievers, R.E. Preparation of Active Proteins, Vaccines and Pharmaceuticals as Fine Powders Using Supercritical or Near-Critical Fluids. *Pharmaceut. Res.* **2008,** *25* (9), 1967–1990.
25. Castro, C.; Zuluaga, R.; Álvarez, C.; Putaux, J. -L.; Caro, G.; Rojas, O. J.; Mondragon, I.; Gañán, P. Bacterial Cellulose Produced by a New Acid-Resistant Strain of Gluconacetobacter Genus. *Carbohydr. Polym.* **2012,** *89* (4), 1033–1037.
26. Castro, I.; Teixeira, J.; Salengke, S.; Sastry, S.; Vicente, A. Ohmic Heating of Strawberry Products: Electrical Conductivity Measurements and Ascorbic Acid Degradation Kinetics. *Innovat. Food Sci. Emerg. Technol.* **2004,** *5* (1), 27–36.
27. International Commission on Education for the Twenty-first Century, *Learning the Treasure Within: Report to UNESCO of the International Commission on Education for the Twenty-First Century: Highlights*. UNESCO Pub. **1996**.
28. Chen, K. G.; Valencia, J. C.; Gillet, J. P.; Hearing, V. J.; Gottesman, M.M. Involvement of ABC Transporters in Melanogenesis and the Development of Multidrug Resistance of Melanoma. *Pigment Cell Melanoma Res.* **2009,** *22* (6), 740–749.
29. Cheung, Y. -C.; Siu, K. -C.; Liu, Y. -S.; Wu, J. -Y. Molecular Properties and Antioxidant Activities of Polysaccharide–Protein Complexes from Selected Mushrooms by Ultrasound-Assisted Extraction. *Process Biochem.* **2012,** *47* (5), 892–895.
30. Cheynier, V. Polyphenols in Foods Are More Complex Than Often Thought. *Am. J. Clinical Nutr.* **2005,** *81* (1), 223S–229S.
31. Clement, R.; Hao, C. Liquid–Liquid Extraction: Basic Principles and Automation. *Comprehensive Sampling and Sample Preparation; Academic Press: Oxford,* 2012; pp 51–63.
32. Corrales, M.; García, A. F.; Butz, P.; Tauscher, B. Extraction of Anthocyanins from Grape Skins Assisted by High Hydrostatic Pressure. *J. Food Eng.* **2009,** *90* (4), 415–421.
33. Corrales, M.; Toepfl, S.; Butz, P.; Knorr, D.; Tauscher, B. Extraction of Anthocyanins from Grape By-Products Assisted by Ultrasonics, High Hydrostatic Pressure or Pulsed Electric Fields: A Comparison. *Innovat. Food Sci. Emerg. Technol.* **2008,** *9* (1), 85–91.
34. Cravotto, G.; Boffa, L.; Mantegna, S.; Perego, P.; Avogadro, M.; Cintas, P. Improved Extraction of Vegetable Oils Under High-Intensity Ultrasound and/or Microwaves. *Ultrason. Sonochem.* **2008,** *15* (5), 898–902.
35. Crespo, M. P.; Yusty, M. L. Comparison of Supercritical Fluid Extraction and Soxhlet Extraction for the Determination of Aliphatic Hydrocarbons in Seaweed Samples. *Ecotoxicol. Environ. Safety.* **2006,** *64* (3), 400–405.
36. Croft, K. D. The Chemistry and Biological Effects of Flavonoids and Phenolic Acids. *Ann. NY Acad. Sci.* **1998,** *854* (1), 435–442.
37. Croteau, R.; Ketchum, R. E.; Long, R. M.; Kaspera, R.; Wildung, M. R. Taxol Biosynthesis and Molecular Genetics. *Phytochem. Rev.* **2006,** *5* (1), 75–97.
38. Da Porto, C.; Porretto, E.; Decorti, D. Comparison of Ultrasound-Assisted Extraction with Conventional Extraction Methods of Oil and Polyphenols from Grape (*Vitis vinifera* L.) seeds. *Ultrason. Sonochem.* **2013,** *20* (4), 1076–1080.
39. Daintree, L.; Kordikowski, A.; York, P. Separation Processes for Organic Molecules Using Scf Technologies. *Adv. Drug Delivery Rev.* **2008,** *60* (3), 351–372.

40. Dal Pra, V.; Lunelli, F. C.; Vendruscolo, R. G.; Martins, R.; Wagner, R.; Lazzaretti, A. P., Jr.; Freire, D. M.; Alexandri, M.; Koutinas, A.; Mazutti, M. A., et al. Ultrasound-Assisted Extraction of Bioactive Compounds from Palm Pressed Fiber with High Antioxidant and Photoprotective Activities. *Ultrason. Sonochem.* **2017,** *36,* 362–366.
41. Day, L.; Seymour, R. B.; Pitts, K. F.; Konczak, I.; Lundin, L. Incorporation of Functional Ingredients into Foods. *Trends Food Sci. Technol.* **2009,** *20* (9), 388–395.
42. De Castro, M. L.; Priego-Capote, F. Soxhlet Extraction: Past and Present Panacea. *J. Chromatogr. A* **2010,** *1217* (16), 2383–2389.
43. Dey, S.; Rathod, V. K. Ultrasound Assisted Extraction of Beta-Carotene from Spirulina Platensis. *Ultrason. Sonochemi.* **2013,** *20* (1), 271–276.
44. Donsì, F.; Ferrari, G.; Pataro, G. Applications of Pulsed Electric Field Treatments for the Enhancement of Mass Transfer from Vegetable Tissue. *Food Eng. Rev.* **2010,** *2* (2), 109–130.
45. Duraipandiyan, V.; Ayyanar, M.; Ignacimuthu, S. Antimicrobial Activity of Some Ethnomedicinal Plants Used by Paliyar Tribe from Tamil Nadu, india. *BMC Complement. Alternat. Med.* **2006,** *6* (1), 35.
46. El Darra, N.; Grimi, N.; Vorobiev, E.; Louka, N.; Maroun, R. Extraction of polyphenols from Red Grape Pomace Assisted by Pulsed Ohmic Heating. *Food Bioprocess Technol.* **2013,** *6* (5), 1281–1289.
47. Esclapez, M.; García-Pérez, J.; Mulet, A.; Cárcel, J. Ultrasound-Assisted Extraction of Natural Products. *Food Eng. Rev.* **2011,** *3* (2), 108.
48. Farnsworth, N. R.; Akerele, O.; Bingel, A. S.; Soejarto, D. D.; Guo, Z. Medicinal Plants in Therapy. *Bull. World Health Organiz.* **1985,** *63* (6), 965.
49. Fincan, M.; Dejmek, P. In situ Visualization of the Effect of a Pulsed Electric Field on Plant Tissue. *J. Food Eng.* **2002,** *55* (3), 223–230.
50. Firn, R. *Nature's Chemicals: The Natural Products That Shaped Our World.* Oxford University Press: Oxford, United Kingdom, 2010.
51. Gachovska, T.; Cassada, D.; Subbiah, J.; Hanna, M.; Thippareddi, H.; Snow, D. Enhanced Anthocyanin Extraction from Red Cabbage Using Pulsed Electric Field Processing. *J. Food Sci.* **2010,** *75* (6), E323–329.
52. Gani, A.; Wani, S.; Masoodi, F.; Hameed, G. Whole-Grain Cereal Bioactive Compounds and Their Health Benefits: A Review. *J. Food Process. Technol.* **2012,** *3* (3), 146–156.
53. Ganzler, K.; Salgo, A. Microwave-Extraction-A New Method Superseding Traditional Soxhlet Extraction. *Z. Lebensmitteluntersuchung Und-Forschung A.* **1987,** *184* (4), 274–276.
54. Ganzler, K.; Salgo, A.; Valko, K. Microwave extraction: A Novel Sample Preparation Method for Chromatography. *J. Chromatogr. A.* **1986,** *371,* 299–306.
55. Gao, T.; Zhang, M.; Fang, Z.; Zhong, Q. Optimization of Microwave-Assisted Extraction of Flavonoids from Young Barley Leaves. *Int. Agrophys.* **2017,** *31* (1), 45–52.
56. Garcia-Ayuso, L.; Sánchez, M.; Fernández de Alba, A.; Luque de Castro, M. Focused Microwave-Assisted Soxhlet: An Advantageous Tool for Sample Extraction. *Anal. Chem.* **1998,** *70* (11), 2426–2431.
57. Gaur, A.; Jewell, D. A.; Liang, Y.; Ridzon, D.; Moore, J. H.; Chen, C.; Ambros, V. R.; Israel, M. A. Characterization of Microrna Expression Levels and Their Biological Correlates in Human Cancer Cell Lines. *Cancer Res.* **2007,** *67* (6), 2456–2468.

58. Ghafoor, A.; Hay, I. D.; Rehm, B. H. Role of Exopolysaccharides in Pseudomonas Aeruginosa Biofilm Formation and Architecture. *Appl. Environ. Microbiol.* **2011,** *77* (15), 5238–5246.
59. Ghafoor, K.; AL-Juhaimi, F. Y.; Choi, Y. H. Supercritical Fluid Extraction of Phenolic Compounds and Antioxidants from Grape (*Vitis labrusca* b.) Seeds. *Plant Foods Human Nutr.* **2012,** *67* (4), 407–414.
60. Govindarajan, V.; Stahl, W. H. Turmeric—Chemistry, Technology, and Quality. *Crit. Rev. Food Sci. Nutr.* **1980,** *12* (3), 199–301.
61. Grigonis, D.; Venskutonis, P.; Sivik, B.; Sandahl, M.; Eskilsson, C.S. Comparison of Different Extraction Techniques for Isolation of Antioxidants from Sweet Grass (*Hierochloe odorata*). *J. Supercritical Fluids.* **2005,** *33* (3), 223–233.
62. Hekmat, S.; Soltani, H.; Reid, G. Growth and Survival of *Lactobacillus reuteri* rc-14 and *Lactobacillus Rhamnosus* gr-1 in Yogurt for Use as a Functional Food. *Innovat. Food Sci. Emerg. Technol.* **2009,** *10* (2), 293–296.
63. Hernández-Alonso, M. D.; Fresno, F.; Suárez, S.; Coronado, J. M. Development of Alternative Photocatalysts to Tio 2: Challenges and Opportunities. *Energy Environ. Sci.* **2009,** *2* (12), 1231–1257.
64. Hii, T. M.; Lee, H. K. Liquid–Liquid Extraction in Environmental Analysis. In *Handbook of Sample Preparation;* 2010; Hoboken, New Jersey, pp 39–51.
65. Hooper, L.; Cassidy, A. A Review of the Health Care Potential of Bioactive Compounds. *J. Sci. Food Agricult.* **2006**, *86* (12), 1805–1813.
66. Huang, Z.; Wang, B.; Eaves, D. H.; Shikany, J. M.; Pace, R.D. Phenolic Compound Profile of Selected Vegetables Frequently Consumed by African Americans in the Southeast United States. *Food Chem.* **2007**, *103* (4), 1395–1402.
67. Inczedy, J.; Lengyel, T.; Ure, A. M.; Gelencsér, A.; Hulanicki, A. Compendium of Analytical Nomenclature. *The Orange Book,* 3rd edn.; Blackwell Science: Hoboken, New Jersey, 1998.
68. Ingle, R.; Magar, D. Heterocyclic Chemistry of Benzimidazoles and Potential Activities of Derivatives. *Int. J. Drug Res. Technol.* **2017,** *1* (1), 7.
69. Jabbar, S.; Abid, M.; Wu, T.; Hashim, M. M.; Saeeduddin, M.; Hu, B.; Lei, S.; Zeng, X. Ultrasound-Assisted Extraction of Bioactive Compounds and Antioxidants from Carrot Pomace: A Response Surface Approach. *J. Food Process. Preserv.* **2015,** *39* (6), 1878–1888.
70. Jain, T.; Jain, V.; Pandey, R.; Vyas, A.; Shukla, S. Microwave Assisted Extraction for Phytoconstituents–An Overview. *Asian J. Res. Chem.* **2009,** *2* (1), 19–25.
71. James, C. *Global Status of Commercialized Biotech/gm Crops: 2010.* International Service for the Acquisition of Agri-Biotech Applications (ISAAA) Ithaca, **2010**.
72. Joana Gil-Chávez, G.; Villa, J. A.; Fernando Ayala-Zavala, J.; Basilio Heredia, J.; Sepulveda, D.; Yahia, E. M.; González-Aguilar, G. A. Technologies for Extraction and Production of Bioactive compounds to be Used as Nutraceuticals and Food Ingredients: An Overview. *Comprehens. Rev. Food Sci. Food Saf.* **2013,** *12* (1), 5–23.
73. Jun, X.; Shuo, Z.; Bingbing, L.; Rui, Z.; Ye, L.; Deji, S.; Guofeng, Z. Separation of Major Catechins from Green Tea by Ultrahigh Pressure Extraction. *Int. J. Pharmaceut.* **2010,** *386* (1), 229–231.

74. Kahkonen, M . P.; Hopia, A. I.; Vuorela, H.J.; Rauha, J. -P.; Pihlaja, K.; Kujala, T. S.; Heinonen, M. Antioxidant Activity of Plant Extracts Containing Phenolic Compounds. *J. Agricul. Food Chem.* **1999**, *47* (10), 3954–3962.
75. Karki, B.; Lamsal, B. P.; Jung, S.; van Leeuwen, J. H.; Pometto, A. L.; Grewell, D.; Khanal, S.K. Enhancing Protein and Sugar Release from Defatted Soy Flakes Using Ultrasound Technology. *J. Food Eng.* **2010,** *96* (2), 270–278.
76. Kaufmann, B.; Christen, P. Recent Extraction Techniques for Natural Products: Microwave-Assisted Extraction and Pressurised Solvent Extraction. *Phytochem. Anal.* **2002,** *13* (2), 105–113.
77. Kaur, R.; Arora, S. Alkaloids-Important Therapeutic Secondary Metabolites of Plant Origin. *J. Crit. Rev.* **2015,** *2* (3), 1–8.
78. Khosravi-Darani, K. Research Activities on Supercritical Fluid Science in Food Biotechnology. *Crit. Rev. Food Sci. Nutr.* **2010,** *50* (6), 479–488.
79. Kitts, D. D.; Weiler, K. Bioactive Proteins and Peptides from Food Sources. Applications of Bioprocesses Used in Isolation and Recovery. *Curr. Pharmaceut. Des.* **2003,** *9* (16), 1309–1323.
80. Knirsch, M. C.; Dos Santos, C. A.; de Oliveira Soares, A. A. M.; Penna, T. C. V. Ohmic Heating–A Review. *Trends Food Sci. Technol.* **2010,** *21* (9), 436–441.
81. Knorr, D.; Froehling, A.; Jaeger, H.; Reineke, K.; Schlueter, O.; Schoessler, K. Emerging Technologies in Food Processing. *Ann. Rev. Food Sci. Technol.* **2011,** *2*203–235.
82. Kolak, M.; Yki-Järvinen, H.; Kannisto, K.; Tiikkainen, M.; Hamsten, A.; Eriksson, P.; Fisher, R. M. Effects of Chronic Rosiglitazone Therapy on Gene Expression in Human Adipose Tissue In Vivo in Patients with Type 2 Diabetes. *J. Clin. Endocrinol. Metabol.* **2006,** *92* (2), 720–724.
83. Kris-Etherton, P.; Lefevre, M.; Beecher, G.; Gross, M.; Keen, C.; Etherton, T. Bioactive Compounds in Nutrition and Health-Research Methodologies for Establishing Biological Function: The Antioxidant and Anti-Inflammatory Effects of Flavonoids on Atherosclerosis. *Ann. Rev. Nutr.* **2004,** *24,* 511–538.
84. Lakkakula, N. R.; Lima, M.; Walker, T. Rice Bran Stabilization and Rice Bran Oil Extraction Using Ohmic Heating. *Biores. Technol.* **2004,** *92* (2), 157–161.
85. Lange, B. M.; Ahkami, A. Metabolic Engineering of Plant Monoterpenes, Sesquiterpenes and Diterpenes—Current Status and Future Opportunities. *Plant Biotechnol. J.* **2013,** *11* (2), 169–196.
86. Lebovka, N.; Bazhal, M.; Vorobiev, E. Estimation of Characteristic Damage Time of Food Materials in Pulsed-Electric Fields. *J. Food Eng.* **2002,** *54* (4), 337–346.
87. Lee, N. Y.; Yunus, M. A. C.; Idham, Z.; Ruslan, M. S. H.; Aziz, A. H. A.; Irwansyah, N. Extraction and Identification of Bioactive Compounds from Agarwood Leaves. *IOP Conference Series: Mater. Sci. Eng.* **2016,** *162* (1), 012028.
88. Leighton, T. G. What Is Ultrasound? *Progr. Biophys. Mol. Biol.* **2007**, *93* (1), 3–83.
89. Letellier, M.; Budzinski, H. Microwave Assisted Extraction of Organic Compounds. *Analusis.* **1999,** *27* (3), 259–270.
90. Lima, G. P. P.; Vianello, F.; Corrêa, C. R.; Campos, R. A.d. S.; Borguini, M. G. Polyphenols in Fruits and Vegetables and Its Effect on Human Health. *Food Nutr. Sci.* **2014**, *05* (11), 1065–1082.

91. Lin, H. S.; Yue, B. D.; Ho, P. C. Determination of Pterostilbene in Rat Plasma by a Simple HPLC-UV Method and Its Application in Pre-Clinical Pharmacokinetic Study. *Biomed. Chromatogr.* **2009,** *23* (12), 1308–1315.
92. Liu, C. -T.; Wu, C. -Y.; Weng, Y. -M.; Tseng, C. -Y. Ultrasound-Assisted Extraction Methodology as a Tool to Improve the Antioxidant Properties of Herbal Drug Xiao-Chia-Hu-Tang. *J. Ethnopharmacol.* **2005,** *99* (2), 293–300.
93. López-Fandiño, R. High Pressure-Induced Changes in Milk Proteins and Possible Applications in Dairy Technology. *Int. Dairy J.* **2006,** *16* (10), 1119–1131.
94. López, N.; Puértolas, E.; Condón, S.; Raso, J.; Alvarez, I. Enhancement of the Extraction of Betanine from Red Beetroot by Pulsed Electric Fields. *J. Food Eng.* **2009,** *90* (1), 60–66.
95. Loypimai, P.; Moonggarm, A.; Chottanom, P. Effects of Ohmic Heating on Lipase Activity, Bioactive Compounds and Antioxidant Activity of Rice Bran. *Austr. J. Basic Applied Sci.* **2009,** *3* (4), 3642–3652.
96. Loypimai, P.; Moongngarm, A.; Chottanom, P.; Moontree, T. Ohmic Heating-Assisted Extraction of Anthocyanins from Black Rice Bran to Prepare a Natural Food Colourant. *Innovative Food Sci. Emerg. Technol.* **2015,** *27,* 102–110.
97. Lusas, E.; Rhee, K.; Watkins, L. Processing Methods to Modify Soybean Components. **1988**.
98. Manach, C.; Scalbert, A.; Morand, C.; Rémésy, C.; Jiménez, L. Polyphenols: Food Sources and Bioavailability. *Am. J. Clin. Nutr.* **2004**, *79* (5), 727–747.
99. Manach, C.; Williamson, G.; Morand, C.; Scalbert, A.; Rémésy, C. Bioavailability and Bioefficacy of Polyphenols in Humans. I. Review of 97 Bioavailability Studies. *American J. Clin. Nutr.* **2005,** *81* (1), 230S–242S.
100. Marella, A.; Tanwar, O. P.; Saha, R.; Ali, M. R.; Srivastava, S.; Akhter, M.; Shaquiquzzaman, M.; Alam, M. M. Quinoline: A Versatile Heterocyclic. *Saudi Pharmaceut. J.* **2013,** *21* (1), 1–12.
101. Marette, S.; Roosen, J.; Blanchemanche, S.; Feinblatt-Mélèze, E. Functional Food, Uncertainty and Consumers' Choices: A Lab Experiment with Enriched Yoghurts for Lowering Cholesterol. *Food Policy* **2010,** *35* (5), 419–428.
102. Martinez, F. O.; Sica, A.; Mantovani, A.; Locati, M. Macrophage Activation and Polarization. *Front. Biosci.: J. Virt. Libr.* **2008,** *13,* 453–461.
103. Md Othman, S. N. A.; Hassan, M. A.; Nahar, L.; Basar, N.; Jamil, S.; Sarker, S. D. Essential Oils from the Malaysian Citrus (Rutaceae) Medicinal Plants. *Medicines* **2016,** *3* (2), 13.
104. Medina-Meza, I. G.; Barbosa-Cánovas, G. V. Assisted Extraction of Bioactive Compounds from Plum and Grape Peels by Ultrasonics and Pulsed Electric Fields. *J. Food Eng.* **2015,** *166,* 268–275.
105. Merry, A. H.; Schouten, L. J.; Goldbohm, R. A.; van den Brandt, P. A. Body Mass Index, Height and Risk of Adenocarcinoma of the Oesophagus and Gastric Cardia: A Prospective Cohort Study. *Gut* **2007**, *56* (11), 1503–1511
106. Mertz, C.; Gancel, A. L.; Gunata, Z.; Alter, P.; Dhuique-Mayer, C.; Vaillant, F.; Perez, A. M.; Ruales, J.; Brat, P. Phenolic Compounds, Carotenoids and Antioxidant Activity of Three Tropical Fruits. *J. Food Compos. Anal.* **2009,** *22* (5), 381–387.

107. Miró, M.; Hartwell, S. K.; Jakmunee, J.; Grudpan, K.; Hansen, E. H. Recent Developments in Automatic Solid-Phase Extraction with Renewable Surfaces Exploiting Flow-Based Approaches. *TrAC Trends Anal. Chem.* **2008,** *27* (9), 749–761.
108. Möller, N. P.; Scholz-Ahrens, K. E.; Roos, N.; Schrezenmeir, J. Bioactive Peptides and Proteins from Foods: Indication for Health Effects. *Eur. J. Nutr.* **2008**, *47* (4), 171–182.
109. Nair, D. P.; Podgórski, M.; Chatani, S.; Gong, T.; Xi, W.; Fenoli, C. R.; Bowman, C. N. The Thiol-Michael Addition Click Reaction: A Powerful and Widely Used Tool in Materials Chemistry. *Chem. Mater.* **2014,** *26* (1), 724–744.
110. Ncube, N.; Afolayan, A.; Okoh, A. Assessment Techniques of Antimicrobial Properties of Natural Compounds of Plant Origin: Current Methods and Future Trends. *Afr. J. Biotechnol.* **2008,** *7* (12), 1797–1806.
111. Nobili, S.; Lippi, D.; Witort, E.; Donnini, M.; Bausi, L.; Mini, E.; Capaccioli, S. Natural Compounds for Cancer Treatment and Prevention. *Pharmacolog. Res.* **2009,** *59* (6), 365–378.
112. Ohtani, K.; Mizutani, K.; Kasai, R.; Tanaka, O. Selective Cleavage of Ester Type Glycoside-Linkages and Its Application to Structure Determination of Natural Oligoglycosides. *Tetrahed. Lett.* **1984,** *25* (40), 4537–4540.
113. Ollanketo, M.; Peltoketo, A.; Hartonen, K.; Hiltunen, R.; Riekkola, M. -L. Extraction of Sage (salvia officinalis l.) by Pressurized Hot Water and Conventional Methods: Antioxidant Activity of the Extracts. *Eur. Food Res. Technol.* **2002,** *215* (2), 158–163.
114. Palaniappan, S.; Sastry, S. K. Electrical Conductivities of Selected Solid Foods During Ohmic Heating. *J. Food Process Eng.* **1991,** *14* (3), 221–236.
115. Paré, J. J.; Bélanger, J. M.; Stafford, S. S. Microwave-Assisted Process (Map™): A New Tool for the Analytical Laboratory. *TrAC Trends Anal. Chem.* **1994,** *13* (4), 176–184.
116. Patil, P.; Shettigar, R. An Advancement of Analytical Techniques in Herbal Research. *J. Adv. Sci. Res.* **2010,** *1* (1), 8–14.
117. Pereira, R. N.; Rodrigues, R. M.; Genisheva, Z.; Oliveira, H.; de Freitas, V.; Teixeira, J. A.; Vicente, A. A. Effects of Ohmic Heating on Extraction of Food-Grade Phytochemicals from Colored Potato. *LWT - Food Sci. Technol.* **2016,** *74,* 493–503.
118. Plaza, M.; Cifuentes, A.; Ibáñez, E. In the Search of New Functional Food Ingredients from Algae. *Trends Food Sci. Technol.* **2008,** *19* (1), 31–39.
119. Pollier, J.; Goossens, A. Oleanolic Acid. *Phytochemistry* **2012,** *77,* 10–15.
120. Przygoda, K.; Wejnerowska, G. Extraction of Tocopherol-Enriched Oils from Quinoa Seeds by Supercritical Fluid Extraction. *Industr. Crops Prod.* **2015,** *63,* 41–47.
121. Puertolas, E.; Cregenzan, O.; Luengo, E.; Alvarez, I.; Raso, J. Pulsed-Electric-Field-Assisted Extraction of Anthocyanins from Purple-Fleshed Potato. *Food Chem.* **2013,** *136* (3–4), 1330–1336.
122. Puértolas, E.; López, N.; Condón, S.; Álvarez, I.; Raso, J. Potential Applications of Pef to Improve Red Wine Quality. *Trends Food Sci. Technol.* **2010,** *21* (5), 247–255.
123. Puértolas, E.; Luengo, E.; Álvarez, I.; Raso, J. Improving Mass Transfer to Soften Tissues by Pulsed Electric Fields: Fundamentals and Applications. *Ann. Rev. Food Sci. Technol.* **2012,** *3,* 263–282.

124. Qu, W.; Ma, H.; Jia, J.; He, R.; Luo, L.; Pan, Z. Enzymolysis Kinetics and Activities of Ace Inhibitory Peptides from Wheat Germ Protein Prepared with SFP Ultrasound-Assisted Processing. *Ultrason. Sonochem.* **2012,** *19* (5), 1021–1026.
125. Rassem, H.; Nour, A.; Yunus, R. Gc-ms Analysis of Bioactive Constituents of Hibiscus Flower. *Austral. J. Basic Appl. Sci.* **2017**, *11,* 91–97.
126. Rios, J.; Recio, M. Medicinal Plants and Antimicrobial Activity. *J. Ethnopharmacol.* **2005,** *100* (1), 80–84.
127. Ruttarattanamongkol, K.; Siebenhandl-Ehn, S.; Schreiner, M.; Petrasch, A. M. Pilot-Scale Supercritical Carbon Dioxide Extraction, Physico-Chemical Properties and Profile Characterization of *Moringa Oleifera* Seed Oil in Comparison with Conventional Extraction Methods. *Industr. Crops Prod.* **2014,** *58,* 68–77.
128. Safdar, M. N.; Kausar, T.; Jabbar, S.; Mumtaz, A.; Ahad, K.; Saddozai, A. A. Extraction and Quantification of Polyphenols from Kinnow (*Citrus Reticulate* l.) Peel using Ultrasound and Maceration Techniques. *J. Food Drug Anal.* **2017,** *25* (3), 488–500.
129. Sahena, F.; Zaidul, I.; Jinap, S.; Karim, A.; Abbas, K.; Norulaini, N.; Omar, A. Application of Supercritical CO_2 in Lipid Extraction–A Review. *J. Food Eng.* **2009,** *95* (2), 240–253.
130. Şahin, S.; Samli, R.; Tan, A. S. B.; Barba, F. J.; Chemat, F.; Cravotto, G.; Lorenzo, J. M. Solvent-Free Microwave-Assisted Extraction of Polyphenols From Olive Tree Leaves: Antioxidant and Antimicrobial Properties. *Molecules.* **2017,** *22* (7), 1056.
131. Sakr, M.; Liu, S. A Comprehensive Review on Applications of Ohmic Heating (oh). *Renew. Sustain. Energy Rev.* **2014,** *39,* 262–269.
132. Sánchez-Moreno, C.; Cano, M. P.; de Ancos, B.; Plaza, L.; Olmedilla, B.; Granado, F.; Martín, A. Consumption of High-Pressurized Vegetable Soup Increases Plasma Vitamin C and Decreases Oxidative Stress and Inflammatory Biomarkers in Healthy Humans. *J. Nutr.* **2004,** *134* (11), 3021–3025.
133. Sastry, S. Ohmic Heating and Moderate Electric Field Processing. *Revista Agaroquimica Tecnol. Alimentos* **2008,** *14* (5), 419–422.
134. Sastry, S. K.; Barach, J. T. Ohmic and Inductive Heating. *J. Food Sci.* **2000,** *65* (s8), 42–46.
135. Silva, L.; Nelson, D.; Drummond, M.; Dufossé, L.; Glória, M. Comparison of Hydrodistillation Methods for the Deodorization of Turmeric. *Food Res. Int.* **2005,** *38* (8), 1087–1096.
136. Singh, B.; Sharma, R. A. Plant terpenes: Defense Responses, Phylogenetic Analysis, Regulation and Clinical Applications. *3 Biotech* **2015,** *5* (2), 129–151.
137. Siro, I.; Kápolna, E.; Kápolna, B.; Lugasi, A. Functional Food. Product Development, Marketing and Consumer Acceptance—A Review. *Appetite* **2008**, *51* (3), 456–467.
138. Soria, A. C.; Villamiel, M. Effect of Ultrasound on the Technological Properties and Bioactivity of Food: A Review. *Trends Food Sci. Technol.* **2010**, *21* (7), 323–331.
139. Spring, A.; Hungerbuhler, S.; Sanwald, M.; Bossart, E. Extractor for Soxhlet Solid–Liquid Extraction. Google Patents; 1998.
140. Strati, I. F.; Gogou, E.; Oreopoulou, V. Enzyme and High Pressure Assisted Extraction of Carotenoids from Tomato Waste. *Food Bioprod. Process.* **2015,** *94,* 668–674.
141. Taiz, L.; Zeiger, E.; Møller, I. M.; Murphy, A. *Plant Physiology and Development.* Sinauer Associates, Incorporated. *Nat. Sci.* **2016,** *8,* 216–219

142. Tatke, P.; Rajan, M. Comparison of Conventional and Novel Extraction Techniques for the Extraction of Scopoletin from Convolvulus Pluricaulis. *Ind. J. Pharmaceut. Educat. Res.* **2014,** *48* (1), 27–31.
143. Tiwari, S. Plants: A Rich Source of Herbal Medicine. *J. Nat. Prod.* **2008,** *1,* 27–35.
144. Toepfl, S.; Mathys, A.; Heinz, V.; Knorr, D. Potential of High Hydrostatic Pressure and Pulsed Electric Fields for Energy Efficient and Environmentally Friendly Food Processing. *Food Rev. Int.* **2006**, *22* (4), 405–423.
145. Upadhyay, R.; Ramalakshmi, K.; Rao, L. J. M. Microwave-Assisted Extraction of Chlorogenic Acids from Green Coffee Beans. *Food Chem.* **2012,** *130* (1), 184–188.
146. Vankar, P. S. Essential Oils and Fragrances from Natural Sources. *Resonance* **2004**, *9* (4), 30–41.
147. Vilkhu, K.; Mawson, R.; Simons, L.; Bates, D. Applications and Opportunities for Ultrasound Assisted Extraction in the Food Industry—A Review. *Innovat. Food Sci. Emerg. Technol.* **2008,** *9* (2), 161–169.
148. Vinatoru, M.; Toma, M.; Radu, O.; Filip, P.; Lazurca, D.; Mason, T. The Use of Ultrasound for the Extraction of Bioactive Principles from Plant Materials. *Ultrason. Sonochem.* **1997,** *4* (2), 135–139.
149. Vorobiev, E.; Jemai, A. B.; Bouzrara, H.; Lebovka, N.; Bazhal, M. Pulsed Electric Field Assisted Extraction of Juice from Food Plants. *Novel Food Process. Technol.* **2005,** 105–130.
150. Vorobiev, E.; Lebovka, N. Extraction of Intercellular Components by Pulsed Electric Fields. *Pulsed Electric Fields Technology for the Food Industry;* Springer: Boston, MA, 2006; pp 153–193.
151. Vorobiev, E.; Lebovka, N. Pulsed-electric-fields-induced effects in plant tissues: Fundamental Aspects and Perspectives of Applications. In *Electrotechnologies for Extraction from Food Plants and Biomaterials*; Springer: Verlag, New York, 2009; pp 39–81.
152. Vranová, E.; Coman, D.; Gruissem, W. Structure and Dynamics of the Isoprenoid Pathway Network. *Mol. Plant.* **2012,** *5* (2), 318–333.
153. Wang, L. -J.; Li, D.; Tatsumi, E.; Liu, Z. -S.; Chen, X. D.; Li, L. -T. Application of Two-Stage Ohmic Heating to Tofu Processing. *Chem. Eng. Proc.: Proc. Intensif.* **2007,** *46* (5), 486–490.
154. Wang, L.; Wang, X.; Wang, P.; Xiao, Y.; Liu, Q. Optimization of Supercritical Carbon Dioxide Extraction, Physicochemical and Cytotoxicity Properties of Gynostemma Pentaphyllum Seed Oil: A Potential Source of Conjugated Linolenic Acids. *Separat. Purif. Technol.* **2016,** *159,* 147–156.
155. Wang, L.; Weller, C. L. Recent Advances in Extraction of Nutraceuticals from Plants. *Trends Food Sci. Technol.* **2006,** *17* (6), 300–312.
156. Wang, L.; Yang, B.; Du, X.; Yi, C. Optimization of Supercritical Fluid Extraction of Flavonoids from Pueraria Lobata. *Food Chem.* **2008**, *108* (2), 737–741.
157. Wang, W.; Meng, B.; Lu, X.; Liu, Y.; Tao, S. Extraction of Polycyclic Aromatic Hydrocarbons and Organochlorine pesticides from Soils: A Comparison Between Soxhlet Extraction, Microwave-Assisted Extraction and Accelerated Solvent Extraction Techniques. *Anal. Chim. Acta* **2007,** *602* (2), 211–222.
158. Wynn, T. A. Common and Unique Mechanisms Regulate Fibrosis in Various Fibroproliferative Diseases. *J. Clin. Invest.* **2007,** *117* (3), 524.

159. Xi, J.; Shen, D.; Zhao, S.; Lu, B.; Li, Y.; Zhang, R. Characterization of Polyphenols from Green Tea Leaves using a High Hydrostatic Pressure Extraction. *Int. J. Pharmaceut.* **2009,** *382* (1), 139–143.
160. Xiao, J. Dietary Flavonoid Aglycones and Their Glycosides: Which Show Better Biological Significance? *Crit. Rev. Food Sci. Nutr.* **2017,** *57* (9), 1874–1905.
161. Yang, F.; Kubota, F.; Baba, Y.; Kamiya, N.; Goto, M. Selective Extraction and Recovery of Rare Earth Metals from Phosphor Powders in Waste Fluorescent Lamps Using an Ionic Liquid System. *J. Hazardous Mater.* **2013,** *254,* 79–88.
162. Yeh, G. Y.; Eisenberg, D. M.; Kaptchuk, T. J.; Phillips, R. S. Systematic Review of Herbs and Dietary Supplements for Glycemic Control in Diabetes. *Diabetes Care* **2003,** *26* (4), 1277–1294.
163. Zeidan, S.; Hijazi, A.; Rammal, H.; Kobaissi, A.; Badran, B. Extraction of Phenolic Compounds and Flavonoids from Eryngiumcreticum l. By Conventional and Non-Conventional Extraction Techniques. *World J. Pharm. Pharmaceut. Sci.* **2014,** *3* (17), 1889–1898.
164. Zhang, D.; Hamauzu, Y. Phenolics, Ascorbic Acid, Carotenoids and Antioxidant Activity of Broccoli and Their Changes During Conventional and Microwave Cooking. *Food Chem.* **2004,** *88* (4), 503–509.
165. Zhang, Q.; Worsnop, D.; Canagaratna, M.; Jimenez, J. Hydrocarbon-like and Oxygenated Organic Aerosols in Pittsburgh: Insights into Sources and Processes of Organic Aerosols. *Atmos. Chem. Phys.* **2005,** *5* (12), 3289–3311.

CHAPTER 2

EXTRACTION OF BIOACTIVE COMPOUNDS: CONVENTIONAL AND GREEN EXTRACTION TECHNIQUES

AFKA DEEN, RIZLIYA VISVANATHAN, and RUVINI LIYANAGE*

Division of Nutritional Biochemistry, National Institute of Fundamental Studies, Hantana Road, Kandy 20000, Sri Lanka

Corresponding author. E-mail: ruvini@ifs.ac.lk

ABSTRACT

The increasing demand for bioactive compounds in plants encourages the discovery of novel extraction methods. Especially, the usage of phytochemicals in different industries such as pharmaceutical, chemical, and food further reinforces the importance of extraction. Moreover, the choice of extraction method is an essential factor which decides the quality and yield of the extraction. Both traditional and nonconventional extraction methods can be used to extract phytochemicals from plant matrix. However, time consumption and high volume of hazardous solvent usage make conventional methods less popular among researchers. Furthermore, the concern of environment has led to a green approach of extraction by utilizing novel green extraction methods which result better yield in a short period of time with minimal solvent usage. This chapter mainly deals with principles of different extraction methods along with their strengths and limitations, and their comparative usage in bioactive compound extraction.

2.1 INTRODUCTION

Natural products, such as plant material extracts, open a new horizon for many discoveries which lead to tremendous changes in the field of medicine and human health.[24] Plant material usage is as old as the start of mankind. Earlier, plants were used for their nutritional value, yet later on the discovery of the medicinal aspects of plants led to the advancement in human health along with research studies.[2] Bioactive compounds are mainly identified as the essential source in plant materials.[2] They are produced as secondary metabolites.[15,40] Plants have been evolved for millions of years to withstand bacteria, insects, fungi, and tough weather that led to the production of diverse secondary metabolites.[20] According to the World Health Organization, 80% of people still depend on plant-based traditional medicine.[20,21] With the advancement in science, these compounds are isolated and used in medicines and for that the active compounds should be extracted from the plant material. Hence, the extraction and separation from coextractive compounds are essential.

Extraction is considered as an important and essential step since it affects the eventual output of the study.[9,35] Extraction is a standard procedure where medicinally active ingredients of plant (and animal) tissues are separated using selective solvents.[49] Extraction history begins from the discovery of fire. Extraction and distillation methods were invented to manufacture perfumes, cosmetics, and certain food items.[12] Since then, the evolution of extraction started with the advancement of technology. The main conventional extraction methods are Soxhlet extraction, maceration, and hydrodistillation.[33] For the past 126 years, Soxhlet was the most used conventional techniques among the others.[11] Traditional methods of extraction mainly depend on the solvent type. In addition, heat/agitation process increases the efficiency of conventional methods.[47] However certain major disadvantages, such as time consumption, high solvent requirement, and less extraction yield, have made traditional methods less popular among industry-based researchers. Hence, the replacement of conventional extraction methods by green extraction methods has opened up a new research approach on advancement of alternative analytical techniques.[49]

Most of the novel extraction methods are referred as "green extraction methods" which mainly focus on reducing energy requirements and usage of nonhazardous alternative solvents while ensuring high-quality extracts. Ultrasound-assisted extraction (UAE), pressurized liquid extraction (PLE), microwave-assisted extraction (MAE), enzyme-assisted extraction, supercritical fluid extraction (SFE), and pulse electric field (PEF) extraction are few examples of green extraction methods. Green extraction technologies are considered as an innovative approach to improve the outcome and quality of extraction.[15] Owing to the strength and limitations of both traditional and nonconventional methods, it is still a question mark on finding a universal extraction method. Overall, the main aim of this chapter is to summarize and give an overview of conventional and nonconventional extraction methods.

2.2 EXTRACTION METHODS

There is an increasing demand for developing suitable techniques for more efficient and effective extraction from the plant materials. Researchers believe that extraction is an important part in any analytical study required for the precession of final outcome. Extraction is also referred as "sample preparation technique" which has a significant impact on the quality and quantity of the final yield.[9,46] Moreover, a fine technique of extraction and a screen-out process is required to find the specific compound which exerts the health benefits among the broad spectrum of bioactive compounds available in different plant materials.[9,21]

The major objective of all the extraction methods is to separate the soluble phytochemicals from the insoluble part of the plant matrix.[37] Extraction of bioactive compounds in plant materials is carried out by various methods. In early days, commonly used traditional methods were Soxhlet, maceration, and hydrodistillation.[9] Along with the conventional methods, the advancement of science and technology has led to the invention of novel methods, for example, solid-phase microextraction, supercritical fluid extraction, PLE, MAE, solid-phase extraction, and surfactant-mediated techniques are a few examples for innovative extraction methods.[9] These novel methods are specially designed to omit the limitations of conventional methods. Moreover, due to the decrease of synthetic and organic solvent usage, nonconventional methods are referred as green extraction methods. In addition, these methods reduce the time of extraction leading to a better yield and high-quality extract.[15] However, a suitable extraction method for a particular study is basically decided on the objective of the study, type of the plant material, and the compound of interest.

2.2.1 CONVENTIONAL EXTRACTION METHODS

Soaking, maceration, boiling, grinding, magnetic stirrer, water percolation, and heat reflux are considered to be some of the main conventional techniques.[60] Moreover, commonly used conventional methods are Soxhlet, maceration, and hydrodistillation, and out of all these methods, Soxhlet is considered as the oldest traditional method. The efficacy of a traditional method depends on the power of the solvent used and the physical treatment (heat and/or mixing) that is carried out in the process. The fact mentioned above points out the significance of solvent choice on the efficiency of a conventional method. Table 2.1 depicts the different bioactive compounds extracted to a particular solvent.

TABLE 2.1 Types of Solvents Used for Different Phytochemicals.

Water	Ethanol	Methanol	Chloroform	Dichloro-methanol	Ether	Acetone
Anthocyanin	Tannins	Antho-cyanins	Terpenoids	Terpenoids	Alkaloid	Flavonoid
Tannin	Polyphenol	Terpenoids	Flavonoid		Terpenoid	
Saponnin	Flavonol	Saponins				
Terpenoids	Terpenoids	Tannins				
	Alkaloid	Flavones				
		Polyphenols				

2.2.1.1 SOXHLET EXTRACTION (HOT CONTINUOUS EXTRACTION)

Soxhlet is a laboratory apparatus invented in 1879 by a German agricultural chemist Franz von Soxhlet.[26] Furthermore Soxhlet extraction is considered as a special tool in preperative step since the analyte of interest can be conceretrated as a whole or it can be separated from the interfering matrix.[36] It was originally discovered for the purpose of fat extraction. Nevertheless, now it has been extensively used to extract numerous bioactive compouds from various natural sources and pesticides from soil samples.[9,26,50] In addition, Soxhlet has been used as a comparison model to indicate the advancement of novel extarction methods.

2.2.1.1.1 Principle and Studies

In Soxhlet extraction, the sample is repeatadly brought into contact with the solvent.[50] Generally, a thimble containing a small amount of sample is fixed to a distillation flask with the solvent of intrest. The extraction solvent in the distillation flask is heated, which is evaporated into the thimble, then again condenses in the condenser and fills up the thimble. When the liquid content reaches the siphon arm, the liquid content is emptied into the bottom flask and again the process is continued.[37] Filtration is not required after the extraction process.[50]

Due to certain limitations, as stated below, many attempts were made to improve the efficacy of the Soxhlet extractor.[50] Supporting the statement, a study done by Subramanian et al.[50] on the extraction of piperine from *Pipernigrum* demonstrated that a modified, double bypasses sidearm Soxhlet apparatus has the ability to reduce the extraction time comparing to general Soxhlet extraction. Yet another study done to extract flavonoids from rape bee pollen (*Brassica campestris*) using a hybrid Soxhlet technique demonstrated that the method is effective for flavonoid extraction.[32] This hybrid Soxhlet study emphasizes the key significance of separating the targeted compound from the plant matrix.

2.2.1.1.2 Strengths and Limitations

When focusing on the positive side of Soxhlet, it indicates that this extraction method requires little training due to simple methodology and can extract more sample compared to many modern techniques.[32,33] However, this technique is associated with certain adverse characteristics such as use of large volumes of hazardous and flammable organic solvents and at the same time, it requires highly pure solvents, which are costly. Yet the requirement of solvent is less in comparison with maceration.[32] Use of large quantity of hazardous solvent makes it a less environmental-friendly method compared to SFE.[36,37] Furthermore, it is a time-consuming technique.[28,36]

2.2.1.2 MACERATION

Maceration was a method used in wine making. But later on, it was utilized in medicinal plant extraction.

2.2.1.2.1 Principle and Studies

This method mainly involves soaking of plant materials in a solvent. Coarsely ground material is inserted in a stoppered container with the solvent (menstruum) and kept with frequent agitation for 3–4 days.[15,21,23] Use of finely powdered material increases the surface area for contact with solvent. Also, frequent shaking in maceration increases the diffusion and removes the concentrated solvent from the surface of the sample by transferring a new solvent to the sample surface. This eventually increases the extraction rate.[9]

2.2.1.2.2 Strengths and Limitations

This is the simplest, inexpensive, and easiest method of extraction. However, a large volume of solvent is required which again creates environmental issue unless a proper waste management is utilized. But, selecting a proper temperature and solvent can reduce the volume needed for extraction.[61]

2.2.1.3 HYDRODISTILLATION

Hydrodistillation is a conventional extraction method used for essential oil and bioactive compound extraction from plant materials. The specific features in this extraction method are extraction before dehydration of plant material and solvent-free extraction.[58] Water distillation, water and steam distillation, and direct steam distillation are the three main hydrodistillation techniques.[51,58]

2.2.1.3.1 *Principle and Studies*

In this extraction method, plant material is packed in a still compartment and sufficient amount of water is added. Then water is boiled and steam is directed into the plant materials. Contact of water vapor with plant materials increases the release of bioactive compounds from plant tissues. Isolation of essential oil from plant materials is also conducted by the same three methods stated above. Generally, the separation of extraction into water and oil is done by indirect cooling. Moreover, the condensed mixture is separated to oil and bioactive compounds by directing to a separator. A study by Ma et al.[32] to extract essential oil from agarwood showed that the highest amount of compound extraction was obtained using acetone. Furthermore, this study demonstrated that there is no significant difference between laboratory-scale and industrial-scale hydrodistillation.

2.2.1.3.2 *Strengths and Limitations*

Hydrodistillation is considered as an absolute environment-friendly method since it is an organic solvent-free technique. Yet, it is a slow process, and the longer distillation time leads to the problem of high-fuel consumption. Hence, this method is regarded as a noneconomical process. Also this method is not suitable for high boiling point, hard roots, and woody plants. Furthermore, this method is not recommended for thermally-labile compounds (Table 2.2).[9]

TABLE 2.2 A Brief Summary of the Experimental Conditions for Main Conventional Methods Used in Extraction of Active Components in Plant.

Parameters	Soxhlet	Maceration	Hydrodistillation
Commonly used solvent	Methanol, ethanol, or mixture of alcohol and water	Methanol, ethanol, or mixture of alcohol and water	Water
Temperature (°C)	Depends on the solvent	Room temperature	Boiling temperature
Time required	3–18 h	3–4 days	4–5 h
Required solvent volume (mL)	150–200	Depends on the quantity of the sample	No solvent required
References	[9, 36, 37, 50]	[28, 37, 52]	[9, 52, 58]

2.2.2 NONCONVENTIONAL METHOD

The consistent demand for plant extraction has led to the invention of novel extraction methods. Especially, the advancement in chromatography technique and environment safety factors have influenced the discovery of nonconventional extraction methods. The decline in the use of synthetic organic solvents, reduction in time of extraction, and quality and quantity of extract make the nonconventional methods more desirable than conventional methods.[9] Moreover, the type of solvent, ratio between solvent to solid, size of particles, and temperature and time of extraction are the certain parameters that increase the yield of nonconventional methods.[13] The most promising nonconventional methods in use are UAE, PLE, MAE, enzyme-assisted extraction, SFE, and PEF extraction.

2.2.2.1 PRESSURIZED LIQUID EXTRACTION METHOD

PLE is known as enhanced solvent extraction or accelerated solvent extraction. This novel extraction technique uses organic solvents at high pressure and temperature above their normal boiling point.[4] This method is also known as pressurized hot water extraction when water becomes the extraction solvent. In comparison with Soxhlet, PLE technique has a shorter extraction time, requires less solvent, and has high penetrability into the sample. The usage of less organic solvent makes PLE a green extraction method.[25] Also, this method was successfully employed for extraction of agrochemicals, various pollutants, and pharmaceuticals.[10]

2.2.2.1.1 Principle and Studies

In PLE method, the sample is packed in an extraction cell. Then, the extraction is carried out using a suitable solvent under high pressure (500–3000 psi) and temperature (40–200°C) conditions for a short period of time (5–15 min). The process ends with collection of extraction using a compressed gas.[4,51]

PLE is a famous method utilized for extraction of environmental pollutants such as persistent organic pollutants (POPs). Chemicals that persist for a long period of time are known as POPs. It is important to monitor the amount of POPs to analyze the environmental conditions. Owing to the facts above, Suchan et al.[51] have utilized the PLE to analyze polychlorinated biphenyls and organochlorine pesticides in fish samples which demonstrated that this technique is much better than the traditional Soxhlet method. Yet another study done to extract derivatives of isoflavone from freeze-dried soybeans depicted that isoflavones were extracted without degradation using the optimized PLE extraction conditions. Hence, these findings emphasize the importance of PLE technique in numerous extraction studies.

2.2.2.1.2 Strengths and Limitations

PLE requires 15 mL of solvent in comparison with Soxhlet extraction which requires about 500 mL. This decline in solvent amount is mainly due to the efficient usage of solvent. Moreover, the whole extraction and clean-up procedure can be completed within 30 min, whereas traditional methods consume around 10–16 h. However, optimized parameters were recommended for especial thermolabile compounds since they can degrade. Yet another drawback of using pressurized fluid technology is that the method requires expensive equipment to withstand the high pressure utilized.[25]

2.2.2.2 MICROWAVE-ASSISTED EXTRACTION

MAE technique combines both microwave and traditional solvent extraction.[63] The usage of MAE was started in 1980s and gained popularity due to its tremendous advantages over traditional extraction methods. The frequency that ranges from 300 MHz to 300 GHz in the electromagnetic spectra is known as microwaves. The usage of microwave for heating the solvents and plant tissue has shown an increase in the efficacy of the extraction.[44,63] As a modern technique MAE has a shorter time of extraction in comparison with traditional Soxhlet or sonication methods; nevertheless, the method operates at a higher temperature and pressure.[16] Moreover, MAE is subjected to several advancement to make the method more efficient such as pressurized microwave-assisted extraction and solvent-free microwave-assisted extraction.

2.2.2.2.1 Principle and Studies

The mechanism of microwave extraction consists of three main steps: (1) separation of bioactive compounds from the active sites of the plant tissue with increased temperature and pressure, (2) the process of diffusion of solvent into plant tissue, and (3) the release of targeted bioactive compounds to the solvent.[6] In MAE samples, which are enclosed in teflon vessel with solvent, are heated under controlled temperature. Basically, the solvent choice for MAE depends on the solubility of the target compound and the interaction of microwave with the solvent.[16]

Application of MAE method on medicinal plants to extract bioactive compounds is found to be easy and effective. In addition, this method is utilized in flavonoid extraction and analytical quantification.[31] Over the past 15 years, researchers and food manufactures show an interest toward phenolic compounds due to their incredible health benefits. A study by Alupului et al.[6] focused on the extraction of phenolic compounds in *Cynarascolymus* leaves, which is a herb that is rich in flavonoids and phenolic acids; it demonstrated that MAE provides higher purity in extraction compared with classical extraction methods.

2.2.2.2.2 Strengths and Limitations

MAE technique has many advantages such as smaller extraction time, less solvent requirement, higher extraction rate, and lower cost in comparison to traditional methods. One of the main novel approaches is that making MAE as a solvent-free extraction method. This makes MAE a green extraction method especially for essential oils and some volatile natural methods.[9,63]

2.2.2.3 ENZYME-ASSISTED EXTRACTION

Enzyme-assisted extraction method is also gaining attention due to the need of green extraction methods.[42] Certain bioactive compounds in the plants are distributed in the cytosol and some are constrained in the polysaccharide–lignin complex through hydrogen or hydrophilic bonds. The latter type of phytochemicals are not easy to extract with a general solvent extraction method.[9] Thus, the treatment of enzymes is considered as an effective way to increase the extraction yield. Generally, cellulose, hemicellulose, and pectinase are the enzymes used to hydrolyze the cell wall.[42] Bacteria, fungi, animal organs, vegetable, and fruit extracts are some of the well-known sources of enzymes. It is important to have an idea on the catalytic property, mode of action, and optimal conditions when utilizing enzymes in extraction.[42]

2.2.2.3.1 Principle and Studies

Enzyme-assisted aqueous extraction (EAAE) and enzyme-assisted cold pressing (EACP) are the two main approaches in enzyme extraction method. It is found that in EAAE method, the seed cell wall is degraded and polysaccharide–protein colloid is ruptured which results in an emulsion formation and low yield. In comparison with EAAE, in EACP enzyme facilitates the hydrolysis of seed cell wall, since there is no polysaccharide–protein colloid in nonaqueous system.[9,30] Recently, enzymes were combined with ultrasound, microwave, and supercritical fluid to increase the efficacy of extraction. The combination of enzyme and ultrasound has been studied and applied in many plants for the extraction of natural product. Ultrasonic wave is found to improve the capability of enzyme and at the same time, it results an even spreading of the enzyme.[13] Moreover, the enzyme-mediating reactions are conducted at very low temperatures (i.e., around 15–45°C), which explains the usefulness of this method in the extraction of thermo-labile compounds. Normally, it is found that at high temperature (above 60°C) the enzyme activity is irreversibly altered. This is explained in the means of protein degradation that occurs due to the change in the shape, folding, and cross-linking in polypeptide chain.[7]

Compounds such as polysaccharides, oils, natural pigments, flavors, and medicinal compounds are found to be extracted in large amounts using enzyme-assisted extraction.[30] Hence, the application of enzyme has improved the quality as well as the yield of natural product extraction including oils, flavorings, and bioactive compounds from plant materials. Thi et al.[54] focused on improving the oil yield extract from gac fruit aril (*Momordica cochinchinensis Spreng*) using enzyme-assisted extraction method. The study demonstrated that application of Viscozyme L helped to increase the oil recovery yield and total carotenoid content in oil. Generally, it is known that viruses cause life-threatening diseases such as hepatitis, influenza, HIV/AIDS, ebola, and herpes. Yet another study based on enzyme-assisted hydrolysis on extracting antiviral agents from seaweeds depicted an improvement in bioactive materials in the extraction. *Codium fragile* and *Chondrus crispus,* green and red seaweeds, respectively, are found in the North Atlantic coasts. This study demonstrated that the usage of synthetic commercial enzymes such as proteases and carbohydrates increased the yield of extraction compared to water extraction (Fig. 2.1).[29]

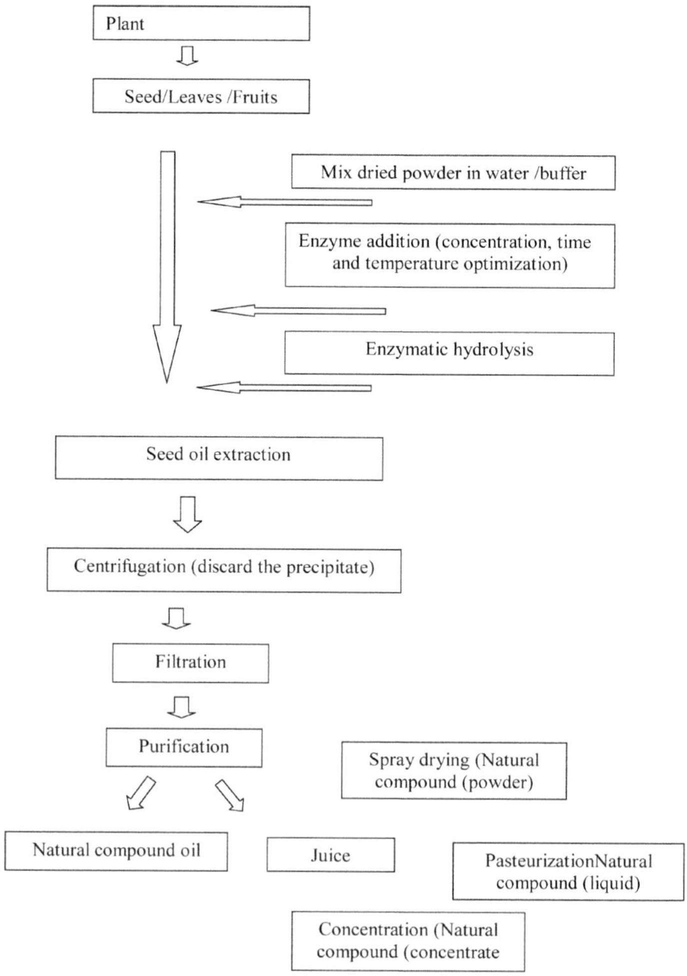

FIGURE 2.1 Enzyme-assisted extraction of phytochemicals from a plant source.

2.2.2.3.2 *Strengths and Limitations*

Enzyme-assisted technology has gained attention among many researchers due to several specific advantages. The method has a high bioactive yielding technology by which cell walls are broken down and desired bioactive products are released. Thus, this method has a lower impact on the environment. Enzyme-assisted extraction leads to a reduction in extraction time, solvent volume, and energy consumption when compared to nonenzymatic methods.[42] Also, this technique results in high-quality product.[13] However, the method is relatively expensive for large industrial production since the enzyme activity is limited by environmental conditions as well as, the available enzyme cannot break down the plant cell walls completely.[13]

2.2.2.4 *SUPERCRITICAL FLUID EXTRACTION*

Discovery of supercritical fluid by Hannay and Hogarth in 1879 led to the invention of SFE. Moreover, Zosel, the patent holder for decaffeination of coffee using SFE, also has an equal contribution toward SFE method discovery.[3,9,64] Since then, the SFE method has improved tremendously. Especially, this method is gaining attention as an alternative green extraction method in natural product extraction.[3]

2.2.2.4.1 *Principle and Studies*

Supercritical fluid is a substance which lies above its critical point temperature and pressure.[43] Under this condition, the fluid behaves like both liquid and gas. A supercritical fluid has a density of a liquid and viscosity similar to gas. Due to their high diffusivity and low viscosity, supercritical fluids penetrate into solid materials better than normal liquids which eventually results in high extraction yield.[23] Carbon dioxide and water are the most commonly used supercritical fluids.[43] However, since CO_2 is nonpolar molecule, it is ideal for fat, lipid, and nonpolar substances, yet it has certain difficulties in extracting polar substances such as pharmaceuticals and drug samples from natural matrices.[9] In order to overcome this problem, modifiers (also called cosolvents) are commonly used. A small amount of modifier can result a significant change in the polarity of

CO_2. For example, 0.5 mL of dichloromethane (CH_2Cl_2) can enhance the extraction which consumes 4 h hydrodistillation.[9,23,49] Especially due to the low critical temperature of carbon dioxide (31.1°C), thermally-labile food products can be extracted using this method. As well as a solvent-free extraction is obtained due to the gaseous nature of CO_2.[49] Other than CO_2, propane and ethane can be used as supercritical fluids to extract natural products. These fluids have a high solvating power compared to CO_2, yet these solvents are flammable and costly.[3] Generally, there are certain major factors such as temperature, pressure, particle size, and moisture content of the material that should be taken into account to increase the efficiency of this method.[9]

A general SFE system must contain a tank for the mobile phase (CO_2), a pump, a pressure cell to contain the sample, a collecting vessel, and a way of controlling the pressure of the system.[9] The liquid pumped to the heating zone is heated to supercritical condition. Then, it is passed to an extraction vessel where it diffuses into the plant material and dissolves it. After that, the extracted material is directed toward a separator under low-pressure conditions, and the extracted material is separately obtained. Finally, the CO_2 will be cooled, recompressed and recycled, or discharged to the atmosphere.[54] Generally, supercritical fluids have been used in numerous applications such as essential oil extraction, metal cation extraction polymer synthesis, and particle nucleation.[39,49] A study done by Giannuzzo et al. (2003) to extract sunflower oil using SC-CO_2 extraction resulted in maximum amount of sunflower oil (54.37 wt%). Hence, this method is considered as one of the efficient method to extract oil. Atil[8] extracted limonoid glucosides from grapefruit molasses using SFE. The results of extraction showed that SFE has a practical significance in commercial production of limonoid glucosides. It has been reported that supercritical fluid extraction has gained much attention in industrial usage, especially in palm oil industry.[3] Use of CO_2 as a solvent depicts certain advantages such as nontoxicity, nonflammablility, and nonpolluting supercritical fluid. SFE has shown 100% efficiency in oil extraction, and it is considered as a safe method due to organic solvent-free extraction.[3]

2.2.2.4.2 *Strengths and Limitations*

The significance of using SFE technique is indicated by following advantages such as the use of supercritical fluids with different physiochemical properties (density, diffusivity, viscosity, and dielectric constant). The low viscosity and relative high diffusivity help the fluid to penetrate inside the matrix as compared to liquids which result in high extraction yield.[49] The higher efficiency of the method reduces the extraction time.[49]

2.2.2.5 ULTRASOUND-ASSISTED EXTRACTIONOR SONICATION EXTRACTION METHOD

Ultrasound wave lies beyond human hearing within the frequency range of 20 kHz to 100 MHz. Ultrasound-assisted extraction has incorporated with other classical methods since it can increase the efficacy of the respective technique. As an instance, ultrasound-assisted Soxhlet extraction, ultrasound-assisted Clevenger distillation, and continuous UAE.[9] Furthermore, there are techniques which combine UAE with MAE and SFE methods.[46] The usage of ultrasound disrupts the plant cell wall thereby enhances the solvent penetration and leads to an effective extraction.[56]

2.2.2.5.1 Principle and Studies

Generally, waves move forward through a medium by expansion and compression which eventually result into formation of bubbles. The conversion of the kinetic energy of motion to heat the bubble produces a large amount of energy. Hence, the bubbles possess high temperature and pressure. Based on this principle, UAE method has been invented. UAE method deals with two physical phenomena such as the diffusion of solvent through the cell wall and washing out of the cell content after disrupting the cell wall.[9] Furthermore, the size of the sample, content of moisture in the sample, degree of milling, and solvent play a vital role in the efficiency of the method.

Samaram et al.[46] demonstrated that UAE is a convenient technique for recovering oil from the papaya seed. Furthermore, the study revealed that papaya seed oil was obtained in desirable fatty acid profile with shorter extraction time and mild conditions. Soxhlet, maceration, heat reflux, and MAE are well-known methods for phenolic compound extraction from medicinal plants. Another study by Altemimi et al.[5] to extract phenolic compound from pumpkins and peaches showed that UAE is an appropriate alternative method for phenolic compound extraction. Yet, another study by Zhang et al.[62] to extract phenolic compounds from *Inulahelenium* using UAE resulted in a high amount of phenolics which again proves that UAE is a suitable method for phenolic compound extraction.

2.2.2.5.2 Strengths and Limitations

UAE method is used as an alternative extraction method due to several advantages such as easy use, inexpensiveness compared to other nonconventional methods, and reduced amount of solvent required, temperature, and time.[9,46,62] The most specific advantage is that this method can be used for thermo-sensitive and unstable compounds.[55,62] However, use of ultrasound energy more than 20 kHz may have an inappropriate effect on phytochemicals, which is formed through free radicals.[9]

2.2.2.6 PULSE ELECTRIC FIELD EXTRACTION METHOD

PEF extraction method is gaining attention as one of the most important green extraction methods due to its contribution toward food, agriculture, and pharmaceutical industries.[60] Generally, PEF is considered as a nonthermal technique. Moreover, electric field with short pulses is used to block the activity of most of the microorganisms and some of the enzymes at the room temperature. Hence electric field with short pulses aids in preserving and improving the standard of medicinal materials and foods.[60]

2.2.2.6.1 Principle and Studies

In PEF extraction method, a direct current with high-voltage pulses is applied for a material placed in between two electrodes for microsecond to millisecond time range. Exposure of electric field on plant tissue enhances the extraction of the intercellular materials by increasing the porosity of cell wall.[1,59] Abenoza et al.[1] extracted olive oil using PEF which demonstrates that PEF can be used as a better alternative method to reduce the current and temperature in olive oil extraction industry. Yet another study done to extract anthocyanin from purple-fleshed potato using PEF revealed that this method is suitable to improve anthocyanin extraction yield, and can be used to reduce the extraction temperature. In addition, the method can eliminate or reduce the use of organic solvents used in the recovery.[1]

2.2.2.6.2 *Strengths and Limitations*

PEF technique increases mass transfer by destroying membrane structure of plant materials. This eventually reduces the extraction time and increase the extraction yield.[9] PEF treatment is carried out at a modest electric field (500 and 1000 V/cm for 10^{-4}–10^{-2} s) with the small increase in the temperature. Thus, this method is considered as a suitable method for thermolabile compound extraction from plant tissues. However, low capacity of treatment, combination with other extraction methods, and the breakage in continuous process make this a less acceptable method in industries.

2.3 COMPARISON OF VARIOUS EXTRACTION METHODS

A suitable extraction method depends on both extraction yield and extraction purity. Hence, there are several studies conducted to compare the efficacy of different extraction methods on several important bioactive compounds in plant materials (Table 2.3). For example, Boryana and Trusheva[11] extracted phenolic compounds and flavonoids from propolis using various methods such as conventional maceration extraction, UAE, and MAE. According to this study, UAE method gave the maximum yield of phenolics and MAE method was found to result a high yield of nonphenolic and nonflavonoid material compared to maceration technique.

TABLE 2.3 Phytochemicals Extracted from Plants Using Conventional and Nonconventional Methods.

Plant material	Secondary metabolites	Extraction methods	Performances	Reference
Osbeckia parvifolia	Phenolic compounds, tannins, and flavonoids	Maceration, Soxhlet	The study reveals that Soxhlet method has a higher recovery of targeted samples than maceration	[11]

TABLE 2.3 *(Continued)*

Plant material	Secondary metabolites	Extraction methods	Performances	Reference
Aerial parts of *Haplophyllum robustum*	Monoterpenes such as 1,8-cineole, camphor, and terpinen-4-ol	HD, MAHD, and SFME	MAHD and SFME extraction percentages are 21.1% and 21.0%, respectively, in comparison with HD (19.0%). A similar extraction yield was obtained at significantly shorter extraction time for MAHD and SFME in comparison with HD. Also MAHD and SFME are environment friendly than HD	[34]
Phyllanthus niruri, Orthosiphon stamins, Labisia pumila	Tannin and flavonoid	Soxhlet extraction, cold maceration	The study has documented that Soxhlet extraction is a preferable technique compared to cold maceration	[27]
Citrus paradisi	Naringin, the major flavonoid from the peel	Maceration, reflux Soxhlet, SFE	The maximum outcome (14.4 g/kg) was obtained by SFE method. SFE requires less solvent and it has a shorter extraction time with comparison to conventional extraction methods	[22]
Roots of *Inulahelenium* L	Alantolactone and isoalantolactone	Nonconventional methods: UAE, conventional methods: maceration, infusion, microsteam, and distillation extraction	Amounts of a lantolactone and isoalantolactone obtained from UAE for 30 min was same as the compounds obtained from 24 h of maceration. Hence, this study clearly shows that nonconventional methods are more efficient in comparison to conventional method	[57]

TABLE 2.3 *(Continued)*

Plant material	Secondary metabolites	Extraction methods	Performances	Reference
Pterocarpus marsupium	Pterostilbene (3',5'dimethoxy-4-stilbenol)	Infusion, decoction, maceration percolation, UAE, and MAE	Among the conventional extraction methods, percolation gave the highest yield The MAE method showed a highest extraction yield compared to UAE	[18]
Rubus idaeus L	Anthocyanins	Conventional extraction method UAE processes	Amount of total anthocyanin extraction is approximately equal in both extraction methods. Yet time required for UAE is less than the conventional method	[38]
Mucuna pruriens	L-DOPA	Refluxing, UAE, and MAE	Among all the three extraction methods, MAE method provides the best results for yield and quality in a short period of time	[19]
Convolvulus pluricaulis	Scopoletin	Soxhlet extraction, reflux extraction, SFE, UAE, and MAE	Nonconventional methods are better than traditional extraction methods due to less time consumption, less solvent consumption and high yield of scopoletin. Among the methods used in this study, microwave extraction method was found to be the most efficient method	[53]

HD, hydrodistillation; L-DOPA, 3,4-dihydroxyphenylalanine; MAE, microwave-assisted extraction; MAHD, microwave-assisted hydrodistillation; SFE, supercritical fluid extraction; SFME, solvent-free microwave extraction; UAE, ultrasound-assisted extraction.

2.4 SUMMARY

The widespread application of bioactive compounds in pharmaceutical, food, and chemical industries signifies the importance of a standard extraction method. Owing to the fact above, an ideal extraction method should be swift and the extract should be obtained without degradation. Yet, it is difficult to find a universal extraction method since each and every extraction method holds a unique approach toward the plant of interest. The efficacy of the classical and nonconventional extraction methods depends on the scientific skills, plant type, and chemical nature of bioactive compounds. Hence, it is also important to focus on above parameters prior to the selection of an extraction method.

KEYWORDS

- **extraction method**
- **phytochemicals**
- **conventional extraction methods**
- **nonconventional extraction methods**
- **Soxhlet**
- **maceration**
- **hydrodistillation**

REFERENCES

1. Abenoza, M. Effects of Pulsed Electric Field on Yield Extraction and Quality of Olive Oil. *Food Bioprocess. Technol.* **2013,** *6* (6), 1367–1373.
2. The Norwegian Academy of Science and Letters. In *Bioactive Compounds in Plants—Benefits and Risks for Man and Animals*, Bernhoft A, Ed.; Proceedings from a Symposium Held at The Norwegian Academy of Science and Letters, Oslo, November 13–14, 2008.
3. Akanda, M. J. H. Applications of Supercritical Fluid Extraction (SFE) of Palm Oil and Oil from Natural Sources. *Molecules* **2012,** *17* (2), 1764–1794.
4. Alonso-Salces, R. M. Pressurized Liquid Extraction for the Determination of Polyphenols in Apple. *J. Chromatogr. A.* **2001,** *933* (1–2), 37–43.
5. Altemimi, A. Ultrasound Assisted Extraction of Phenolic Compounds From Peaches and Pumpkins. *PLoS One* **2016,** *11* (2), 1–20.

6. Alupului, A.; Călinescu, I.; Lavric, V. Microwave Extraction of Active Principles From Medicinal Plants. *UPB Sci. Bull., Ser. B: Chem. Mater. Sci.* **2012,** *74* (2), 129–142.
7. Sowbhagya, H. B.; Chitra, V. N. Enzyme-Assisted Extraction of Flavorings and Colorants From Plant Materials. *Crit. Rev. Food Sci. Nutr.* **2010,** *50* (2), 146–161.
8. Yu, J.; Dandekar, D. V.; Toledo, R. T.; Singh, R. K.; Patil, B. S. Supercritical Fluid Extraction of Limonoid Glucosides from Grapefruit Molasses. **2006,** *54* (16), 6041–6045.
9. Azmir, J. Techniques for Extraction of Bioactive Compounds From Plant Materials: A Review. *J. Food Eng.* **2013,** *117* (4), 426–436.
10. Benthin, B.; Danz, H.; Hamburger, M. Pressurized Liquid Extraction of Medicinal Plants. *J. Chromatogr. A* **1999,** *837* (1–2), 211–219.
11. Trusheva, B.; Trunkova, D.; Bankova, V. Different Extraction Methods of Biologically Active Components From Propolis: A Preliminary Study. *Chem. Cent. J.* **2007,** *1* (13), 1050–1074.
12. Chemat, F. Ultrasound Assisted Extraction of Food and Natural products. Mechanisms, Techniques, Combinations, Protocols and Applications. A Review. *Ultrason. Sonochem.* **2017,** *34*, 540–560.
13. Cheng, X. In *Advances in Enzyme Assisted Extraction of Natural Products*, 3rd International Conference on Material, Mechanical and Manufacturing Engineering, Guangzhou, China, (June 27-28) 2015, 371–375.
14. Conservation, R. *Methods Optimization in Accelerated Solvent Extraction*, Technical Note 208, 2013, 1–4.
15. Dar, N. G. Evaluation of Different Techniques for Extraction of Antioxidants as Bioactive Compounds from Citrus Peels (Industrial By Products). *Am-Eur J. Agric. Environ. Sci.* **2015,** *15* (4), 676–682.
16. Delazar, A. Natural Products Isolation. **2012a,** 864.
17. Delazar, A. Natural Products Isolation. **2012b,** 864.
18. Devgun, M.; Nanda, A.; Ansari, S. H. Comparison of Convetional and Non Conventional Methods of Extraction of Heartwood of *Pterocarpus Marsupium* Roxb. *Acta Poloniae Pharmaceutica* **2012,** 69 (3), 475–485.
19. Dhanani, T. Comparison of Green Extraction Methods with Conventional Extraction Method for Extract Yield, L-DOPA Concentration and Antioxidant Activity of *Mucuna Pruriens* Seed. *Green Chem. Lett. Rev.* **2015,** *8* (2), 43–48.
20. Dias, D. A.,; Urban, S.; Roessner, U. A. Historical Overview of Natural Products in Drug Discovery. *Metabolites* **2012,** *2* (4), 303–336.
21. Farnsworth, N. R. Medicinal Plants in Therapy. *Bull. World Health Organ.* **1985,** *63* (6), 965–981.
22. Giannuzzo, A. N. Supercritical Fluid Extraction of Naringin from the Peel of *Citrus Paradisi. Phytochem. Anal.* **2013,** *14* (4), 221–223.
23. Herrero, M.; Cifuentes, A.; Ibanez, E. Sub- and Supercritical Fluid Extraction of Functional Ingredients From Different Natural Sources: Plants, Food-By-Products, Algae and Microalgae—A Review. *Food Chem.* **2006,** *98* (1), 136–148.
24. Ingle, K. P. Phytochemicals: Extraction Methods, Identification and Detection of Bioactive Compounds From Plant Extracts. *Afr. J. Tradit. Complement Altern. Med.* **2017,** *6* (1), 32–36.

25. Janghel, A. Supercritical Fluid Extraction (SFE) Techniques as an Innovative Green Technologies for the Effective Extraction of the Active Phytopharmaceuticals. *Res. J. Pharm. Technol.* **2015,** *8* (6), 775–786.
26. Jensen, W. B. The Origin of the Soxhlet Extractor. *J. Chem. Educ.* **2007,** *84* (12), 1913.
27. Kamarudin, N. A.; Markom, M.; Latip, J. Effects of Solvents and Extraction Methods on Herbal Plants *Phyllanthus Niruri*, *Orthosiphon Stamineus* and *Labisia Pumila*, *Indian J. Sci. Technol.* **2016,** *9*, 3–7.
28. Kostova, I. Natural Product Chemistry For Drug Discovery. *J. Nat. Prod.* **2010,** *5* (8), 440.
29. Kulshreshtha, G. Enzyme-Assisted Extraction of Bioactive Material From *Chondrus Crispus* and *Codium Fragile* and Its Effect on Herpes Simplex Virus (HSV-1). *Mar. Drugs* **2015,** *13* (1), 558–580.
30. Latif, S.; Anwar, F. Physicochemical Studies of Hemp (*Cannabis sativa*) Seed Oil Using Enzyme-Assisted Cold-Pressing. *Eur. J. Lipid Sci. Technol.* **2009,** *111* (10), 1042–1048.
31. Lavric, V. Microwave Extraction of Active Principles, **2012**, *74*.
32. Ma, S. Soxhlet-Assisted Matrix Solid Phase Dispersion to Extract Flavonoids From Rape (*Brassica campestris*) Bee Pollen. *J. Chromatogr. B: Anal. Technol. Biomed. Life Sci.* **2015**, 1005, 17–22.
33. Shahid, M.; Yusuf, M.; Mohammad, F. Plant Phenolics : A Review on Modern Extraction Techniques. **2015**.
34. Moradalizadeh, M.; Samadi, N.; and Rajaei, P. Comparison of Hydrodistillation, Microwave Hydrodistillation and Solvent Free Microwave Methods in Analysis of the Essential Oils from Aerial Parts of *Haplophyllum robustum* Bge. By GC/MS Method. *Int. J. Adv. Biol. Biomed. Res.* **2013,** *1* (9), 1058–1067.
35. Murugan, R.; Parimelazhagan, T. Comparative Evaluation of Different Extraction Methods for Antioxidant and Anti-inflammatory Properties From Osbeckia Parvifolia Arn—An In Vitro Approach. *J. King Saud Univ.—Science* **2014,** *26* (4), 267–275.
36. Naudé, Y. Comparison of Supercritical Fluid Extraction and Soxhlet Extraction for the Determination of DDT, DDD, and DDE in Sediment. *Water SA* **1998,** *24* (3), 205–214.
37. Azwanida, N. N. Review on the Extraction Methods Use in Medicinal Plants, Principle, Strength and Limitation. *Med. Aromat. Plants* **2015,** *4* (3), 3–8.
38. Oancea, S. Conventional and Non-Conventional Extraction of Anthocyanins From *Rubus Idaeus* L. Grown in Romania. **2002,** 247–251.
39. Pourmortazavi, S. M.; Hajimirsadeghi, S. S. Supercritical Fluid Extraction in Plant Essential and Volatile Oil Analysis. *J. Chromatogr. A* **2007,** *1163* (1–2), 2–24.
40. Alara, O. R.; Olalere, O. A. A Critical Overview on the Extraction of Bioactive Compounds from Phaleria macrocarpa (Thymelaceae). *Nat. Prod. Chem. Res.* **2016,** *4,* 232. DOI:10.4172/2329-6836.1000232.
41. Puértolas, E. Pulsed-Electric-Field-Assisted Extraction of Anthocyanins From Purple-Fleshed Potato. *Food Chem.* **2013,** *136* (3–4), 1330–1336.
42. Puri, M.; Sharma, D.; Barrow, C. J. Enzyme-Assisted Extraction of Bioactives From Plants. *Trends Biotechnol.* **2012,** *30* (1), 37–44.

43. Rai, A.; Mohanty, B.; Bhargava, R. Supercritical Extraction of Sunflower Oil: A Central Composite Design for Extraction Variables. *Food Chem.* **2016,** *192*, 647–659.
44. Routray, W.; Orsat, V. Microwave-Assisted Extraction of Flavonoids: A Review. *Food Bioprocess Technol.* **2012,** *5* (2), 409–424.
45. Rostagno, M. A.; Palma, M.; Barroso, C. G. Pressurized Liquid Extraction of Isoflavones From Soybeans. *Analytica Chimica Acta* **2004,** *522,* 169–177.
46. Samaram, S. Ultrasound-Assisted Extraction (UAE) and Solvent Extraction of Papaya Seed Oil: Yield, Fatty Acid Composition and Triacylglycerol Profile. *Molecules* **2013**, *18* (10), 12474–12487.
47. Sasidharan, S. Extraction, Isolation and Characterization of Bioactive Compounds From Plants' Extracts. *Afr. J. Tradit. Complement. Altern. Med.* **2011,** *8* (1), 1–10.
48. Shams, K. A. Green Technology: Economically and Environmentally Innovative Methods for Extraction of Medicinal and Aromatic Plants (MAP) in Egypt. *J. Chem. Pharm. Res.* **2015,** *7* (5), 1050–1074.
49. da Silva, R. P. F. F.; Rocha-Santos, T. A. P.; Duarte, A. C. Supercritical Fluid Extraction of Bioactive Compounds. *TrAC—Trends Anal. Chem.* **2016,** *76*, 40–51.
50. Subramanian, R. Double Bypasses Soxhlet Apparatus for Extraction of Piperine From *Piper nigrum*. *Arab. J. Chem.* **2016,** *9*, S537–S540.
51. Suchan, P. Pressurized Liquid Extraction in Determination of Polychlorinated Biphenyls and Organochlorine Pesticides in Fish Samples. *Anal. Chim. Acta* **2004,** *520* (1–2), 193–200.
52. Swami Handa, S. Extraction Techniques of Medicinal Plants. *Extr. Technol. Med. Aromat. Plants* **2008,** 1–10.
53. Tatke, P.; Jaiswal, Y. An Overview of Microwave Assisted Extraction and Its Applications in Herbal Drug Research. *Res. J. Med. Plant* **2011,** 21–31.
54. Thi, T.; Nhi, Y.; Tuan, D. Q. Enzyme Assisted Extraction of GAC Oil (*Momordica cochinchinensis* Spreng) From Dried Aril. **2016,** *4* (1), 1–6.
55. Tiwari, B. K. Ultrasound: A Clean, Green Extraction Technology. *TrAC—Trends Anal. Chem.* **2015,** *71*, 100–109.
56. Trendafilova, A.; Chanev, C.; Todorova, M. Ultrasound-Assisted Extraction of Alantolactone and Isoalantolactone From Inula Helenium Roots. *Pharmacogn. Mag.* **2010,** *6* (23), 234.
57. Trusheva, B.; Trunkova, D.; Bankova, V. Different Extraction Methods of Biologically Active Components From Propolis: A Preliminary Study. *Chem. Cent. J.* **2007,** *1* (1), p. 13.
58. Vankar, P. S. Essential Oils and Fragrances From Natural Sources. *Resonance* **2004,** *9* (4), 30–41.
59. Xue, D.; Farid, M. M. Pulsed Electric Field Extraction of Valuable Compounds From White Button Mushroom (*Agaricus bisporus*). *Innov. Food Sci. Emerg. Technol.* **2015,** *29*, 178–186.
60. Yan, L.-G.; He, L.; Xi, J. High Intensity Pulsed Electric Field as an Innovative Technique for Extraction of Bioactive Compounds—A review. *Crit. Rev. Food Sci. Nutr.* **2017,** *57* (13), 2877–2888.
61. Yung, O. H.; Maskat, M. Y.; Wan Mustapha, W. A. Kesan Pengekstrakan Terhadap Kandungan Polifenol, Aktiviti Antipengoksida Dan pH Ekstrak Pegaga (*Centella asiatica*). *Sains Malays.* **2010,** *39* (5), 747–752.

62. Zhang, D.-S. Ultrasound-Assisted Extraction of Total Flavonoids From *Inula helenium*. *Pharmacog. Mag.* **2012,** *8* (30), 166.
63. Zhang, H. F.; Yang, X. H.; Wang, Y. Microwave Assisted Extraction of Secondary Metabolites From Plants: Current Status and Future Directions. *Trends Food Sci. Technol.* **2011,** *22* (12), 672–688.
64. Zhang, J. Extraction Methods of Natural Products from Traditional Chinese Medicines. *1263*, 177–185.

CHAPTER 3

BIOACTIVE COMPONENTS IN FRUITS AND VEGETABLES

AAMIR SHEHZAD[1,*], NIDA KANWAL[1], ANAM LATIF[1], ASNA ZAHID[1], and HAFIZ ANSAR RASUL SULERIA[2]

[1]*Faculty of Food, Nutrition and Home Sciences, National Institute of Food Science and Technology, University of Agriculture, Faisalabad 38040, Pakistan*

[2]*School of Life and Environmental Sciences, Faculty of Science, Engineering and Built Environment, Centre for Chemistry and Biotechnology, Deakin University, 75 Pigdons Road, Geelong, VIC 3216, Australia*

[*]*Corresponding author. E-mail: draamir@uaf.edu.pk*

ABSTRACT

Consumption of fruits and vegetables in daily diet is substantially escalating as they provide high amount of biologically active components or phytochemicals that bestow benefits to human health beyond basic nutritional needs. The biologically active compounds present in fruits and vegetables are diverse ranging from polyphenols, fibers, conjugated isomers of linoleic acid, gallate, D-limolene, epigallocatechin, soy protein, vitamins A, B, C, and E, isoflavones, selenium, calcium, alipharin, chlorophyllin, catechin, sulfides, tetrahydrocurecumin, uric acid, seasaminol, glutathione, indoles, thiocyanates, protease inhibitors, etc. These phytochemicals have been categorized into various broad classes such as phenolics, N-containing compounds, alkaloids, organosulfur compounds, and phytosterols. The intake of dietary phytochemicals is significantly related to lower mortality from coronary heart diseases and cancer. They prevent the low-density lipoprotein oxidation and also have anti-ischemic, antioxidative, antiangiogenic, antihypercholesteroemic, and anti-inflammatory properties. Likewise, they also inhibit the cancer progression by altering cellular signal transduction pathway and regulating apoptosis. Phytochemicals also have angiotensin-converting enzyme inhibiting and hypoglycemic role. In order to attain the maximum benefits, these phytochemicals are extracted by using conventional or modern techniques.

3.1 INTRODUCTION

Plants, particularly those with ethnopharmacological usages, have always been the prime basis of medicines for initial drug discovery. It has been shown by an analysis that 80% of plant-derived drugs have ethno-pharmacological purposes. That is why World Health Organization is reassuring the use of herbal products or medicines in developing countries to treat numerous chronic ailments.[69]

The use of fruits and vegetables is increasing at a much higher rate as they contain higher quantities of biologically active components that bestow benefits to human health beyond basic nutritional needs.[23] Fruits and vegetables deliver an optimum combination of antioxidants, such as polyphenols, vitamins C and E, and carotenoids, accompanying complex carbohydrates and fiber. When consumed by humans, antioxidants perform various functions from controlling blood sugar, lowering high blood pressure, acting as anticarcinogenic, and lowering cholesterol to support the immunity. They have antifungal, antibacterial, antiviral, anti-inflammatory, and antithrombic properties.[14]

The most meticulously examined dietary bioactive compounds in vegetables and fruits serving as antioxidants are polyphenols, fibers, conjugated isomers of linoleic acid, gallate, D-limolene, epigallocatechin, soy protein, vitamins A, B, C, and E, isoflavones, selenium, calcium, alipharin, chlorophyllin, catechin, sulfides, tetrahydrocurecumin, uric acid, seasaminol, glutathione, indoles, thiocyanates, and protease inhibitors.[23] Bioactive components are present in both, the foods of natural or synthetic origin having definite physiological or metabolic actions provided with proven safety for human consumption.[61] The levels of bioactive components in all fruits and vegetables are determined by factors like the growing conditions, the cultivar, transport, and storage conditions.[11]

The bioactive compounds are extracted to attain their maximum benefits. Their extraction is dependent on many other factors such as extraction technique, extraction solvent used, and raw materials.[64] The extraction techniques are categorized into conventional and non-conventional techniques. Conventional techniques necessitate the application of organic solvents, agitation, and temperature. Soxhlet, hydrodistillation, and maceration are the examples of conventional technique. Non-conventional techniques, also known as modern techniques, are clean or green techniques as they require less energy and less application of organic solvents, providing benefit to the atmosphere.[56]

3.2 PHYTOCHEMICALS IN FRUITS AND VEGETABLES

Phytochemicals are nonnutritive plant-derived components with various bioactive activities; they are credited to guard the plants against pathogens, predators, competitors, and other environmental offenses, and boost human health. At present, approximately 20,000 phytochemicals have been certified in around 7000 edible plants. These phytochemicals have been categorized into various broad classes such as phenolics, N-containing compounds, alkaloids, organosulfur compounds, and phytosterols.[39]

TABLE 3.1 Categorization of Phytochemicals Present in Fruits and Vegetables.

Classification	Subgroups	Bioactive components	Main sources
Phenolics	Anthocyanins	Cyanidin Delphinidin	Apple, peach, berries, red grape, red cabbage, carrot
	Flavonols	Quercetin Kaempferol	Apple, citrus fruits, apricot, onions, berries
	Flavanols	Catechin Epicatechin	Apple, purple/red grapes peach, berries, apricot Green tea
	Flavones	Apigenin	Carrot, parsley, celery
	Hydroxybenzoic acids	Gallic acid	Blackberry, mango, cherry, purple/red grapes
	Lignan	Pinoresinol	Banana, orange, cranberry, broccoli, peach
	Coumarins	Coumarins	Tonka bean, mullein, vanilla grass

TABLE 3.1 *(Continued)*

Classification	Subgroups	Bioactive components	Main sources
Alkaloids	Betacyanins	Betanin	Red beet
N-containing compounds	Benzylamines	Capsaicin	Chili peppers
	Tryptamines	Psilocybin	Mushrooms
	Aliphatic glucosinulates	Sulforaphane	Broccoli, cabbage, Brussels, cauliflower
Organic acids	Aldonic acids	Ascorbic acid	Pepper, berries, kiwi, tomato, citrus fruits
Steroids	Sterols	Campesterol β-sitosterol Stigmastrero	Apples, banana, cruciferous vegetables, onions, asparagus

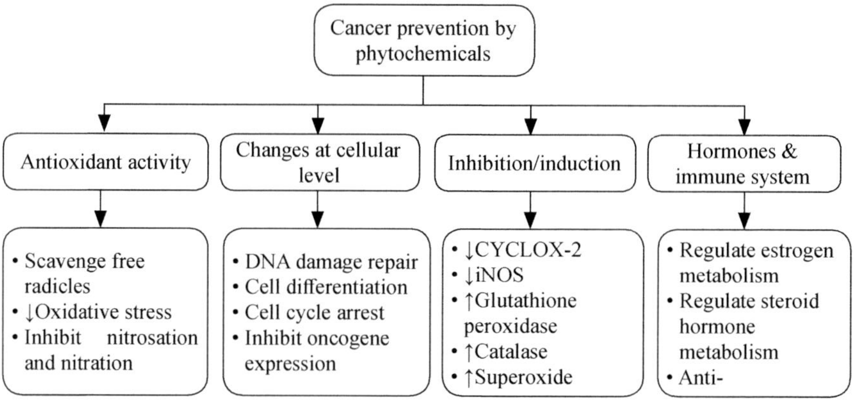

FIGURE 3.1 Mechanism of cancer prevention by phytochemicals.

3.3.3 ANGIOTENSIN CONVERTING ENZYME INHIBITORS

Hypertension is one of the main public health problems arising across the globe. The main contributing factor in developing hypertension is the overactivation of renin–angiotensin aldosterone system (RAAS). Angiotensinogen converting enzyme (ACE) performs an important role in RAAS. It converts the angiotensin I to angiotensin II, a peptide accountable for increasing the blood pressure.[7] Thus, the foremost therapeutic tool in regulating high blood pressure is the inhibition of angiotensin converting enzyme (ACE). The investigation on ACE inhibitors is intensifying largely and most of them are on natural plant-derived products as polyphenols and peptides.[6]

A research was carried out to investigate the effect of plant flavonoids as ACI inhibitor in controlling high blood pressure. The ACE inhibition of different doses of plant flavonoids were in range of 43–75%. Thus, it was concluded that certain dietary flavonoids from fruits and vegetables possess hypertension regulation properties.[6]

3.3.4 ANTIDIABETIC PROPERTIES

Diabetes is one of the most prevalent noncommunicable disease worldwide,[76] and nowadays, known as one of the major threats to human well-being.[19] The prevalence of diabetes is rapidly increasing worldwide with about 382 million affected people, and this figure is expected to escalate to 552 million by 2035.[33] The management of this disease necessitates rigorous efforts. Application of dietary therapy in the treatment of this disease is one of the options aimed at controlling diabetes-induced hyperglycaemia. The mechanism by which antidiabetic functional foods help in managing diabetes is by their ability to act as either insulin-secreting, -sensitizing, or -mimetic agents. Foods that increase insulin secretion by the β-cell of pancreas are termed as insulin-secreting foods, whereas those which improve the response of insulin toward glucose are labeled as insulin-sensitizing. Insulin-mimetic foods are those which have the ability to mimic insulin action.[36]

Numerous studies observed dietary designs and occurrence of type 2 diabetes and indicated that consumption of certain fruits and vegetables lowers the risk of diabetes.[31,44,66] Furthermore, intake of green leafy vegetables has potential to reduce the occurrence of type 2 diabetes as indicated by many researchers.[8,20,40,71] Antioxidants, for instance, polyphenols in fruits and vegetables are responsible for protective effects against diabetes.[2] Hypoglycemic effects of fruits and vegetables are attributed to their insulin-like activity,[63] or due to bioactive components such as anthocyanidins and anthocyanins present in them.[34,74]

Some plant components comprising carotenoids, steroids, flavonoids, alkaloids, amino acids, triterpenoids, guanidine, glycopeptides, peptidoglycans, polysaccharides, galactomannan gum, and inorganic ions have shown antidiabetic activity.[24,43] Alkaloids show antidiabetic activity by inhibition of α-glucosidase and thereby, lowering transportation of glucose over intestinal epithelium.[49] Steroidal glycosides, triterpenoids, and polysaccharides improve serum insulin levels, lower blood glucose, and increase glucose tolerance[54]; however, flavonoids increase activity of hepatic glucokinase thereby enhancing insulin release from the pancreatic islets.[70]

A number of fruits and vegetables have shown antidiabetic activity including avocado, barberry, bitter gourd, garlic, ginger, grape, guava, jackfruit, lychee, onion, pomegranate, papaya, pumpkin, watermelon, and chicory.[9,36] The future research must aim at isolation of these bioactive components responsible for hypoglycemic or antidiabetic effects for nutraceuticals as well as functional food ingredients.

3.4 EXTRACTION TECHNIQUES

Fruits and vegetables are rich sources of bioactive compounds which can be extracted and have positive effects on human health[61] Extraction of bioactive components from fruits and vegetables depends upon various factors including nature of raw material, extraction technique, and type of solvent applied to extract the bioactive compounds.[64] Up till now, several techniques have been used for extracting bioactive components from fruits and vegetables and they were classified into two categories by many authors, that is, conventional and nonconventional or modern techniques. Recently, many researchers/studies have used the word "green techniques" or "green technologies" for modern or nonconventional techniques as they implement the use of organic solvent, are environment-friendly, and require minimum energy.[56]

Figure 3.2 shows an outline of conventional and nonconventional extraction techniques. Generally, bioactive compounds from fruits and vegetables or any other natural source are extracted following well-defined protocols. This section provides a clear summary/overview of techniques used to extract bioactive components from the fruits and vegetables.

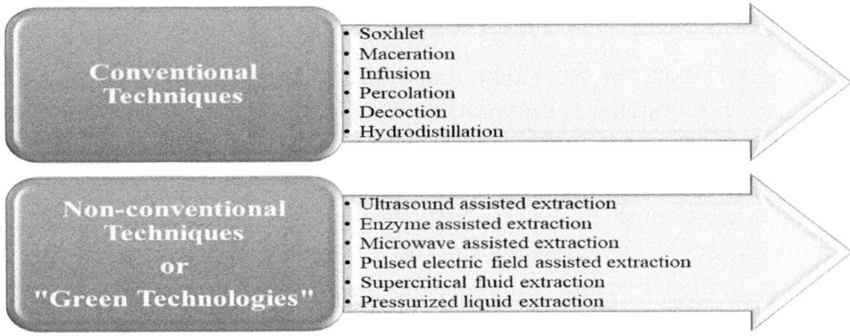

FIGURE 3.2 Techniques used to extract bioactive compounds.

3.4.1 CONVENTIONAL EXTRACTION TECHNIQUES

For extracting bioactive compounds from fruits and vegetables, various classical extraction techniques have been employed. The extraction efficiency of these techniques lies behind the extraction power of solvent and the application of heat.[4] Conventional extraction techniques, Soxhlet, maceration, infusion, percolation, decoction, and hydrodistillation, are mainly applied for extracting bioactive compounds including[4,5,15]:

3.4.1.1 SOXHLET EXTRACTION

Extraction of food components using Soxhlet method dates back to 1879 when German chemist Franz Ritter Von Soxhlet proposed a Soxhlet extractor for the first time.[62] It was intended to be used primarily for lipid extraction at that time but with passage of time, it has been extensively used to extract valuable components from natural sources. This technique is used as a reference or standard for comparative evaluation of alternative new extraction techniques. Commonly, a lesser quantity of dried sample is kept in thimble; which is positioned in distillation flask containing solvent of certain interest. When solvent approaches to a run-off level, a siphon aspirates thimble holder solution, which is unloaded back to distillation flask. The bulk liquid in distillation flask has extracted solute and solvent goes back to solid bed of plant. The process is carried out until complete extraction of the desired compound.[4] However, this method involves long extraction time and bulk quantity of solvent.[29]

3.4.1.2 MACERATION

Maceration is one of the oldest techniques, and from a long time, it has been used for homemade preparation of tonics and medicinal formulations. To get bioactive components from plant materials, maceration was considered as a popular and inexpensive way for small-scale extractions. Maceration process comprises several steps: particle size of sample is reduced by grinding process, which also increases the surface area accessible by the solvent. A solvent named as menstruum is added in a closed container or vessel along with powdered sample and kept at room temperature with the sample for minimum 3 days under repeated agitation process. The soluble bioactive components are released during this process, whereas impurities are removed through filtration.[25] Extraction process using maceration technique is facilitated by random shaking, which improves extraction efficiency by two ways (1) increased diffusion and (2) removing concentrated solution from the surface of sample carrying fresh solvent toward the surface for improving the extraction yield.[4,5]

3.4.1.3 DECOCTION AND INFUSION

The basic principle for extraction of bioactive components using decoction and infusion technique is same as of maceration. The raw material is soaked in boiling or cold water. The only difference of decoction from maceration and infusion is that it requires shorter time period and specified volume of water (i.e., 1:4 or 1:16) and is only appropriate for extraction of heat-stable components and hard plant materials.[5] This is an appropriate method for extraction of bioactive compounds that are soluble in water. This processes do not require any expensive equipment.[12]

3.4.1.4 PERCOLATION

This technique employs use of the unique equipment known as percolator and follows same fundamental principle of maceration. Finely ground sample is filled in the percolator; then, boiling water is applied and macerated for 2 h. The process is accomplished at moderate rate (six drops per minute) until complete extraction. Finally, evaporation is done to get extracts in concentrated form.[55]

3.4.1.5 HYDRODISTILLATION

Hydrodistillation is a traditional method for extraction of plants metabolites that does not use organic solvents. The three types of hydrodistillation are: (1) water distillation, (2) water and steam distillation, and (3) direct steam distillation.[68] For the extraction of bioactive compounds using hydrodistillation, the raw material is filled in a compartment or vessel; sufficient amount of water is added and mixture is brought to boiling. Otherwise, direct steam can also be injected into the plant sample. The three main physicochemical procedures involved in hydrodistillation are: (1) hydrodiffusion, (2) hydrolysis, and (3) decomposition by heat. One of the major drawbacks of this technique is that it causes destruction of some volatile components at a higher temperature which limits its use for extraction of heat-sensitive compounds.[59]

3.4.2 NONCONVENTIONAL TECHNIQUES OR GREEN TECHNOLOGIES

The challenges associated with conventional extraction techniques directed to the introduction of new and promising extraction techniques, termed as nonconventional extraction techniques.[61] Some of the propitious extraction techniques include: ultrasound-assisted extraction (UAE), enzyme-assisted extraction (EAE), microwave-assisted extraction (MAE), pulsed electric field-assisted extraction, supercritical fluid extraction (SFE), and pressurized liquid extraction (PLE). These methods were referred to as "green techniques" or "green technologies" because of their compliance with standards established by Environmental Protection Agency (EPA), USA.[4]

3.4.2.1 ULTRASOUND-ASSISTED EXTRACTION

UAE involves use of special sound waves in range of 20 kHz to 100 MHz.[25] UAE works on principle of cavitation effect which is only produced by liquids and liquid comprising solid material.[30] Permeability of cell walls as well as surface contact between solvent and sample is increased because of mechanical effect of acoustic cavitation produced by the ultrasound waves. The ultrasound waves cause disruption of plant cell wall and alter physicochemical properties of plant materials, which consequently facilitate release of bioactive compounds and increase bulk transportation of solvents toward plant cell wall.[16,30] This technique employs simple and comparatively low-cost technology that is suitable for extraction of bioactive components both at small and large scale.

3.4.2.2 ENZYME-ASSISTED EXTRACTION

EAE implicates use of particular enzymes like pectinase, α-amylase as well as cellulose for extracting bioactive components from plants. The procedure involves pretreatment with specific enzymes which is reflected as new and efficacious approach to liberate bounded components as well as to enhance overall production.[57] This technique is particularly effective in releasing phytochemicals from plant matrices that are dispersed in hydrophobic or hydrogen bonding in polysaccharide–lignin complex or retained in cell cytoplasm. In a routine extraction process, the bounded or dispersed phytochemicals are not available with a solvent and therefore, addition of specific enzymes during extraction process improves recovery by cell wall breakage and hydrolysis of structural polysaccharides and lipid moieties.[60] EAE comprises two methods: (1) enzyme-assisted cold pressing and (2) enzyme-assisted aqueous extraction.[38] Extraction efficiency of this technique is influenced by many factors including moisture content of plant materials, enzyme concentration and composition, hydrolysis time, solid to water ratio, and particle size of plant materials.[18,47] The technique is accepted as environmental friendly for extracting bioactive components as it involves water usage as solvent instead of organic elements.[53]

3.4.2.3 MICROWAVE-ASSISTED EXTRACTION

MAE employs use of microwaves in the frequency ranging from 300 MHz to 300 GHz and is recognized as an innovative technique for extraction of soluble components into a fluid by means of microwave energy.[50] The three successive stages involved in the mechanism lying behind MAE are: (1) separation of bioactive components from sample milieu at elevated pressure and temperature, (2) solvent is diffused through sample matrix, and

(3) the solute or bioactive compounds are released from sample medium to the solvent.[1]

3.4.2.4 PULSED-ELECTRIC FIELD EXTRACTION

During last decade, pulsed electric field was accepted as an appropriate method for enhancing the drying, pressing, diffusion, and extraction efficiency.[3,72,73] The principle mechanism involved in pulsed-electric field extraction (PEF) is its ability to destroy the structure of cell membrane and thereby improving extraction. Generally, a simple circuit having exponential decay pulses is employed for PEF treatment of plant samples. It comprises a treatment chamber having two electrodes for placement of sample. The PEF process can be run either in batch or continuous mode depending on the chamber design.[52] The extraction efficiency of PEF technique relies upon field strength, pulse number, temperature, energy input, and properties of the sample.[28] This technique is also appropriate as a pretreatment process on plant materials before any conventional extraction method is applied to decrease the energy required for extraction process.[42]

3.4.2.5 SUPERCRITICAL FLUID EXTRACTION

The extraction of bioactive components using SFE dates back to 1879 with the discovery of supercritical fluid by Hannay and Hogarth.[26] Supercritical fluid has solvating properties of a liquid and works more alike gas. An excellent example is CO_2 that is converted to supercritical fluid above 7380 kPa and 31.1°C. The importance of CO_2 for being used in SFE was developed because of its ability to act as excellent solvent for nonpolar solutes and also its availability at low cost with little toxicity. Addition of small quantity of methanol or ethanol enables extraction of polar compounds using CO_2 as supercritical fluid. However, one of the main limitations of this technique is high initial cost of equipment.[5]

3.4.2.6 PRESSURIZED LIQUID EXTRACTION

PLE was developed in 1996 for the first time and currently, it is known by a number of names including accelerated fluid extraction, pressurized fluid extraction, high-pressure solvent extraction, and enhanced solvent extraction.[46] The main concept lying behind this extraction technique is the use of high pressure to keep solvent liquid above its regular boiling point. The extraction process is facilitated by application of high pressure and it requires small quantity of solvent as faster extraction is accomplished by combination of high pressure and temperature.[32] Recently, PLE has been accepted as a possible alternative to SFE for extraction of polar compounds.[35]

3.4.3 *COMBINATION TECHNIQUES*

With the advancement and development of extraction approaches, recently, combination techniques were introduced to save resources and energy. The strategy used in these techniques is combination of analytical techniques with different sample preparation methods, for example, extraction of bioactive compounds using ultrasound could be combined with other extraction techniques such as supercritical CO_2, extraction heat-reflux, and microwave.[45] Many authors used combination techniques and established that they can be used to improve extraction efficiency and amount of bioactive components and to reduce processing duration.[17,22,51,75]

3.5 SUMMARY

Conclusively, phytochemicals in fruits and vegetables can be regarded as natural remedies to a number of ailments owing to their health-promoting potentials. The ability of these phytochemicals to prevent health-related disorders is attributed to the presence of diverse biologically active components. Keeping in view, their disease-preventive abilities, filling up human diet with rich diversity of fruits and vegetables on a daily basis will improve the health, appearance, and performance. These components can be efficiently extracted using specific conventional and nonconventional techniques. However, isolation of specific phytochemicals, their structure, and specificity for biological activity are the major challenges to be addressed in this field. It is expected that the development of specific plants with higher levels of phytochemicals through bioengineering will not only help to overcome such concerns but also incorporate enough phytochemicals in our food.

KEYWORDS

- **phytochemicals**
- **anticarcinogenic**
- **hypoglycemic**
- **conventional techniques**
- **non-conventional techniquesantioxidants**
- **nutraceuticals and functional foods**

REFERENCES

1. Alupului, A. Microwave Extraction of Active Principles from Medicinal Plants. *U. P. B. Sci. Bull. Series B.* **2012,** *74* (2), 129–142.
2. Anderson, R. A.; Broadhurst, C. L.; Polansky, M. M.; Schmidt, W. F.; Khan, A.; Flanagan, V. P.; Schoene, N. W.; Graves, D. J. Isolation and Characterization of Polyphenol Type-A Polymers from Cinnamon with Insulin-like Biological Activity. *J. Agricult. Food Chem.* **2004,** *52*, 65–70.
3. Angersbach, A.; Heinz, V.; Knorr, D. Effects of Pulsed Electric Fields on Cell Membranes in Real Food Systems. *Innov. Food Sci. Emerg. Technol.* **2000,** *1* (2), 135–149.
4. Azmir, J.; Zaidul, I. S. M.; Rahman, M. M.; Sharif, K. M.; Mohamed, A.; Sahena, F.; Jahurul, M. H. A.; Ghafoor, K.; Norulaini, N. A. N.; Omar, A. K. M. Techniques for Extraction of Bioactive Compounds from Plant Materials: A Review. *J. Food Eng.* **2013,** *117*, 426–436.
5. Azwanida, N. N. A Review on the Extraction Methods Use in Medicinal Plants, Principle, Strength and Limitation. *Med. Aromatic Plants*. **2015,** *4*, 1–6.
6. Balasuriya, B. N.; Rupasinghe, H. V. Plant Flavonoids as Angiotensin Converting Enzyme Inhibitors in Regulation of Hypertension. *Funct. Foods Health Dis.* **2011,** *1* (5), 172–188.
7. Balasuriya, N.; Rupasinghe, H. V. Antihypertensive Properties of Flavonoid-Rich Apple Peel Extract. *Food Chem.* **2012,** *135* (4), 2320–2325.
8. Bazzano, L. A.; Li, T. Y.; Joshipura, K. J.; Hu, F. B. Intake of Fruit, Vegetables, and Fruit Juices and Risk of Diabetes in Women. *Diabetes Care* **2008,** *31*, 1311–1317.
9. Beidokhti, M. N.; Jäger, A. K. Review of Antidiabetic Fruits, Vegetables, Beverages, Oils and Spices Commonly Consumed in the Diet. *J. Ethnopharmacol.* **2017,** *201*, 26–41.

10. Benetou, V.; Orfanos, P.; Lagiou, P.; Trichopoulos, D.; Boffetta, P.; Trichopoulos, A. Vegetables and Fruits in Relation to Cancer Risk: Evidence from the Greek EPIC Cohort Study. *Cancer Epidemiol. Biomark. Prevent.* **2008,** *17* (2), 387–392.
11. Bennett, L. E.; Jegasothy, H.; Konczak, I.; Frank, D.; Sudharmarajan, S.; Clingeleffer, P. R. Total Polyphenolics and Anti-Oxidant Properties of Selected Dried Fruits and Relationships to Drying Conditions. *J. Funct. Foods* **2011,** *3* (2), 115–124.
12. Bimakr, M.; Rahman, R. A.; Taip, S. F.; Ganjloo, A.; MdSalleh, L.; Selamat, J.; Hamid, A.; Zaidul, I. S. M. Comparison of Different Extraction Methods for the Extraction of Major Bioactive Flavonoid Compounds from Spearmint (*Mentha spicata* L.) Leaves. *Food Bioprod. Process.* **2011,** *89*, 67–72.
13. Boffetta, P.; Couto, E.; Wichmann, J.; Ferrari, P.; Trichopoulos, D.; Bueno-de-Mesquita, H. B.; van Duijnhoven, F. J.; Büchner, F. L.; Key, T.; Boeing, H.; Nöthlings, U. Fruit and Vegetable Intake and Overall Cancer Risk in the European Prospective Investigation into Cancer and Nutrition (EPIC). *J. Natl. Cancer Inst.* **2010,** *102* (8), 529–537.
14. Bub, A.; Watzl, B.; Blockhaus, M.; Briviba, K.; Liegibel, U.; Müller, H.; PoolZobel, B. L.; Rechkemmer, G. Fruit Juice Consumption Modulates Antioxidative Status, Immune Status and DNA Damage. *J. Nutr. Biochem.* **2003,** *14* (2), 90–98.
15. Bubalo, M. C.; Vidovic′, S.; Redovnikovic′, I. R.; Jokic′, S. New Perspective in Extraction of Plant Biologically Active Compounds by Green Solvents. *Food Bioprod. Process.* **2018**, *109*, 52–73.
16. Dhanani, T.; Shah, S.; Gajbhiye, N. A.; Kumar, S. Effect of Extraction Methods on Yield, Phytochemical Constituents and Antioxidant Activity of *Withania somnifera. Arab. J. Chem.* **2017,** *10*, S1193–S1199.
17. Dias, A. L. B.; Sergio, C. S. A.; Santos, P.; Barbero, G. F.; Rezende, C. A.; Martínez, J. Effect of Ultrasound on the Supercritical CO_2 Extraction of Bioactive Compounds from Dedo de Moça Pepper (*Capsicum baccatum* L. var. pendulum). *Ultrason. Sonochem.* **2016,** *31*, 284–294.
18. Dominguez, H.; Ntiiiez, M. J.; Lema, J. M. Enzyme-Assisted Hexane Extraction of Soybean Oil. *Food Chem.* **1995,** *54* (2), 223–231.
19. Ezuruike, U. F.; Prieto, J. M. The Use of Plants in the Traditional Management of Diabetes in Nigeria: Pharmacological and Toxicological Considerations. *J. Ethnopharmacol.* **2014,** *155*, 857–924.
20. Ford, E. S.; Mokdad, A. H. Fruit and Vegetable Consumption and Diabetes Mellitus Incidence Among US Adults. *Prevent. Med.* **2001,** *32,* 33–39.
21. Freedman, N. D.; Park, Y.; Subar, A. F.; Hollenbeck, A. R.; Leitzmann, M. F.; Schatzkin, A.; Abnet, C. C. Fruit and Vegetable Intake and Head and Neck Cancer Risk in a Large United States Prospective Cohort Study. *Int. J. Cancer*. **2008,** *122* (10), 2330–2336.
22. Garcia-Mendoza, M. P.; Paula, J. T.; Paviani, L. C.; Cabral, F. A.; MartinezCorrea, H. A. Extracts from Mango Peel by-Product Obtained by Supercritical CO_2 and Pressurized Solvent Processes. *LWT-Food Sci. Technol.* **2015,** *62* (1), 131–137.
23. Gonzalez-Aguilar, G.; Robles-Sánchez, R. M.; Martínez-Téllez, M. A.; Olivas, G. I.; Alvarez-Parrilla, E.; De La Rosa, L. A. Bioactive Compounds in Fruits: Health Benefits and Effect of Storage Conditions. *Stewart Postharvest Rev.* **2008,** *4* (3), 1–10.

24. Grover, J. K.; Yadav, S.; Vats, V. Medicinal Plants of India with Anti-Diabetic Potential. *J. Ethnopharmacol.* **2002,** *81*, 81–100.
25. Handa, S. S.; Khanuja, S. P. S.; Longo, G.; Rakesh, D. D. *Extraction Technologies for Medicinal and Aromatic Plants*, 1st ed.; Earth, Environmental and Marine Sciences and Technologies: Trieste, Italy, 2008; p 260.
26. Hannay, J. B.; Hogarth, J. On the Solubility of Solid in Gases. *Proc. R. Soc. Lond.* **1879,** *29*, 324–326.
27. Heidemann, C.; Schulze, M. B.; Franco, O. H.; van Dam, R. M.; Mantzoros, C. S.; Hu, F. B. Dietary Patterns and Risk of Mortality from Cardiovascular Disease, Cancer, and All-Causes in a Prospective Cohort of Women. *Circulation* **2008,** *118* (3), 230–237.
28. Heinz, V.; Toepfl, S.; Knorr, D. Impact of Temperature on Lethality and Energy Efficiency of Apple Juice Pasteurization by Pulsed Electric Fields Treatment. *Innovat. Food Sci. Emerg. Technol.* **2003,** *4* (2), 167–175.
29. Heleno, S. A.; Diz, P.; Prieto, M. A.; Barros, L.; Rodrigues, A.; Barreiro, M. F.; Ferreira, I. C. Optimization of Ultrasound-Assisted Extraction to Obtain Mycosterols from *Agaricus bisporus* L. by Response Surface Methodology and Comparison with Conventional Soxhlet Extraction. *Food Chem.* **2016,** *197*, 1054–1063.
30. Herrera, M. C.; Luque de Castro, M. D. Ultrasound-Assisted Extraction of Phenolic Compounds from Strawberries Prior to Liquid Chromatographic Separation and Photodiode Array Ultraviolet Detection. *J. Chromatogr.* **2005,** *1100* (1), 1–7.
31. Hodge, A. M.; English, D. R.; O'Dea, K.; Giles, G. G. Dietary Patterns and Diabetes Incidence in the Melbourne Collaborative Cohort Study. *Am. J. Epidemiol.* **2007,** *165*, 603–610.
32. Ibanez, E.; Herrero, M.; Mendiola, J. A.; Castro-Puyana, M. Extraction and Characterization of Bioactive Compounds with Health Benefits from Marine Resources: Macro and Micro Algae, Cyanobacteria, and Invertebrates. In Hayes, M., Ed., *Marine Bioactive Compounds: Sources, Characterization and Applications*; Springer: New York, 2012; pp 55–98.
33. International Diabetes Federation (IDF). *Global Diabetes Scorecard-Tracking Progress for Action*; Brussels, Belgium 2014; p 150.
34. Jayaprakasam, B.; Vareed, S. K.; Olson, L. K.; Nair, M. G. Insulin Secretion by Bioactive Anthocyanins and Anthocyanidins Present in Fruits. *J. Agricult. Food Chem.* **2005,** *53*, 28–31.
35. Kaufmann, B.; Christen, P. Recent Extraction Techniques for Natural Products: Microwave-Assisted Extraction and Pressurized Solvent Extraction. *Phytochem. Anal.* **2002,** *13* (2), 105–113.
36. Kazeem, M. I.; Davies, T. C. Anti-Diabetic Functional Foods as Sources of Insulin Secreting, Insulin Sensitizing and Insulin Mimetic Agents. *J. Funct. Foods*. **2016,** *20*, 122–138.
37. Khoo, H. E.; Prasad, K. N.; Kong, K. W.; Jiang, Y.; Ismail, A. 2011. Carotenoids and Their Isomers: Color Pigments in Fruits and Vegetables. *Molecules* **2011,** *16* (2), 1710–1738.
38. Latif, S.; Anwar, F. Physicochemical Studies of Hemp (*Cannabis sativa*) Seed Oil Using Enzyme-Assisted Cold-Pressing. *Eur. J. Lipid Sci. Technol.* **2009,** *111* (10), 1042–1048.

39. Liu, R. H. Potential Synergy of Phytochemicals in Cancer Prevention: Mechanism of Action. *J. Nutr.* **2004a,** *134* (12), 3479–3485.
40. Liu, S.; Serdula, M.; Janket, S. J.; Cook, N. R.; Sesso, H. D.; Willett, W. C.; Manson, J. E.; Buring, J. E. A Prospective Study of Fruit and Vegetable Intake and the Risk of Type 2 Diabetes in Women. *Diabetes Care* **2004b,** *27*, 2993–2996.
41. Liu, R. H. Dietary Bioactive Compounds and Their Health Implications. *J. Food Sci.* **2013,** *78* (1), 18–25.
42. Lopez, N.; Puertolas, E.; Condon, S.; Raso, J.; Alvarez, I. Enhancement of the Extraction of Betanine from Red Beetroot by Pulsed Electric Fields. *J. Food Eng.* **2009,** *90* (1), 60–66.
43. Malviya, N.; Jain, S.; Malviya, S. A. P. N. A. Antidiabetic Potential of Medicinal Plants. *Acta Poloniae Pharmaceut.* **2010,** *67*, 113–118.
44. Montonen, J.; Knekt, P.; Härkänen, T.; Järvinen, R.; Heliövaara, M.; Aromaa, A.; Reunanen, A. Dietary Patterns and the Incidence of Type 2 Diabetes. *Am. J. Epidemiol.* **2005,** *161*, 219–227.
45. Mustafa, A.; Turner, C. Pressurized Liquid Extraction as a Green Approach in Food and Herbal Plants Extraction: A Review. *Anal. Chim. Acta.* **2011,** *703* (1), 8–18.
46. Nieto, A.; Borrull, F.; Pocurull, E.; Marcé, R. M. Pressurized Liquid Extraction: A Useful Technique to Extract Pharmaceuticals and Personal-Care Products from Sewage Sludge. *TrAC Trends Anal. Chem.* **2010,** *29* (7), 752–764.
47. Niranjan, K.; Hanmoungjai, P. Enzyme-Aided Aquous Extraction. In Dunford, N. T., Dunford, H. B., Eds., *Nutritionally Enhanced Edible Oil Processing*; AOCS Publishing: Urbana, Illinois, 2004; p 314.
48. Oude Griep, L. M.; Geleijnse, J. M.; Kromhout, D.; Ocké, M. C.; Verschuren, W. M. M. Raw and Processed Fruit and Vegetable Consumption and 10-Year Coronary Heart Disease Incidence in a Population-Based Cohort Study in The Netherlands. *PLoS One* **2010,** *5* (10), e13609.
49. Pan, G. Y.; Huang, Z. J.; Wang, G. J.; Fawcett, J. P.; Liu, X. D.; Zhao, X. C.; Sun, J. G.; Xie, Y. Y. The Antihyperglycaemic Activity of Berberine Arises from a Decrease of Glucose Absorption. *Planta Med.* **2003,** *69*, 632–636.
50. Paré, J. J. R.; Bélanger, J. M. R.; Stafford, S. S. Microwave-Assisted Process (MAP™): A New Tool for the Analytical Laboratory. *TrAC Trends Anal. Chem.* **1994,** *13* (4), 176–184.
51. Parniakov, O.; Barba, F. J.; Grimi, N.; Lebovka, N.; Vorobiev, E. Extraction Assisted by Pulsed Electric Energy as a Potential Tool for Green and Sustainable Recovery of Nutritionally Valuable Compounds from Mango Peels. *Food Chem.* **2016,** *192*, 842–848.
52. Puértolas, E.; López, N.; Saldaña, G.; Álvarez, I.; Raso, J. Evaluation of Phenolic Extraction During Fermentation of Red Grapes Treated by a Continuous Pulsed Electric Fields Process at Pilot-Plant Scale. *J. Food Eng.* **2010,** *119* (3), 1063–1070.
53. Puri, M.; Sharma, D.; Barrow, C. J. Enzyme-Assisted Extraction of Bioactives from Plants. *Trends Biotechnol.* **2012,** *30* (1), 37–44.
54. Quanhong, L.; Caili, F.; Yukui, R.; Guanghui, H.; Tongyi, C. Effects of Protein Bound Polysaccharide Isolated from Pumpkin on Insulin in Diabetic Rats. *Plant Foods Human Nutr.* **2005,** *60* (1), 13–16.

55. Rathi, B. S.; Bodhankar, S. L.; Baheti, A. M. Evaluation of Aqueous Leaves Extract of *Moringa oleifera* Linn for Wound Healing in Albino Rats. *Ind. J. Exp. Biol.* **2006,** *44*, 898–901.
56. Rodríguez-Pérez, C.; Quirantes-Piné, R.; Fernández-Gutiérrez, A.; Segura Carretero, A. Optimization of Extraction Method to Obtain a Phenolic Compounds-Rich Extract from *Moringa Oleifera* Lam Leaves. *Industr. Crops Prod.* **2015,** *66*, 246–254.
57. Rosenthal, A.; Pyle, D. L.; Niranjan, K. Aqueous and Enzymatic Processes for Edible Oil Extraction. *Enzyme Microb. Technol.* **1996,** *19* (6), 402–420.
58. Scalbert, A.; Andres-Lacueva, C.; Arita, M.; Kroon, P.; Manach, C.; Urpi-Sarda, M.; Wishart, D. Databases on Food Phytochemicals and Their Health-Promoting Effects. *J. Agricult. Food Chem.* **2011,** *59*, 4331.
59. Silva, L. V.; Nelson, D. L.; Drummond, M. F. B.; Dufosse, L.; Gloria, M. B. A. Comparison of Hydrodistillation Methods for the Deodorization of Turmeric. *Food Res. Int.* **2005,** *38* (8–9), 1087–1096.
60. Singh, R. K.; Sarker, B. C.; Kumbhar, B. K.; Agrawal, Y. C.; Kulshreshtha, M. K. Response Surface Analysis of Enzyme-Assisted oil Extraction Factors for Sesame, Groundnut, and Sunflower Seeds. *J. Food Sci. Technol.* **1999,** *36* (6), 511–514.
61. Soquetta, M. B., Terra, L. D. M.; Bastos, C. P. Green Technologies for the Extraction of Bioactive Compounds in Fruits and Vegetables. *CyTA-J. Food.* **2018,** *16* (1), 400–412.
62. Soxhlet. F. Die gewichtsanalytische Bestimmung des Milchfettes. *Digitalisier. Polytechnisc. J.* **1879,** *232*, 461–465.
63. Survay, N. S.; Ko, E. Y.; Upadhyay, C. P.; Jang, M.; Park, S. W.; Lee, D. H.; Hong, S. J. Hypoglycemic Effects of Fruits and Vegetables in Hyperglycemic Rats for Prevention of Type-2 Diabetes. *Kor. J. Horticult. Sci. Technol.* **2010,** *28*, 850–856.
64. Tiwari, B. K. Ultrasound: A Clean, Green Extraction Technology. *TrAC—Trends Anal. Chem.* **2015,** *71*, 100–109.
65. Trichopoulos, A.; Costacou, T.; Bamia, C.; Dimitrios, T. Adherence to a Mediterranean Diet and Survival in a Greek Population. *N. Engl. J. Med.* **2003,** *348*, 2599–2608.
66. Van Dam, R. M.; Rimm, E. B.; Willett, W. C.; Stampfer, M. J.; Hu, F. B. Dietary Patterns and Risk for Type 2 Diabetes Mellitus in US Men. *Ann. Internal Med.* **2002,** *136*, 201–209.
67. van Duijnhoven, F. J.; Bueno-De-Mesquita, H. B.; Ferrari, P.; Jenab, M.; Boshuizen, H. C.; Ros, M. M.; Casagrande, C.; Tjønneland, A.; Olsen, A.; Overvad, K.; Thorlacius-Ussing, O. Fruit, Vegetables, and Colorectal Cancer Risk: The European Prospective Investigation into Cancer and Nutrition. *Am. J. Clin. Nutr.* **2009,** *89* (5), 1441–1452.
68. Vankar, P. S. Essential Oils and Fragrances from Natural Sources. *Resonance* **2004,** *9* (4), 30–41.
69. Vasanthi, H. R.; ShriShriMal, N.; Das, D. K. Phytochemicals from Plants to Combat Cardiovascular Disease. *Curr. Med. Chem.* **2012,** *19* (14), 2242–2251.
70. Vessal, M.; Hemmati, M.; Vasei, M. Antidiabetic Effects of Quercetin in Streptozocin-Induced Diabetic Rats. *Compar. Biochem. Physiol. Part C Toxicol. Pharmacol.* **2003,** *135*, 357–364.

71. Villegas, R.; Shu, X. O.; Gao, Y. T.; Yang, G.; Elasy, T.; Li, H.; Zheng, W. Vegetable but not Fruit Consumption Reduces the Risk of Type 2 Diabetes in Chinese Women. *J. Nutr.* **2008,** *138*, 574–580.
72. Vorobiev, E.; Jemai, A. B.; Bouzrara, H.; Lebovka, N. I.; Bazhal, M. I. Pulsed Electric Field Assisted Extraction of Juice from Food Plants. In Barbosa-Canovas, G., Tapia, M. S., Cano, M. P., Eds., *Novel Food Processing Technologies*; CRC Press: New York, 2005; pp 105–130.
73. Vorobiev, E.; Lebovka, N. I. Extraction of Intercellular Components by Pulsed Electric Fields. In Raso, J., Heinz, V., Eds.; *Pulsed Electric field Technology for the Food Industry: Fundamentals and Applications*; Springer: New York,. 2006; pp 153–194.
74. Wedick, N. M.; Pan, A.; Cassidy, A.; Rimm, E. B.; Sampson, L.; Rosner, B.; Willett, W.; Hu, F. B.; Sun, Q.; van Dam, R. M. Dietary Flavonoid Intakes and Risk of Type 2 Diabetes in US Men and Women. *Am. J. Clin. Nutr.* **2012,** *95*, 925–933.
75. Yang, Y. C.; Wei, M. C. Kinetic and Characterization Studies for Three Bioactive Compounds Extracted from Rabdosia Rubescens Using Ultrasound. *Food Bioprod. Process.* **2015,** *94*, 101–113.
76. Zimmet, P.; Alberti, K. G.; Shaw, J. Global and Societal Implications of the Diabetes Epidemic. *Nature* **2001,** *414*, 782–787.

CHAPTER 4

PHYTOCHEMICALS FROM FRUIT WASTES FOR SUSTAINABLE AND SOCIALLY ACCEPTABLE POULTRY PRODUCTION

ZAFAR HAYAT[1], KASHIF AKRAM[2,*], ATHAR MAHMUD[3], UMAR FAROOQ[4], MUHAMMAD UMAR FARID[1], and AFSHAN SHAFI[4]

[1]Department of Animal Sciences, CVAS-Jhang campus, University of Veterinary & Animal Sciences, Lahore, Pakistan

[2]Department of Food Sciences, Cholistan University of Veterinary and Animal Sciences Bahawalpur, Pakistan

[3]Department of Poultry Production, University of Veterinary & Animal Sciences, Lahore, Pakistan

[4]Department of Food Science & Technology, MNS-University of Agriculture Multan, Pakistan

**Corresponding author. E-mail: kashifakramft@gmail.com*

ABSTRACT

Antibiotics have been supplemented in poultry diets as growth promoters for disease prevention and efficient production for the many decades. However, residues of some antibiotics used in poultry feeds as growth promoters have been found in poultry products that pose considerable risks to human health and acceptability. Emergence of antimicrobial resistance in human is the major threat associated with the incessant practice of supplementing the poultry diets with multiple antibiotics. Therefore, the scientific community has great concern on indiscriminate usage of antibiotic growth promoters (AGPs) in poultry feeds and most of the industrialized countries have banned the use of AGPs in the diets of poultry. Nevertheless, intensive poultry production requires alternatives to AGPs for sustainable and efficient poultry farming to feed the ever-increasing human population around the globe. Therefore, search for some potent, cost-effective, natural, and safe antibacterial principles are urgently required. Plants are a well-known source for isolation of a wide array of high-value components and bioactives, especially free and bound phenolics with potential as drug and nutraceutical leads. Agro-wastes are one of the potential sources of high-value bioactives including phenolics, which can be revalorized into useful nutraceuticals and pharmaceutical products. Currently, there has been an increasing interest in the investigation of bioactives obtained from the agro-residues such as fruit wastes. However, despite their significant nutraceutical potential, these agro-residues are often discarded as agro-wastes and thus are treated as underutilized materials. Considerable quantity and variety of bioactive compounds with therapeutic value are reported in fruit wastes. Anthocyanin, anthocyanidins, caffeic acid, catechin, cyanidin, epicatechin, eriocitrin, gallic acid, gallic acid derivatives, and procyanidins are some of the important phytochemicals present in waste materials of different fruits. These bioactive principles possess significant antibacterial capacity along with other health-promoting and disease-preventing potential. The use of fruit waste as a natural growth promoter is certainly considered as safe with low cost and no side effects that can enhance feed efficiency because of positive effects on the nutrient digestibility and better immune response to pathogenic organisms ensuring the efficient poultry production in terms of meat and eggs. The exploration of phytogenic feed additives based on fruit wastes is very helpful in global food and nutritional security and thus is a step forward toward utilization of sustainable natural plant resources.

4.1 INTRODUCTION

The growing human population and increasing demand for safe, value-added, and healthy food are major and continuous challenges in the 21st century. Concerns are especially focused on the animal agriculture, its potential impact on the environment, and role of animal products on human health. Poultry birds are efficient convertor of animal feed to supreme quality products with comparatively low global greenhouse gas emissions. Poultry meat and eggs are nutritious and have a large share toward satisfying the human demands for animal protein. However, sustainable supply of safe, healthy, and socially acceptable poultry products at an affordable price are important questions posed to government, academia, researchers, industry, and farmers around the globe. Antibiotics have been supplemented in poultry diets as growth promoters for disease prevention and efficient production for the last many decades. However, residues of some antibiotics used in poultry feeds as growth promoters have been found in poultry products that pose considerable risks to human health and acceptability. Emergence of antimicrobial resistance (AMR) in human is the major threat associated with the incessant practice of supplementing the poultry diets with multiple antibiotics. Therefore, the scientific community has great concern on indiscriminate usage of antibiotic growth promoters (AGPs) in poultry feeds and most of the industrialized countries have banned the use of AGPs in the diets of poultry. Nevertheless, intensive poultry production requires alternatives to AGPs for sustainable and efficient poultry farming to feed the ever-increasing human population around the globe. Therefore, search for some potent, cost-effective, natural, and safe antibacterial principles are urgently required.

Plants are a well-known source for isolation of a wide array of high-value components and bioactive, especially free and bound form phenolics with potential as drug and nutraceutical leads. These substances, especially phenolic compounds, have ability to donate protons and transfer electrons.[138] Fruit-wastes are one of the potential sources of high-value bioactives including phenolics that can be revalorized into useful nutraceuticals and pharmaceutical products. Currently, there has been an increasing interest in the investigation of bioactives obtained from agro-wastes such as peels, seeds, leaves, flowers, and stem bark of different fruits and vegetables. Despite their significant nutraceutical potential, these agro-residues are often discarded as agro-wastes and thus are treated as underutilized materials. Due to wide-scale production, consumption, and industrialization of the subject food crops, large quantities of agro-wastes in the form of seeds, rind/peel, and skin are thus produced, which not only have disposal problem but also are wasted as underutilized materials. Efficient techniques can be explored for separation and isolation of natural bioactive compounds from such materials. Considerable quantity and variety of bioactive compounds with therapeutic value are reported in fruit wastes. Anthocyanin, anthocyanidins, caffeic acid, catechin, cyanidin, epicatechin, eriocitrin, gallic acid, gallic acid derivatives, and procyanidins are some of the important phytochemicals present in waste materials of different fruits. These bioactive principles possess significant antibacterial capacity along with other health-promoting and disease-preventing potential.

The use of fruit waste as a natural growth promoter is certainly considered as safe with low cost and with no side effects that can enhance feed efficiency because of positive effects on the nutrient digestibility and better immune response to pathogenic organisms ensuring the efficient poultry production in terms of meat and eggs. The exploration of phytogenic feed additives based on fruit wastes would be helpful in national food and nutritional security and thus, a step forward toward sustainable utilization of agro-industrial waste.

This chapter focuses on the potential use of phytochemicals from fruit wastes as alternatives to antibiotics for sustainable and socially acceptable poultry production. Scientific data on existing, new, and/or emerging phytochemicals from fruit wastes that have potential use in poultry feed as growth promoters are discussed. A discussion of in vivo studies on the effects of phytochemicals on poultry nutrition, performance, and health is also included. This chapter also summarizes current experimental knowledge on the classification, sources, and possible mechanisms of action of phytochemicals as feed additives for poultry production.

4.2 ANTIBIOTICS: HISTORY AND DEVELOPMENT

The root of the term "antibiotic" can be traced back to the word "antibiose" to be first utilized as an antonym to symbiosis by Paul Vuillemin in his 1890 publication to portray the adversarial activity between various microorganisms (e.g., bacteria vs. fungi; bacteria vs. protozoa).[13,179] Afterward, antibiotic was characterized as an agent that was naturally produced by an organism or prepared synthetically that retards or kills microorganisms. The term "antibiotic" is derived from two words: "anti" means against and "biota" means life. Hence, in this way, antibiotic can be defined as "anti-life."

In the history of medicine, antibiotics presumably standout among the best types of chemotherapy. It is not important to duplicate here the number of lives they have spared and how fundamentally they played their role in control of infectious diseases that were the major causes of human illness and mortality.[7,179] Irrespective of the common belief that exposure to antibiotics is limited to the modern "antibiotic era," researches have revealed that this is not the situation. For example, remnants of ancient human skeleton of Sudanese Nubia dating back to 350–550 C.E. showed that the hints of tetracycline medication.[12,127] The presence of tetracycline drug in human skeletal bones is only possible after the exposure of tetracycline containing drug in their food of these historic people.

In 1929, the first antibiotic "penicillin" was discovered by Sir Alexander Fleming. Later, Ernst Chain and Howard Florey were able to develop a new technique to separate penicillin which has been used to treat bacterial infections during World War II. The discovery of penicillin was a great breakthrough for public health as it reduces the chances of disease spreading. In 1946, penicillin was first introduced for clinical use and made a significant effect on well-being of general public health. Revolutionary discoveries of French chemist and bacteriologist Louis Pasteur, including the four "Koch postulates" building up a causative relationship among microorganism and disease, impelled bacteriology toward its advanced era.[73,92]

The disclosure of first three antimicrobials, Salvarsan, prontosil, and penicillin, was model, as those investigations set up the standard for future research in drug discovery. The time period between the 1950s and 1970s was surely the golden era of discovery of different antibiotics drugs, with no new classes found from that point forward. However, with the decline in discovery rate, the standard approach for the development of new drugs to combat resistance of pathogens with antibiotics has been the alteration of existing antimicrobial drugs.[13]

4.3 ANTIMICROBIALS IN POULTRY PRODUCTION

Poultry industry has been using antibiotics to enhance production performance and feed conversion ration by reducing infections and disease burden. Antibiotics, together with efficient management, strict hygiene, and stringent biosecurity regimens, have been combating with noxious diseases and supporting the magnificent growth of poultry industry for the last six decades.[14,47,194] Mainly, antibiotics below the therapeutic levels are used for growth promotion in poultry.[47] The possible mechanism of AGPs may be the modification of poultry intestinal microbiome, which can influence health and immunity of birds.[39,97,175] Change in the dynamics of chicken microbiota has been reported by the use of different AGPs, such as penicillin,[159] salinomycin,[56] viginiamycin,[43] tylosin,[100] virginiamycin, or bacitracin methylene disalicylate plus roxarsone,[97] virginiamycin, and tylosin.[34] However, along with in-feed AGPs, the modulation of poultry microbial diversity depends on ingredient and nutrient profile of feed, magnitude of pathogens invasion, housing, and management conditions.[97]

Most of AGPs affect the population of Lactobacillus species, the principal commensal microbes in the chicken intestine to produce bile hydrolase salt. It is postulated that bile hydrolase salt production is decreased with the reduced growth of Lactobacillus spp. by the use of AGPs. Lower activity of bile hydrolase salt may stimulates energy release from lipid metabolism and thus improves body weight through proportional increase in the production of conjugated bile salts.[100] Tetracycline is a major antibiotic used for commercial poultry operations in North America accounting greater than 66% in The United States.[146] Other antibiotics routinely used for poultry production in North America are tylosin, virginiamycin, bambermycin, bacitracin, and salinomycin.[36]

4.4 CONCERNS RELATED TO USE OF ANTIBIOTICS GROWTH PROMOTERS IN POULTRY PRODUCTION

AMR is most common problem related to massive use of AGPs in poultry production.[11,38,52,57] Basically, AMR is the acquired capacity of bacteria to withstand and thrive in the existence of an antibiotic that is usually meant to check or destroy the similar pathogens.[148] World Health Organization (WHO) considered AMR as global public health concern.[186] Although AMR has complex and multidisciplinary perspective.[18] Nevertheless, the connection between AMR in farm workers and use of antibiotics on farms could be traced back to the 1970s.[98] Dibner and Richards (2005) reported the excessive use of antibiotics as growth promoters for intensive poultry farming which is exponentially increasing in the developing countries.[39]

4.4.1 BIORESISTANCE IN PATHOGENIC BACTERIA

Salmonella, Escherichia coli (particularly *E. coli O157:H7*), and *Campylobacter* are the most common pathogenic bacteria of poultry meat and eggs affecting millions of people by foodborn illnesses annually around the world. Only in The United States, Zhao et al.[193] reported around 76 million foodborne infections each year. Food-related deaths and sickness data of The United States revealed that a vast number of fatal and hospitalization cases were caused by *E. coli, Campylobacter*, and *Salmonella species*.[114,193] In Canada, the magnitude of poultry contaminations from *E. coli, Salmonella*, and *Campylobacter* are 96%, 34%, and 25%, respectively.[32] Ironically, use of multiple AGPs in poultry production contributed to the development of bioresistance in these and other pathogenic bacteria from poultry products that is very deleterious for birds and human health.[158,190]

4.4.2 ANTIBIOTICS RESIDUES IN POULTRY PRODUCTS

Residues of in-feed antibiotics have been found in poultry products that pose considerable risks to human health and acceptability. Residues of some antibiotics have been found in food of animal origin[36,146] posing very lethal effects to human nutrition and health.[26,61,95] Prevalence of bioresistance particularly for compromising human and animal health is also attributed to the antibiotics residues in poultry meat and eggs.[37]

4.4.3 IMPACT OF ANTIBIOTICS ON ENVIRONMENT

Antibiotics are also environmental contaminants and pollutants as their use greatly affect the ecosystem. Antibiotics given to animals are metabolized in the body and their metabolites as well as basic compounds are excreted in considerable quantities form urinary and fecal rout[23] and entered into the environment mainly through water disposal and solid waste management systems.[146] Manzetti and Ghisi (2014) estimated the cumulative use of antibiotics for animals as well humans with range of 0.1–0.2 million t, globally.[107] Consequently, the release of huge quantity (approximately 30–90% of antibiotics intake) is expected into environment[23] that is capable of biotransformation and contamination. Bioaccumulation of harmful antibiotics compounds in different environmental channels pose deleterious effects. Aquatic ecosystems are most susceptible to pollutants of antibiotics origin[107] and also are source of disseminating the genes of AMR.[35] Concerns of AMR are aggravated based on the reports that highlighted the residual effects of antibiotics in the environment.[23,146,154]

4.4.4 PUBLIC SENTIMENTS AND REGULATIONS REGARDING THE USE OF AGPs

Due to threats associated with the incessant practice of supplementing the poultry diets with multiple antibiotics, people are avoiding poultry products. Especially in the era of social media, this hype put a great pressure on the regulatory authorities. Scientific community also have great concern on indiscriminate usage of AGP in poultry feeds and most of the industrialized countries have banned the use of AGP in diets of poultry. Recent achievement by the animal industry in The United Kingdom is based on the realization that public sentiment as well as regulatory pressure is against the use of antibiotics in production. To maintain viable export markets, antibiotic-free or strictly controlled administration is necessary. The major UK retailers have imposed standards of antibiotic-free production on their suppliers. Antibiotic residues and drug resistance is one of the most significant topics of concern to the world's broiler industries were reviewed at the 2018 meeting of the International Poultry Council (IPC) in Amsterdam.

4.5 ALTERNATIVES TO ANTIBIOTICS GROWTH PROMOTERS FOR POULTRY

Although the impact of antibiotics supplementation of animal feeds to develop AMR is a controversial topic, a great debate is happening on the effect of in-feed antibiotics that could directly affect human health and presence of antibiotics residues in the poultry products and other animal source foods as well as the negative impacts associated with their residues in human food and excretion into the environment. However, concerns about the negative impact of antimicrobial and the consumers' pressure the European Union has imposed an embargo to exclude antibiotics from the list of feed additives since 2006.[36] European Surveillance of Veterinary Antimicrobial Consumption (ESVAC) has banned the use AGPs in the participating countries.[49]

Nevertheless, intensive poultry production requires alternatives to AGPs for sustainable and efficient poultry farming to feed the ever-increasing human population around the globe. Therefore, search for some potent, cost-effective, natural, and safe antibacterial principles are urgently required. Exploration of alternatives as substitute to antibiotics supplemented to poultry feed is an exciting area of current research worldwide. The objectives behind the development of alternatives to AGPs are to increase poultry performance without compromising animal health, environmental quality, and human safety. To achieve these objectives, core emphasis is on the provision balance nutrition, strengthening bird's immune system, and improvement in gut health.

Several alternatives have been developed to decrease the use of AGPs in poultry production. Main focus of the research and development in this regards remained on the quest of exploring natural, safe, and sustainable compounds with growth promotion effects comparable to AGPs. The objectives of these strategies are to achieve optimum animal production performance while keeping mortality and morbidity at the minimum possible level and protecting consumer health and environmental quality at the same time. Among different alternatives to AGPs, the most common are phytochemicals, organic acids, probiotics, prebiotics, synbiotics, bacteriophages, herbs, enzymes, immunostimulants, bacteriocins, acidifier, phytoncides, nanoparticles, egg antibodies, and essential oils.[58]

4.6 FRUIT WASTES AS IMPORTANT SOURCE OF PHYTOGENIC COMPOUNDS

Plants are a well-known source for isolation of a wide array of high-value components and bioactives, especially free and bound form phenolics with potential as drug and nutraceutical leads.[31] Phytogenic compounds are bioactives components of plant origin including herbs and spices. Different phytogenic compounds showed great effectiveness as alternatives to antibiotics in animal nutrition and feeding.[53,62,89,91,99,173,185]

Fruit-wastes are one of the potential sources of high-value bioactives including phenolics that can be revalorized into useful nutraceuticals and pharmaceutical products. In recent years, there has been an increasing interest in the investigation of bioactives obtained from the agro-wastes such as peels, seeds, leaves, flowers, and stem bark of different fruits and vegetables. Despite their significant nutraceutical potential, these agro-residues are often discarded as agro-wastes and thus are treated as underutilized materials. Due to wide-scale production, consumption, and industrialization of the subject food crops, large quantities of agro-wastes in the form of seeds, rind/peel, and skins are thus produced, which not only have disposal problem, nevertheless, these are wasted as under-utilized materials. Efficient techniques can be explored for separation and isolation of natural bioactive compounds from such materials. Considerable quantity and variety of bioactive compounds with therapeutic value are reported in fruit wastes. Anthocyanin, anthocyanidins, caffeic acid, catechin, cyanidin, epicatechin, eriocitrin, gallic acid, gallic acid derivatives, and procyanidins are some of important phytochemicals present in waste materials of different fruits. These bioactive principles possess significant antibacterial capacity along with other health-promoting and disease-preventing potential. Anticancer activity of botanical drugs, through improvement in immune system, was also reported.[41]

Fruits-processing industries generate a huge quantity of fruit waste, for example, approximately 20–40% of fruit waste are generated which pose great problem of disposal. Ayala-Zavala et al.[9] reported that different types of fresh-cut fruits produced variable amounts of by-products have high bioactive properties and[40,99,122] demonstrated that activities of mango peel is often more than the pulp. It has been testified that the total phenolic compound content of fruit's seed, such as mangoes, avocados, and jackfruits, is higher than that of the edible product, and that they are valuable sources of phytochemicals.[163] Mango pulping and juice plants produced large amount of mango peel waste that is a rich source of bioactive compounds. The shelf life of fresh processed meat and its products, transported over long distances, could be extended by using plant-based extracts in order to control meat-spoiling bacteria.[85]

In recent studies, it has been found that use of fruit waste meal of pineapple has resulted in better performance, carcass quality, and other parameters.[104] Similarly, it has been described that supplementation of mango peels, mango stone, orange waste, and carrot residues has shown that carrot residue resulted in better performance in terms of egg production and egg size, whereas orange waste and mango peels have comparable results with control in layers. However, mango stones have shown adverse effects. Furthermore, it is claimed that citrus peel feeding in broiler has exhibited significantly positive effects on cholesterol and weight gain.[28] Kasapidou et al.[86] also suggested in a review that fruit waste could be used as functional food to improve the product quality.

It has been reported that fruit by-products increase live body weight of animal without any side effects on human health. Supplementation of apple pomace-based diets with α-amylase, hemicellulose, protease, and β-glucanase improved growth as compared with control.[87,112,119,131,132,141,162,191] Fruit waste has been successfully incorporated in layer diets.[6] Siyal et al.[160] studied the effect of orange and banana peel on broiler growth performance and concluded that banana peel in broiler feed ration as a feed additive up to 3% is economical and have positive effect on broiler performance.

Zhang, et al.[192] examined the impact of mango saponin on growth, meat, and carcass data and blood biochemistry of broilers and found that mango saponin could be used as a feed additive in broiler chicks, and the inclusion level of 0.28% mango saponin in diet can improve broilers' production performance, blood biochemistry, and carcass characteristics. Supplementation of 10% Mango Fruit Reject Meal (MFRM) in diet of broiler chicks is safe and help in better growth performance.[133] Plants (*Trigonella* and *Withania somnifera*) also have antistress and immune-modulatory properties.[1,182,195] Many phytochemicals result in improved performance by increasing digestibility of and retention of nutrients, increased secretion of digestive enzymes,[77,78,96,139] mucous production,[76] antiviral, antioxidant, activation of immune response, and anti-anthelmintic properties.[24,42,54,75,130,166,174] World fruit production during the year 2016 is presented in Table 4.1.

4.7 MAJOR FRUIT WASTES AVAILABLE

4.7.1 APPLE

The processing of fruits including fruit juices, flavors, and concentrates is resulting in large quantity of fruit processing by-products, such as pomace mentioned in Figure 4.1. Out of the world production, 30–40% apples are not marketed due to their damaged appearance, while for juice extraction,

TABLE 4.1 World Fruit Production (2016).

	Production (million t)						Percentage share of world production				
	World	Africa	America	Asia	Europe	Oceania	Africa	America	Asia	Europe	Oceania
Apples	89.33	2.72	9.79	58.80	17.30	0.72	3.05	10.96	65.83	19.36	0.81
Apricots	3.88	0.51	0.10	2.41	0.84	0.01	13.25	2.64	62.06	21.75	0.30
Bananas	113.28	21.02	28.63	61.58	0.41	1.64	18.56	25.27	54.36	0.36	1.45
Berries	0.98	–	0.30	0.36	0.21	0.12	0.33	30.11	36.27	21.24	12.05
Cherries	2.32	0.02	0.44	1.12	0.72	0.02	0.90	18.93	48.38	30.91	0.88
Citrus	96.29	15.19	34.01	40.65	6.02	0.43	15.77	35.32	42.21	6.25	0.45
Cranberries	0.68	–	0.67	0.01	–	–	0.03	97.56	2.08	0.33	–
Dates	8.46	3.78	0.05	4.61	0.02	–	44.71	0.55	54.54	0.20	–
Figs	1.05	0.39	0.07	0.51	0.08	–	37.59	6.88	48.31	7.21	0.01
Grapes	77.44	4.88	13.66	28.92	27.80	2.18	6.30	17.64	37.34	35.90	2.82
Kiwi fruit	4.27	–	0.25	2.77	0.82	0.44	–	5.94	64.75	19.11	10.20
Mangoes, mangosteens, guavas	46.51	5.93	5.96	34.56	–	0.05	12.76	12.82	74.31	–	0.10
Melons, other (including cantaloupes)	31.17	1.89	3.62	23.53	1.89	0.24	6.06	11.61	75.49	6.05	0.79
Papayas	13.05	1.37	4.32	7.34	–	0.02	10.50	33.12	56.27	–	0.12
Peaches and nectarines	24.98	0.86	2.16	17.50	4.37	0.08	3.44	8.65	70.06	17.51	0.34
Pears	27.35	0.76	2.03	21.61	2.81	0.13	2.79	7.43	79.01	10.28	0.49
Pineapples	25.81	4.80	9.73	11.17	–	0.11	18.59	37.71	43.27	–	0.43
Plums and sloes	12.05	0.40	0.96	8.03	2.64	0.02	3.31	7.99	66.66	21.88	0.17
Raspberries	0.80	–	0.28	0.02	0.50	–	0.05	35.26	1.93	62.67	0.09
Strawberries	9.12	0.62	2.09	4.68	1.67	0.05	6.81	22.94	51.36	18.32	0.57
Watermelons	117.02	6.13	6.51	98.45	5.75	0.19	5.23	5.56	84.13	4.92	0.16

Source: Adapted with permission from FAOSATAT (2018).

20–30% apples are processed.[191] Apple pomace, which is the residue left after juice extraction, could be used as feed for livestock. The waste parts of apple fruit such as seed, peel, and pomace contain significant amount of bioactive compounds. Therefore, extraction of bioactive compounds can be achieved and can be used for the development of functional foods. Poly-phenols are present in large amounts in pomace and seeds rather than in edible tissues. These bioactive components have potential to reduce the occurrence of cardiovascular diseases as well as possessed anticancer, antioxidant, immune-modulatory, and antimicrobial properties.[80,178] As a raw material, apple pomace shows comparative advantages for lactic acid production due to (1) high content of polysaccharides (mainly starch, cellulose and hemi-cellulose), (2) presence of mono-, di-, and oligosaccharides, citric acid, and malic acid, as they can be metabolized by lactic acid bacteria, and (3) rich in metal ions (Mg, Mn, and Fe), as they could be helpful in limiting the price of nutrient enrichment for fermentation media.[69]

4.7.1. COMPOSITION OF APPLE POMACE AND SEEDS

Apple pomace consists of apple skin, seed, and stem with approximate composition of 95%, 2–4%, and 1%, respectively. About 70–75% moisture content and high biodegradable load (high BOD and COD) are present in apple pomace. Apple pomace contain a perishable amount of various phytochemicals such as sugars, dietary fibers, pectin, and tocopherols along with its basic nutrients.[16] Apple peel have high dietary fiber content (0.91%) having contribution of soluble and insoluble dietary fiber is 0.46% and 0.43%, respectively. Apple pomace comprises arabinose (14–43%), galacturonic acid (49–64%), and galactose (6–15%), whereas rhamnose, xylose, and glucose are present in traces.[115] Moreover, proteins, vitamins (A and C), and minerals (P 0.07–0.076%, K 0.43–0.95%, Mn 3.96–9.00 mg/kg, Ca 0.06–0.10%, Mg 0.02–0.36%, and Fe 31.80–38.30%) are also present in apple pomace.[71,88] On a dry weight basis, approximately 10–15% pectin is present in apple pomace.[46,180] Polyphenolics having strong antioxidant activity of quercetin glycosides, phloridzin; its oxidative products are present in apple pomace and also in seeds.[70,102,155] Fatty acids (80%) are present in the seeds, mainly in the form of linoleic and oleic acid. The phenolic compound contents are majorly dependent on the apple cultivar and different practices of cultivation.[117,196] Apple extracts having polyphenols have shown to inhibit proliferation of tumor cell I in vitro.[44] Flavanol catechins, epicatechins, and procyanidins present in apples are good for maintaining heart health, Flavanols neutralize cell-damaging free radicals and bolster cellular antioxidant defenses. Urinary tract and heart health can be maintained by the contribution of proanthocyanidins. Many dietary phytonutrients such as phenolics, carotenoids, and vitamins having strong antioxidant capacities are present in apples, protecting against free radicals. Phenolic compounds, which assist in chronic disease prevention, are present in high concentrations in the apple peel. Phenolic compounds also exhibit strong antioxidant properties. Along with antioxidant activity, phenolic compounds also perform anti-tumoral, antiviral, antibacterial, and antimutagenic activities. The production of conventional apple juice (straight apple pulp pressing or after pulp enzyming) has resulted in a juice which is poor in phenolics and exhibited only 3–10% antioxidant activity. Protective properties of polyphenols include antioxidant, antimicrobial, anticancer, and cardiovascular-protective activities. Antiradical activities of apple pomace are contributed by total phenolics, flavonoids, flavan -3-ols, etc.[81]

Zafar (2005) reported that in broiler feed, 20% maize can be safely replaced by the processed damaged apple, decreasing feed cost, and increasing broiler production without any harmful effect.[191] Matoo et al.[112] have reported that broiler feed, enriched with a specific enzyme formulation (α-amylase hemicellulase, protease, and β-glucanase), has contributed to better performance of broilers than those that were fed on nonsupplemented feed.[110]

4.7.2 BANANA (*Musa acuminata*)

In the history of human agriculture, banana (family: Musaceae) is one of the earliest crops.[134] Two types of bananas are present: (1) consumed as fruit when ripe and (2) "plantains" used for making beer, chips, vinegar, or starch. Type (1) represent 56% and type (2) represent 44% of the world banana production. In industries producing banana-based products, banana or plantain peels equivalent to 40% of fresh banana weight are produced as a waste product.[124] Banana waste can be fed to livestock. Banana wastes comprise small-sized bananas, spoiled bananas, peels, leaves, young stalks, and pseudo stems of bananas.

FIGURE 4.1 **(See color insert.)** Fruit-processing waste.

4.7.2.1 COMPOSITION OF BANANA WASTE

The nutrients are present in more concentrated form as the dried banana flaks contain only 3% water. Banana flaks can be used in broiler diets as each gram of it contains 340 calories of energy, 28 g of iron, and 760 IU of vitamin A. In our environment, there is high yield of orange and banana fruits, but their peels were wasted. It was revealed that addition of banana peel into the broiler ration as a feed additive up to level of 3.0% is economical and has positive effects on growth performance.[74,160] Banana peel is produced in a large quantity by the pulp industry and on examination, it was observed that it comprises various biologically active components/phytochemicals such as polyphenols, flavonoids, carotenoids, and others compounds that are advantageous in upgraded heath status of living organisms including humans and animals.[120,170] Emaga et al.[45] reported the nutrient contents in banana peel, which contains 40–50% dietary fiber, 8–11% proteins, and 2.2–10.95% lipids. According to Maini and Sethi (2000), banana peel has 79.2 g/100 g moisture content, 2.11 g/100 g minerals, 1.72/100 g fiber, and carbohydrates 5 g/100 g,[105] whereas Negesse et al. (2009) indicated the composition of banana peel in such a way that crude protein was 163 g/kg DM, ash 48 g/kg DM, ether extract 9.3 MJ/kg DM, and metabolizable energy 124 g/kg.[126] Banana peels are inherently rich in polymers such as lignin, cellulose, hemicellulose, and pectin.[3,45] Gorinstein et al. (2001a) reported that the phenolic compounds in the banana pulp were about 232 mg/100 g DM, which is just 25% of that found in peel. Higher amounts of catecholamines, dopamine, and 1-dopamine were also suggested along with phenolic compounds in banana peels.[65,67]

Maximum inclusion rates of 7.5% and 10% dried banana peels have been suggested for broiler diets.[17] The presence of tannins in the banana peel cause difficulty in digestion, so the inclusion rate of banana peel is kept limited.[183] Weekly weight grains are being decreased by replacing maize grains with dried plantain peelings in broiler diet at the rate of more than 7.5%.[172] Linearly, the feed intake increased as the level of peels increased to 10%, resulting in decreased growth.[149] Feed efficiency was not affected when 10% standard conventional diet of broilers was replaced by the dried plantain leaves.[108]

4.7.3 CITRUS FRUITS

To make juice, about 30% of the citrus fruit production is principally processed. Large amount of oil, pectin, and variety of by-products are present in citrus wastes.

4.7.3.1 CITRUS PULP

Citrus pulp can be defined as "the residues remaining after extraction of juice." Its concentration in whole fruit is 50–70% by weight. Its composition involves 60–65% peel, 10% seeds, whereas remaining 30–35% consist of internal tissues. Oranges (60%), grapefruits, and lemons are the main source of citrus pulp. It contains 5–10% CP, 6.2% ether extract , 54% water-soluble sugars, and 10–40% soluble fiber (pectin). By addition of lime, 1–2% calcium and 0.1% phosphorus were observed.[10] Depending on citrus species, variety and the harvesting season, the composition of dried citrus pulp vary. The occurrence of limonin (which is toxic to monogastrics) and fiber content in the seeds makes them less valuable for pigs and poultry.[63] Phenolics and flavonoid contents were reported to be higher in the by-products as compared to the final products. These compounds may result in the free radical inhibition activity. Higher levels of phytochemical compounds were reported in peel and seeds with respect to the edible tissues. Gorinstein et al.[67] reported that the phenolic compounds were 15% higher in the peels of lemon, oranges, and grapefruits than that of the pulps of these fruits. Lutein and zeaxanthin present in citrus fruits contribute to the maintenance of healthy vision.

Food safety and quality of fresh food and food products was improved by essential oils from citrus that offer the potential for all natural antimicrobial.[82,142] Subba et al.[168] determined antibacterial effects of orange and lemon oil on Salmonella and other food-borne organisms. Fisher and Phillips[50] found that Gram-negative bacteria are less sensitive than Gram positive. Disc diffusion assay technique was used to screen various citrus essential oils for their antibacterial activity against 11 serotypes/strains of Salmonella. Inhibitory activity against the *Salmonella* spp. was found to be exhibited by orange terpenes, singles-folded d-limonene, and orange essence terpenes. Orange terpenes and d-limonene both showed MICs of 1% on the disc diffusion assay.

4.7.3.2 CITRUS PEEL

The major by-product obtained from the citrus-processing industry is the citrus peel, which contains significant quantity of phenols, especially flavonoids and vitamins E and C.[66,94,106,125,147] The bioactive compounds in citrus peel have antibacterial, antifungal, and anti-inflammatory properties and improve immune system through increase in production of antibodies.[116,140] Extract of hexane, butanol, methanol, and benzene obtained from citrus plant peels showed their antimicrobial activity against *Bacillus subtilis, Bacillus cereus, Staphylococcus, E. coli*, and *Salmonella enteritidis*. Attempts were made to use sweet orange peel extract as a natural feed additive, and even as a medicinal supplement for animals. Oil extract from *Citrus reticulata* Blanco and *Citrus acida* Roxb caused destruction of various infection-causing bacteria, for example, *E. coli, pseudo, putida, S. enterica, L. innocua, B. subtilis, Staph. aureus, Lactobacillus lactis, Sacharo. cerevisiae, E. coli, Staphylococuss aureus, Staphylococuss epidermidis, Enterococcus faecalis, S. typhimurium, Enterobacter cloacae, B subtilis, B. cereus, Enterobacter aerogenes, Salmonella typhimurium, Aspergillus ficuum, Aspergillus niger, Aspergillus fumigatus, Aspergillus flavus, Fusarium saloni, Fusarium digitatum,* and *Candida utilis*.[27,128] Due to the presence of nonstarch polysaccharides, which is responsible for impaired growth rate, lowered feed efficiency, and reduced carcass yields, the level of citrus pulp should not exceed 5–10% in the diets of poultry.[123] However, the amount within limits (up to 10%) showed no adverse effect in laying hens.[188] Erhan et al.[48] showed the positive effects of 3 mL/kg orange peel oil on the performance of jejunum microflora and jejunum morphology. In broilers feed, it could be advised to use 3 mL/kg orange peel oil as feed additives.

4.7.4 PINEAPPLE

During the processing of pineapple, before peeling, the crown and stem are cut off and then core is removed. The wastes (peel, core, stem, crown, and leaves) account for 50% of the pineapple weight.[21,153] Valuable components including sucrose, glucose, fructose, and other nutrients are present in pineapple wastes.

4.7.4.1 COMPOSITION OF PINEAPPLE WASTE

In pineapple, waste moisture content is 91.35%; ash, 0.04 mg/100 g; sugar, 9.75%; crude fiber, 0.60 g/100 g; protein, 10 mg/100 g; and ascorbic acid 26.5 mg/100 g.[15,72] Phenolic content of pineapple was found as 40.4 mg/100g as gallic acid, which is equivalent to the highest bound phenolic compound ethyl acetate,[169] 2.58 as chlorogenic equivalent[68] while the juice had 358 mg/L equivalent amount with the gallic acid.[60]

4.7.5 GRAPES

A huge quantity of residue, mainly in the form of grape pomace, is generated from grapes processing and wine production industries; grapes-processing waste is rich in polyphenols, predominantly proanthocyanidins.[181] Proanthocyanidins, also known as condensed tannins, are carbon–carbon bonds-linked oligomers of monomeric flavan-3-ol units. Epicatechin, epicatechin-3-ogallate, and catechin are most abundant proanthocyanidins in grapes waste.[30,187] Free-radical scavenging activity to terminate oxidative reactions and metal-chelating ability of proanthocyanidins may be responsible for their beneficial effects.[59,65,90] Use of grapes wastes in poultry diets demonstrated the improved antioxidant potential[19,20,25,64]

4.7.6 MANGO

Mango waste consists of seed, seed kernel, and peel of mango fruit. Mango seed composition includes shells (29%), kernel (68%), and testa (3%).[36,129] Mango shells can be utilized for burning purpose as fuel, whereas seed kernel contains a vast variety of phytochemicals. The composition of these phytochemicals depends upon climatic conditions, soil structure, mango fruit variety, etc.[137,144] Mango seed kernel contains 11% fat, 6% protein, 77% carbohydrates, 2% ash, and 2% crude fiber on dry weight basis. A noteworthy quantity of minerals (sodium, calcium, phosphorus, potassium, and magnesium) was present in mango kernel.[118,151] Soong and Barlow (2006) documented that mango seed kernel oil comprises polyunsaturated fatty acids (52–56%) and saturated fatty acids (44–48%).[164]

They also observed a large amount of essential amino acids in mango seed kernel oil. Sterol and vitamin K was the predominant biologically active components present in mango seed kernel oil.[145]

Mango peel's phytochemical composition indicated that it contains total dietary fiber (45–78%).[4] Moreover, mango peel also contains pectin, lipids, cellulose, carotenoids, hemicelluloses, proteins, and polyphenols. Considerable amounts of reducing sugars are present in mango peel, which can be further used for various food-processing operations such as bioenergy production, development of fermented products, and various other value-added products.[161] Rehman et al. (2004) reported huge quantity of pectin (10–15%) in mango peel and suggested that soaking process can increase the yield of pectin up to 21%, if it is performed before pectin extraction.[143] Another study performed on comparison of two types of pectins obtained from citrus peel and mango peel indicated that pectin extracted from mango peel has better gelling characteristic as compared to pectin extracted from citrus peel.[93]

Ahmed et al.[2] measured the phenolic composition of mango seed kernel. Bioactive compounds of mango seed kernel were reported as 20.7 mg tannin, 6 mg gallic acid, 12.6 mg coumarin, 7.7 mg caffeic acid, 20.2 mg vanillin, 4.2 mg mangiferin, 10.4 mg ferulic acid, and 11.2 mg cinnamic acid.[109] It was estimated that about 112 mg GAE/100 g polyphenolic contents are present in mango kernel extract. Similar results were also observed by Ajila et al. (2010) and Palmeira et al. (2012) while working on phytochemistry of mango waste.[5,135]

4.8 PHYTOCHEMICALS OF FRUIT WASTES WITH SIGNIFICANT ANTIMICROBIAL POTENTIAL

Significant antimicrobial activities have been attributed to fruit wastes such as grape seeds and peels, pomegranate peel, and mango seed kernel.[83] Phytochemically, the compounds responsible for antimicrobial properties are inherent constituents of the plant defense system protecting the fruits from pathogenic and spoilage microorganisms. Different phytochemicals found in fruit wastes are summarized in Table 4.2

TABLE 4.2 Phytochemicals Present in Fruit Wastes.

Fruit	Processing waste	Major phytochemicals	References
Apple *(Malus* sp., *Rosaceae)*	Apple pomace	Catechins Hydroxycinnamates Phloretin glycosides Procyanidins Quercetin glycosides	[51, 101, 103, 156, 171]
Banana (*Musa x paradisiaca* L., Musaceae)	Banana peel	Anthocyanin Carotenoids (palmitate or caprate, xanthophylls, laurate)	[167]
	Banana bract	Anthocyanidins Delphinidin Pelargonidin Peonidin Petunidin Malvidin Cyanidin	[136]
Citrus	Peel and solid residues	Eriocitrin Hesperidin Naringin Narirutin	[29, 111, 171]
Cranberry	Leaves	Catechin Epicatechin Myricetin-3-xylopiranoside Procyanidin B1 Quercetin-3-*O*-galactoside Dimethoxymyricetin-hexoside Methoxyquercetin-pentoside	[171]
Kiwi fruit	Peel	Caffeic acid Protocatechuic acid P-coumaric acid	[113, 184]
Guava (*Psidium guajava* L., Myrtaceae)	Guava peel and seed	Anthocyanins Coumarin Lycopene Resveratrol	[33]
Mango (*Mangifera indica* L., Anacardiaceae)	Mango peel and kernel	Ellagic acid Flavonol glycosides Gallates Gallic acid Gallotannins	[8, 157]

TABLE 4.2 *(Continued)*

Fruit	Processing waste	Major phytochemicals	References
Grape (*Vitis* sp., Vitaceae)	Grape pomace	Alcohols Aminoethylthio-flavan-3-ol conjugates Anthocyanins Epicatechin Epigallocatechin catechins Flavonol glycosides Phenolic acids Stilbenes	[156, 165, 176]
	Grape seeds	Catechin Epicatechin Epicatechin-O-gallate Epigallocatechin-O-gallate Gallic acid Gallocatechin Gallocatechin-O-gallate Procyanidin B1 Procyanidin B2 Procyanidins	[55, 79, 84, 150]
	Grape skin	Catechin Epicatechin Epigallocatechin Picatechin gallate	[152, 165, 177]
Papaya (*Carica papaya* L., Caricaceae)		Anthocyanins Lycopene	[33]
Pineapple (*Ananas comosus* L. MERR., Bromeliaceae)		Anthocyanins	[33]

4.9 MODES OF ACTION OF ANTIMICROBIAL COMPOUNDS

Natural antimicrobial compounds act on microorganisms through multiple and complex manner. As phytochemicals from fruit wastes are present in the form of a mixture of different compounds, the synergies of the biological activities make them to exhibit potent and broad-spectrum antimicrobial activities.[81] Bioactive principles can cause lysis of pathogenic agents by penetrating into their cell membranes. Phenolics compounds have ability to stop bacterial spore germination.[22] Change in pH is another way of inhibiting microorganism growth. Bacteria normally grow efficiently on pH of 6.5–7.5 (with tolerable limits of pH 4–9). Reduction in pH lower than the tolerable limit could be detrimental to genetic material of pathogens, their cell membrane, protein synthesis system, and metabolic enzymes.[189] Some important antibacterial modes of action of fruit waste's phytochemicals are illustrated in Figure 4.2.

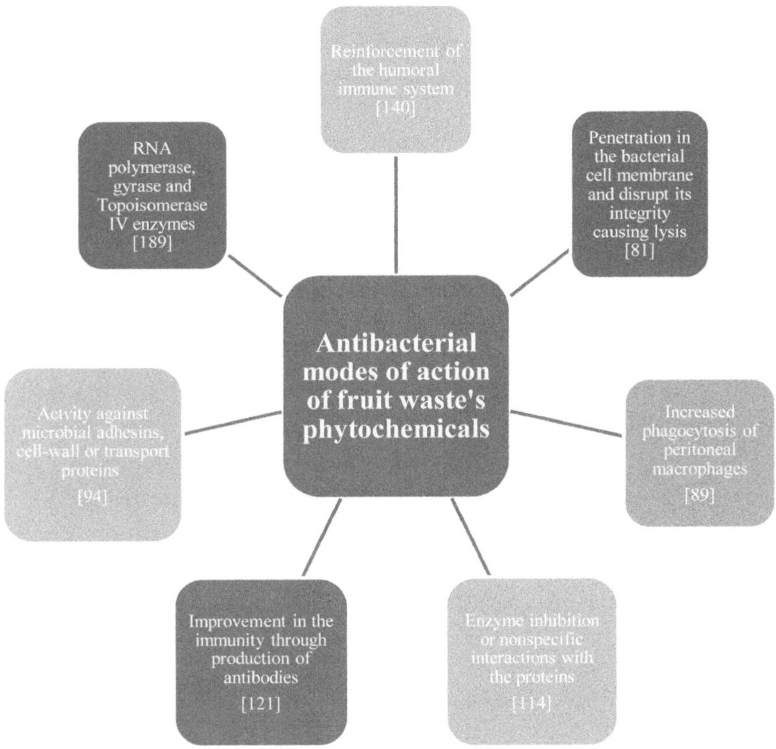

FIGURE 4.2 (See color insert.) Antibacterial modes of action of phytochemicals derived from fruit waste.

4.10 SUMMARY

Antibiotics growth promoters have improved poultry growth and production over the years by reducing infectious diseases and improving feed efficiency. However, their indecent use led to various concerns with respect to animal health, animal welfare, human health, product acceptability, and environmental quality. Fortunately, several alternatives to antibiotics are available and being explored with positive impact on animal growth and productivity while maintaining environmental quality and no side effect on human health and nutrition. Although phytochemicals from fruit wastes are somewhat recent alternatives, they have great opportunity to serve animal industry with safe, economical, and sustainable substitute to AGPs. In the recent years, the quest for these alternatives has increased as consumers are demanding antibiotics-free poultry products. Several studies demonstrated the beneficial effects of phytochemicals from different fruit wastes used in poultry production. Bioactive compounds from fruit wastes have several advantages over AGPs. They are also effective against fungal, viral, and other pathogens, whereas AGPs are only effective against bacterial infections. The use of fruit waste as a natural growth promoter is certainly considered as safe with low cost and no side effects that can enhance feed efficiency because of positive effects on the nutrient digestibility and better immune response to pathogenic organisms ensuring the efficient poultry production in terms of meat and eggs. The exploration of phytogenic feed additives based on fruit wastes would be helpful in national food and nutritional security and thus a step forward toward sustainable utilization of agro-industrial waste. However, research on standardization of extraction/preparation techniques specially based on green chemistry to develop different products from fruit wastes is warranted. There is also a need to explore effective combinations of phytochemicals in different fruit wastes for optimum alternatives for antibiotics-free poultry production.

KEYWORDS

- **antimicrobial resistance**
- **bioactive compounds**
- **fruit waste**
- **growth promoters**
- **nutraceutical**
- **pathogenic organisms**
- **phytochemicals and poultry**

REFERENCES

1. Agarwal, R.; Diwanay, S.; Patki, P.; Patwardhan, B. Studies on Immunomodulatory Activity of Withania Somnifera (Ashwagandha) Extracts in Experimental Immune Inflammation. *J. Ethnopharmacol.* **1999,** *67* (1), 27–35.
2. Ahmed, A.; Saeid, D.; Eman, A.; Reham, E. Egyptian Mango By-product. Compositional Quality of Mango Seed Kernel. *Food Chem.* **2007,** *103*, 1141–1152.
3. Ahnwange, B. A. Chemical Composition of *Musa sapientum* (Banana) Peels. *J. Food Technol.* **2008,** *6*, 263–266.
4. Ajila, C. M.; Bhat, S. G.; Rao, U. J. S. Valuable Components of Raw and Ripe Peels from Two Indian Mango Varieties. *Food Chem.* **2007,** *102*, 1006–1011.
5. Ajila, C. M.; Jaganmohan, R. L.; Rao, U. J. S. Characterization of Bioactive Compounds from Raw and Ripe *Mangifera indica* L. Peel Extracts. *Food Chem. Toxicol.* **2010,** *48*, 3406–3411.
6. Ali, S.; Khan, A. D.; Shah, F. H. Utilisation of Fruit and Vegetable Wastes in Layers' Diet. *J. Sci. Food Agric.* **1994,** *65* (4), 381–383.
7. Aminov, R. I. A Brief History of the Antibiotic Era: Lessons Learned and Challenges for the Future. *Front. Microbiol.* **2010,** *1,* 1–7.
8. Arogba, S. S. Mango (*Mangifera indica*) Kernel: Chromatographic Analysis of the Tannin, and Stability Study of the Associated Polyphenol Oxidase Activity. *J. Food Compos. Anal.* **2000,** *13* (2), 149–156.
9. Ayala-Zavala, J. F.; Rosas-Domínguez, C.; Vega-Vega, V; González-Aguilar, G. A. Antioxidant Enrichment and Antimicrobial Protection of Fresh-Cut Fruits Using Their Own Byproducts: Looking for Integral Exploitation. *J. Food Sci.* **2010,** *75* (8), 175–181.
10. Bakshi, M. P. S.; Wadhwa, M. Evaluation of Cannery and Fruit Wastes as Livestock Feed. *Indian J. Anim. Sci.* **2013,** *83*, 1198–1202
11. Baquero, F.; Tedim, A. S. P.; Coque, T. M. Antibiotic Resistance Shaping Multi-level Population Biology of Bacteria. *Front. Microbiol.* **2013,** *4*, 1–15.
12. Bassett, E. J.; Keith, M. S.; Armelagos, G. J.; Martin, D. L.; Villanueva, A. R. Tetracycline-labeled Human Bone from Ancient Sudanese Nubia (AD 350). *Science* **1980,** *209* (4464), 1532–1534.
13. Bentley, R.; Bennett, J. W., Eds. *Advances in Applied Microbiology;* Academic Press: Cambridge, MA, 2003; Vol. 52, pp 303–331.
14. Bermudez, A. J. Principles of Disease Prevention: Diagnosis and Control. *Diseases of Poultry*, 13th ed.; David, E., Swayne Laboratory Director, John Wiley & Sons, Inc., 2013; pp 3–60.
15. Bhui, K.; Prasad, S.; George, J.; Shukla, Y. Bromelain Inhibits COX-2 Expression by Blocking the Activation of MAPK Regulated NF-kappa B against Skin Tumor-initiation Triggering Mitochondrial Death Pathway. *Cancer Lett.* **2009,** *282* (2), 167–176.
16. Bhushan, S.; Kalia, K.; Sharma, M.; Singh, B.; Ahuja, P. S. Processing of Apple Pomace for Bioactive Molecules. *Crit. Rev. Biotechnol.* **2008,** *28* (4), 285–296.
17. Blandon, J. C.; Hamady, G. A. A.; Abdel-Moneim, M. A. The Effect of Partial Replacement of Yellow Corn by Banana Peels With and Without Enzymes on Broiler's Performance and Blood Parameters. *J. Anim. Poult. Sci.* **2015,** *4*, 10–19.

18. Braykov, N. P.; Eisenberg, J. N.; Grossman, M.; Zhang, L.; Vasco, K.; Cevallos, W.; Foxman, B. Antibiotic Resistance in Animal and Environmental Samples Associated with Small-Scale Poultry Farming in Northwestern Ecuador. *Msphere* **2016,** *1* (1), 21–15.
19. Brenes, A.; Montoro, A. V.; Cambrodón, I. G.; Centeno, C.; Calixto, F. S.; Arija, I. Effect of Grape Seed Extract on Growth Performance, Protein and Polyphenol Digestibilities, and Antioxidant Activity in Chickens. *Spanish J. Agric. Res.* **2010,** (2), 326–333.
20. Brenes, A.; Viveros, A.; Goni, I.; Centeno, C.; Sáyago-Ayerdy, S.; Arija, I.; Saura-Calixto, F. Effect of Grape Pomace Concentrate and Vitamin E on Digestibility of Polyphenols and Antioxidant Activity in Chickens. *Poult. Sci.* **2008,** *87* (2), 307–316.
21. Buckle, K. A. *Biotechnology and the Food Industry;* Rogers, P. L., Ed.; Breach Science Publishers: New York, 1989; pp 261–277.
22. Burt, S. Essential Oils: Their Antibacterial Properties and Potential Applications in Foods—A Review. *Inter. J. Food Microbiol.* **2004,** *94* (3), 223–253.
23. Carvalho, I. T.; Santos, L. Antibiotics in the Aquatic Environments: A Review of the European Scenario. *Environ. Inter.* **2016,** *94*, 736–757.
24. Cavallito, C. J.; Bailey, J. H. Allicin, the Antibacterial Principle of *Allium sativum*. I. Isolation, Physical Properties and Antibacterial Action. *J. Am. Chem. Soc.* **1944,** *66* (11), 1950–1951.
25. Chamorro, S.; Viveros, A.; Centeno, C.; Romero, C.; Arija, I.; Brenes, A. Effects of Dietary Grape Seed Extract on Growth Performance, Amino Acid Digestibility and Plasma Lipids and Mineral Content in Broiler Chicks. *Animal* **2013,** *7* (4), 555–561.
26. Chan, T. Y. Health Hazards due to Clenbuterol Residues in Food. *J. Toxicol.* **1999**, *37* (4), 517–519.
27. Chanda, S.; Baravalia, Y.; Kaneria, M.; Rakholiya, K. Fruit and Vegetable Peels–Strong Natural Source of Antimicrobics. *Curr. Res. Technol. Edu. Topic Appl. Microbiol. Microb. Biotechnol.* **2010,** *444*, 450.
28. Chaudry, M. A.; Badshah, A.; Bibi, N.; Zeb, A.; Ahmed, T.; Ali, S.; Meulen, U. T. Citrus Waste Utilization in Poultry Rations. *Archiv fur Geflugelkunde* **2004,** *68* (5), 206–210.
29. Coll, M. D.; Coll, L.; Laencina, J.; Tomas-Barberan, F. A. Recovery of Flavanones from Wastes of Industrially Processed Lemons. *Zeitschrift für Lebensmitteluntersuchung und-Forschung A* **1998,** *206* (6), 404–407.
30. Cos, P.; Bruyne, T. D.; Hermans, N.; Apers, S.; Berghe, D. V.; Vlietinck, A. Proanthocyanidins in Health Care: Current and New Trends. *Curr. Med. Chem.* **2004,** *11* (10), 1345–1359.
31. Craig, W. J. Phytochemicals: Guardians of Our Health. *J. Acad. Nutri. Diet.* **1997,** *97* (10), 199–204.
32. CSCRA. Système Canadien De Surveillance De La Résistance Aux Antimicrobiens–Rapport De 2016, Gouvernement du Canada **2016**.
33. da Silva, L. M. R.; de Figueiredo, E. A. T.; Ricardo, N. M. P. S.; Vieira, I. G. P.; de Figueiredo, R. W.; Brasil, I. M.; Gomes, C. L. Quantification of Bioactive Compounds in Pulps and By-products of Tropical Fruits from Brazil. *Food Chem.* **2014,** *143*, 398–404.

34. Danzeisen, J. L.; Kim, H. B.; Isaacson, R. E.; Tu, Z. J.; Johnson, T. J. Modulations of the Chicken Cecal Microbiome and Metagenome in Response to Anticoccidial and Growth Promoter Treatment. *PloS One* **2011,** *6* (11), e27949.
35. Devarajan, N.; Köhler, T.; Sivalingam, P.; Van Delden, C.; Mulaji, C. K.; Mpiana, P. T.; Poté, J. Antibiotic Resistant Pseudomonas spp. in the Aquatic Environment: A Prevalence Study under Tropical and Temperate Climate Conditions. *Water Res.* **2017,** *115*, 256–265.
36. Diarra, M. S.; Malouin, F. Antibiotics in Canadian Poultry Productions and Anticipated Alternatives. *Front. Microbiol.* **2014,** *5*, 282.
37. Diarra, M. S.; Rempel, H.; Champagne, J.; Masson, L.; Pritchard, J.; Topp, E. Distribution of Antimicrobial Resistance and Virulence Genes in Enterococcus spp. and Characterization of Isolates from Broiler Chickens. *Appl. Environ. Microbiol.* **2010,** *76* (24), 8033–8043.
38. Diarra, M. S.; Silversides, F. G.; Diarrassouba, F.; Pritchard, J.; Masson, L.; Brousseau, R.; Topp, E. Impact of Feed Supplementation with Antimicrobial Agents on Growth Performance of Broiler Chickens, *Clostridium perfringens* and Enterococcus Counts, and Antibiotic Resistance Phenotypes and Distribution of Antimicrobial Resistance Determinants in *Escherichia coli* Isolates. *Appl. Environ. Microbiol.* **2007,** *73* (20), 6566–6576.
39. Dibner, J. J.; Richards, J. D. Antibiotic Growth Promoters in Agriculture: History and Mode of Action. *Poul. Sci.* **2005,** *84* (4), 634–643.
40. Dilas, S.; Čanadanović-Brunet, J.; Ćetković, G. By-products of Fruits Processing as a Source of Phytochemicals. *Chem. Ind. Chem. Eng. Quart.* **2009,** *15* (4), 191–202.
41. Diwanay, S.; Chitre, D.; Patwardhan, B. Immunoprotection by Botanical Drugs in Cancer Chemotherapy. *J. Ethnopharmacol.* **2004,** *90* (1), 49–55.
42. Domans, H. J. D.; Deans, S. G. Antimicrobial Agents from Plants: Antibacterial Activity of Plant Volatiles Oils. *J. Appl. Microbiol.* **2000,** *88*, 308–316.
43. Dumonceaux, T. J.; Hill, J. E.; Hemmingsen, S. M.; Van Kessel, A. G. Characterization of Intestinal Microbiota and Response to Dietary Virginiamycin Supplementation in the Broiler Chicken. *Appl. Environ. Microbiol.* **2006,** *72* (4), 2815–2823.
44. Eberhardt, T.; Lee, C. Y.; Liu, R. H. Antioxidant Activity of Fresh Apples. *Nature* **2000,** *405*, 903–904.
45. Emaga, T. H.; Bindelle, J.; Agneesens, R.; Buldgen, A.; Wathelet, B.; Paquot, M. Ripening Influences Banana and Plantain Peels Composition and Energy Content. *Trop. Anim. Health Prod.* **2011,** *43* (1), 171–177.
46. Endress, H. U. High Quality Resulting from Product Integrated Environment Protection (PIUS). *Fruit Proc.* **2000,** *10* (7), 273–277.
47. Engberg, R. M.; Hedemann, M. S.; Leser, T. D.; Jensen, B. B. Effect of Zinc Bacitracin and Salinomycin on Intestinal Microflora and Performance of Broilers. *Poul. Sci.* **2000,** *79* (9), 1311–1319.
48. Erhan, M. K.; Bölükbaşı, Ş. C. Citrus Peel Oils Supplementation in Broiler Diet: Effects on Performance, Jejunum Microflora and Jejunum Morphology. *Revista Brasileira de Ciência Avícola*, **2017,** *19* (SPE), 15–22.
49. ESVAC. *Sales of Veterinary Antimicrobial Agents in 30 European Countries in 2015 Trends from 2010 to 2015*; Seventh Esvac Report, 2017; Ema/184855/2017.

50. Fisher, K.; Phillips, C. A. The Effect of Lemon, Orange and Bergamot Essential Oils and Their Components on the Survival of *Campylobacter Jejuni*, *Escherichia coli* O157, *Listeria monocytogenes*, *Bacillus cereus* and *Staphylococcus aureus* In Vitro and In Food Systems. *J. Appl. Microbiol.* **2006,** *101* (6), 1232–1240.
51. Foo, L. Y.; Lu, Y. Isolation and Identification of Procyanidins in Apple Pomace. *Food Chem.* **1999,** *64* (4), 511–518.
52. Forgetta, V.; Rempel, H.; Malouin, F.; Vaillancourt Jr, R.; Topp, E.; Dewar, K.; Diarra, M. S. Pathogenic and Multidrug-resistant *Escherichia fergusonii* From Broiler Chicken. *Poult. Sci.* **2012,** *91* (2), 512–525.
53. Frankic, T.; Voljč, M.; Salobir, J.; Rezar, V. Use of Herbs and Spices and Their Extracts in Animal Nutrition. *Acta. Agric. Slov.* **2009,** *94* (2), 95–102.
54. Fry, J.; Carr, R. I.; Lee, T. D. G. Echinacea: Stimulation of Mouse Macrophages. *FASEB J.* **1998,** *12* (4), 299.
55. Fuleki, T.; Ricardo da Silva, J. M. Catechin and Procyanidin Composition of Seeds from Grape Cultivars Grown in Ontario. *J. Agricul. Food Chem.* **1997,** *45* (4), 1156–1160.
56. Fung, S.; Rempel, H.; Forgetta, V.; Dewar, E. T. K.; Diarra, M. S. In *Ceca Microbiome of Mature Broiler Chickens Fed With or Without Salinomycin,* The Gut Microbiome: The Effector/Regulatory Immune Network Conference (B3), **2013,** Keystone Symposia on Molecular and Cellular Biology.
57. Furtula, V.; Farrell, E. G.; Diarrassouba, F.; Rempel, H.; Pritchard, J.; Diarra, M. S. Veterinary Pharmaceuticals and Antibiotic Resistance of *Escherichia coli* Isolates in Poultry Litter From Commercial Farms and Controlled Feeding Trials. *Poult. Sci.* **2010,** *89* (1), 180–188.
58. Gadde, U.; Kim, W.; Oh, S.; Lillehoj, H. S. Alternatives to Antibiotics for Maximizing Growth Performance and Feed Efficiency in Poultry: A Review. *Anim. Health Res. Rev.* **2017,** *18* (1), 26–45.
59. Galati, G.; Lin, A.; Sultan, A. M.; O'brien, P. J. Cellular and In Vivo Hepatotoxicity Caused by Green Tea Phenolic Acids and Catechins. *Free Rad. Biol. Med.* **2006,** *40* (4), 570–580.
60. Gardener, P. T.; White, T. A. C.; McPhail, D. B.; Duthie, G. G. The Relative Contributions of Vitamin C, Carotenoids and Phenolics to the Antioxidant Potential of Fruit Juices. *Food Chem.* **2000,** *68*, 471–474.
61. Gassner, B.; Wuethrich, A. Pharmacokinetic and Toxicological Aspects of the Medication of Beef-type Calves with an Oral Formulation of Chloramphenicol Palmitate. *J. Vet. Pharmacol. Therapeut.* **1994,** *17* (4), 279–283.
62. Ghasemi, H. A.; Kasani, N.; Taherpour, K. Effects of Black Cumin Seed (*Nigella Sativa* L.), a Probiotic, a Prebiotic and a Synbiotic on Growth Performance, Immune Response and Blood Characteristics of Male Broilers. *Livest. Sci.* **2014,** *164*, 128–134.
63. Göhl, B. Les Aliments du Bétail Sous les Tropiques. FAO, Division de Production et Santé Animale, Roma, Italy, 1982.
64. Goni, I.; Brenes, A.; Centeno, C.; Viveros, A.; Saura-Calixto, F.; Rebole, A.; Estevez, R. Effect of Dietary Grape Pomace and Vitamin E on Growth Performance, Nutrient Digestibility, and Susceptibility to Meat Lipid Oxidation in Chickens. *Poult. Sci.* **2007,** *86* (3), 508–516.

65. González-Paramás, A. M.; Esteban-Ruano, S.; Santos-Buelga, C.; de Pascual-Teresa, S.; Rivas-Gonzalo, J. C. Flavanol Content and Antioxidant Activity in Winery Byproducts. *J. Agricult. Food Chem.* **2004,** *52* (2), 234–238.
66. Gorinstein, S.; Martın-Belloso, O.; Park, Y. S.; Haruenkit, R.; Lojek, A.; Ĉíž, M.; Trakhtenberg, S. Comparison of Some Biochemical Characteristics of Different Citrus Fruits. *Food Chem.* **2001,** *74* (3), 309–315.
67. Gorinstein, S.; Zachwieja, Z.; Folta, M.; Barton, H.; Piotrowicz, J.; Zemser, M.; Màrtín-Belloso, O. Comparative Contents of Dietary Fiber, Total Phenolics, and Minerals in Persimmons and Apples. *J. Agricult. Food Chem.* **2001,** *49* (2), 952–957.
68. Gorinstein, S.; Zemser, M.; Haruenkit, R.; Chuthakorn, R.; Grauer, F.; Martin-Belloso, O.; Trakhtenberg, S. Comparative Content of Total Polyphenols and Dietary Fiber in Tropical Fruits and Persimmon. *J. Nutri. Biochem.* **1999,** *10* (6), 367–371.
69. Gullón, B.; Falqué, E.; Alonso, J. L.; Parajó, J. C. Evaluation of Apple Pomace as a Raw Material for Alternative Applications in Food Industries. *Food Technol. Biotechnol.* **2007,** *45* (4), 426–433.
70. Guyot, S.; Serrand, S.; Le Quéré, J. M.; Sanoner, P.; Renard, C. M. Enzymatic Synthesis and Physicochemical Characterisation of Phloridzin Oxidation Products (POP), a New Water-soluble Yellow Dye Deriving from Apple. *Innov. Food Sci. Emerg. Technol.* **2007,** *8* (3), 443–450.
71. Hang, Y. D.; Walter, R. H. Treatment and Utilization of Apple-Processing Wastes. In *Processed Apple Products;* Downing, D. L., Eds.; Springer: New York, 1989; pp 356–377.
72. Hemalatha, R.; Anbuselvi, S. Physicochemical Constituents of Pineapple Pulp and Waste. *J. Chem. Pharmaceut. Res.* **2013,** *5* (2), 240–242.
73. Henle, F. G. J. *Pathologische Untersuchungen;* Verlag von August: Hirschfeld, Berlin, Prussia, 1840; pp 1–82.
74. Hernán, B.; Enrique, T.; Ríos, B. Technical-economic Evaluation of the Use of Banana Peel as a Supplement for Cattle Feed. *Postcosecha y Agroindustria del Plátano en el Eje Cafetero de Colombia* **2000,** 257–265.
75. Jamroz, D.; Orda, J.; Kamel, C.; Wiliczkiewicz, A.; Wertelecki, T.; Skorupinska, J. The Influence of Phytogenic Extracts on Performance, Nutrient Digestibility, Carcass Characteristics, and Gut Microbial Status in Broiler Chickens. *J. Anim. Feed Sci.* **2003,** *12* (3), 583–596.
76. Jamroz, D.; Wertelecki, T.; Houszka, M.; Kamel, C. Influence of Diet Type on the Inclusion of Plant Origin Active Substances on Morphological and Histochemical Characteristics of the Stomach and Jejunum Walls in Chicken. *J. Anim. Physiol. Anim. Nutri.* **2006,** *90* (5–6), 255–268.
77. Jamroz, D.; Wiliczkiewicz, A.; Wertelecki, T.; Orda, J.; Skorupińska, J. Use of Active Substances of Plant Origin in Chicken Diets Based on Maize and Locally Grown Cereals. *British Poult. Sci.* **2005,** *46* (4), 485–493.
78. Jang, I. S.; Ko, Y. H.; Yang, H. Y.; Ha, J. S.; Kim, J. Y.; Kim, J. Y.; Lee, C. Y. Influence of Essential Oil Components on Growth Performance and the Functional Activity of the Pancreas and Small Intestine in Broiler Chickens. *Asian Austra. J. Anim. Sci.* **2004,** *17* (3), 394–400.

79. Jayaprakasha, G. K.; Singh, R. P.; Sakariah, K. K. Antioxidant Activity of Grape Seed (*Vitis vinifera*) Extracts on Peroxidation Models In Vitro. *Food Chem.* **2001,** *73* (3), 285–290.
80. Joshi, V. K.; Gupta, K.; Devrajan, A.; Lal, B. B.; Arya, S. P. Production and Evaluation of Fermented Apple Pomace in the Feed of Broilers. *J. Food Sci. Technol.* **2000,** *37* (6), 609–612.
81. Joshi, V. K.; Kumar, A.; Kumar, V. Antimicrobial, Antioxidant and Phyto-chemicals from Fruit and Vegetable Wastes: A Review. *Inter. J. Food Ferment. Technol.* **2012,** *2* (2), 123.
82. Joshi, V. K.; Sharma, R.; Kumar, V. Antimicrobial Activity of Essential Oils: A Review. *Inter. J. Food Ferment. Technol.* **2011,** *1* (2), 161–172.
83. Kabuki, T.; Nakajima, H.; Arai, M.; Ueda, S.; Kuwabara, Y.; Dosako, S. I. Characterization of Novel Antimicrobial Compounds from Mango (*Mangifera indica* L.) Kernel Seeds. *Food Chem.* **2000,** *71* (1), 61–66.
84. Kallithraka, S.; Garcia-Viguera, C.; Bridle, P.; Bakker, J. Survey of Solvents for the Extraction of Grape Seed Phenolics. *Phytochem. Anal.* **1995,** *6* (5), 265–267.
85. Kanatt, S. R.; Chawla, S. P. Shelf-life Extension of Chicken Packed in Active Film Developed with Mango Peel Extract. *J. Food Safety* **2018,** *38* (1), 12385.
86. Kasapidou, E.; Sossidou, E.; Mitlianga, P. Fruit and Vegetable Co-products as Functional Feed Ingredients in Farm Animal Nutrition for Improved Product Quality. *Agricul.* **2015,** *5* (4), 1020–1034.
87. Keithly, J. H.; Taggart, T. Dried Citrus Peel Supplement for Use in Commercial Poultry Feed. US; **2006;** Patent *7*; p 298.
88. Kennedy, M. J. Apple Pomace and Kiwifruit: Processing Options. *Australasian Biotechnol.* **1994,** *4*, 43.
89. Khan, R.; Nikousefat, Z.; Tufarelli, V.; Naz, S.; Javdani, M.; Laudadio, V. Garlic (*Allium sativum*) Supplementation in Poultry Diets: Effect on Production and Physiology. *World's Poult. Sci. J.* **2012,** *68* (3), 417–424.
90. Kim, E.-Y.; Ham, S.-K.; Shigenaga, M. K.; Han, O. Bioactive Dietary Polyphenolic Compounds Reduce Nonheme Iron Transport Across Human Intestinal Cell Monolayers. *J. Nutri.* **2008,** *138* (9), 1647–1651.
91. Kirubakaran, A.; Moorthy, M.; Chitra, R.; Prabakar, G. Influence of Combinations of Fenugreek, Garlic, and Black Pepper Powder on Production Traits of the Broilers. *Vet. World* **2016,** *9* (5), 470.
92. Koch, R. Die Aetiologie der Tuberculose. *Berl. Klin. Wochenschr*. **1882,** *19*, 287–296.
93. Koubala, B. B.; Kansci, G.; Garnier, C.; Thibault, J. F.; Ralet, M. C. Physico-chemical Properties of Dietary Fibres Prepared from Ambarella (*Spondias cytherea)* and Mango (*Mangifera indica*) Peels. *Food Biopr. Technol.* **2013,** *6*, 591–597.
94. Kumar, S.; Pandey , A. K. Chemistry and Biological Activities of Flavonoids: An Overview. *Sci. World J.* **2013**.
95. Kümmerer, K. Antibiotics in the Aquatic Environment—A Review—Part I. *Chemosph.* **2009,** *75* (4), 417–434.
96. Lee, K. W.; Everts, H.; Kappert, H. J.; Frehner, M.; Losa, R.; Beynen, A. C. Effects of Dietary Essential Oil Components on Growth Performance, Digestive Enzymes and Lipid Metabolism in Female Broiler Chickens. *British Poult. Sci.* **2003,** *44* (3), 450–457.

97. Lee, K. W.; Hong, Y. H.; Lee, S. H.; Jang, S. I.; Park, M. S.; Bautista, D. A.; Lillehoj, E. P. Effects of Anticoccidial and Antibiotic Growth Promoter Programs on Broiler Performance and Immune Status. *Res. Vet. Sci.* **2012,** *93* (2), 721–728.
98. Levy, S. B.; FitzGerald, G. B.; Macone, A. B. Changes in Intestinal Flora of Farm Personnel After Introduction of a Tetracycline-supplemented Feed on a Farm. *New Engl. J. Med.* **1976,** *295* (11), 583–588.
99. Li, H.; Zhao, P.; Lei, Y.; Hossain, M.; Kim, I. Phytoncide, Phytogenic Feed Additive as an Alternative to Conventional Antibiotics, Improved Growth Performance and Decreased Excreta Gas Emission Without Adverse Effect on Meat Quality in Broiler Chickens. *Livest. Sci.* **2015,** *181*, 1–6.
100. Lin, J.; Hunkapiller, A. A.; Layton, A. C.; Chang, Y. J.; Robbins, K. R. Response of Intestinal Microbiota to Antibiotic Growth Promoters in Chickens. *Foodb. Pathog. Dis.* **2013,** *10* (4), 331–337.
101. Lommen, A.; Godejohann, M.; Venema, D. P.; Hollman, P. C. H.; Spraul, M. Application of Directly Coupled HPLC−NMR−MS to the Identification and Confirmation of Quercetin Glycosides and Phloretin Glycosides in Apple Peel. *Analy. Chem.* **2000,** *72* (8), 1793–1797.
102. Lu, Y.; Foo, L. Y. Antioxidant and Radical Scavenging Activities of Polyphenols from Apple Pomace. *Food Chem.* **2000,** *68*, 81–85.
103. Lu, Y.; Foo, L. Y. Constitution of Some Chemical Components of Apple Seed. *Food Chem.* **1998,** *61* (1–2), 29–33.
104. Mandey, J. S.; Tulung, B.; Leke, J. R.; Sondakh, B. F. J. Performance and Carcass Quality of Broiler Chickens Fed Diet Containing Pineapple Waste Meal Fermented by "Ragi Tape." *Earth Environ. Sci.* **2018,** *102* (1), 12042.
105. Mani, S. B.; Sethi, V. Utilization of Fruits and Vegetables Processing Waste. In *Post-harvest Technology of Fruits and Vegetables: Handling Processing Fermentation and Waste Management;* Verma, L. R., Joshi, V. K., Eds.; Indus Publishing Co.: New Delhi, 2000; Vol. 2, p 1006.
106. Manthey, J. A. Fractionation of Orange Peel Phenols in Ultrafiltered Molasses and Mass Balance Studies of Their Antioxidant Levels. *J. Agric. Food Chem.* **2004,** *52* (25), 7586–7592.
107. Manzetti, S.; Ghisi, R. The Environmental Release and Fate of Antibiotics. *Marine Poll. Bull.* **2014,** *79* (1–2), 7–15.
108. Marin, A.; Carias, D.; Maria, C. A.; Hevia, P. Nutritive Value of Leaves of Musa *paradisiaca* and *Clitoria ternatea* as Dilutents in Diets for Broilers. *Interciencia* **2003,** *28* (1), 51–56.
109. Masibo, M.; He, Q. Mango Bioactive Compounds and Related Nutraceutical Properties: A Review. *Food Rev. Inter.* **2009,** *25* (4), 346–370.
110. Mason, T. L. Inactivation of Red Beet β-glucan Synthase by Native and Oxidized Phenolic Compounds. *Phytochem.* **1987,** *26* (8), 2197–2202.
111. Matharu, A. S.; de Melo, E. M.; Houghton, J. A. Opportunity for High Value-added Chemicals from Food Supply Chain Wastes. *Biores. Technol.* **2016,** *215*, 123–130.
112. Matoo, F. A.; Beat, G. A.; Banday, M. T.; Ganaie, T. A. S. Performance of Broilers Fed On Apple Pomace Diets Supplemented with Enzyme (S). *Indian J. Anim. Nutr.* **2001,** *18* (4), 349–352.

113. Mattila, P.; Hellström, J.; Törrönen, R. Phenolic Acids in Berries, Fruits, and Beverages. *J. Agricul. Food Chem.* **2006,** *54* (19), 7193–7199.
114. Mead, P. S.; Slutsker, L.; Dietz, V.; McCaig, L. F.; Bresee, J. S.; Shapiro, C.; Tauxe, R. V. Food-related Illness and Death in the United States. *Emerg. Infect. Dis.* **1999,** *5* (5), 607.
115. Mehrländer, K.; Dietrich, H.; Sembries, S.; Dongowski, G.; Will, F. Structural Characterization of Oligosaccharides and Polysaccharides from Apple Juices Produced by Enzymatic Pomace Liquefaction. *J. Agric. Food Chem.* **2002,** *50* (5), 1230–1236.
116. Middleton Jr, E.; Kandaswami, C. Potential Health-promoting Properties of Citrus Flavonoids. *Food Technol.* **1994,** *48,* 115–119.
117. Mikulic-Petkovsek, M.; Slatnar, A.; Stampar, F.; Veberic, R. The Influence of Organic/Integrated Production on the Content of Phenolic Compounds in Apple Leaves and Fruits in Four Different Varieties Over a 2-year Period. *J. Sci. Food Agricul.* **2010,** *90* (14), 2366–2378.
118. Mirghani, M. E. S.; Yosuf, F.; Kabbash, N. A.; Vejayan, J.; Yosuf, Z. B. M. Antibacterial Activity of Mango Kernel Extracts. *J. Appl. Sci.* **2009,** *9*, 3013–3019.
119. Moghazy, M. E. S. E.; Boushy, A. E. Some Neglected Poultry Feedstuffs from Vegetable and Fruit Wastes. *World's Poult. Sci. J.* **1982,** *38* (1), 18–27.
120. Mohapatra, D.; S. Mishra.; N, Sutar. Banana and Its By-product Utilization: An Overview. *J. Sci. Industr. Res.* **2010,** *6*, 323–329.
121. Mona, S, R.; Hanan, A. Effects of Using Dried Egyptian Clover and Orange Peels as Natural Feed Additives on Egg Production Egg Quality and Immune Response of Laying Hens. *Fay. J. Agricult. Res. Develop.* **2007,** *21*, 188–205.
122. Morais, D. R.; Rotta, E. M.; Sargi, S. C.; Schmidt, E. M.; Bonafe, E. G.; Eberlin, M. N.; Visentainer, J. V. Antioxidant Activity, Phenolics and UPLC–ESI–MS of Extracts from Different Tropical Fruits Parts and Processed Peels. *Food Res. Internat.* **2015,** *77*, 392–399.
123. Mourao, J. L.; Pinheiro, V. M.; Prates, J. A. M.; Bessa, R. J. B.; Ferreira, L. M. A., Fontes, C. M. G. A.; Ponte, P. I. P. Effect of Dietary Dehydrated Pasture and Citrus Pulp on the Performance and Meat Quality of Broiler Chickens. *Poult. Sci.* **2008,** *87*, 733–743.
124. Nagarajaiah, S. B.; Prakash, J. Chemical Composition and Antioxidant Potential of Peels from Three Varieties of Banana. *Asian J. Food Agro-Indust.* **2011,** *4*, 31–36.
125. Nagy, S. Vitamin C Contents of Citrus Fruit and Their Products: A Review. *J. Agricult. Food Chem.* **1980,** *28* (1), 8–18.
126. Negesse, T.; Makkar, H. P. S.; Becker, K. Nutritive Value of Some Non-conventional Feed Resources of Ethiopia Determined by Chemical Analyses and an In Vitro Gas Method. *Anim. Feed Sci. Technol.* **2009,** *154* (3–4), 204–217.
127. Nelson, M. L.; Dinardo, A.; Hochberg, J.; Armelagos, G. J. Brief Communication: Mass Spectroscopic Characterization of Tetracycline in the Skeletal Remains of an Ancient Population from Sudanese Nubia 350–550 CE. *Am. J. Phy. Anthropol.* **2010,** *143*, 151–154.
128. O'Bryan, C. A.; Crandall, P. G.; Chalova, V. I.; Ricke, S. C. Orange Essential Oils Antimicrobial Activities Against Salmonella spp. *J. Food Sci.* **2008,** *73* (6), 264–267.

129. Odunsi, A. A. Response of Laying Hens and Growing Broilers to the Dietary Inclusion of Mango (*Mangifera indica* L.) Seed Kernel Meal. *Trop. Anim. Healt. Prod.* **2005,** *37*, 139–150.
130. O'Hara, M.; Kiefer, D.; Farrell, K.; Kemper, K. A Review of 12 Commonly Used Medicinal Herbs. *Archiv. Fam. Med.* **1998,** *7* (6), 523.
131. Oluremi, O. I. A.; Ngi, J.; Andrew, I. A. Phytonutrients in Citrus Fruit Peel Meal and Nutritional Implication for Livestock Production. *Livest. Res. Rur. Devel.* **2007,** *19* (7), 1–5.
132. Oluremi, O. I. A.; Ojighen, V. O.; Ejembi, E. H. The Nutritive Potentials of Sweet Orange (*Citrus sinensis*) Rind in Broiler Production. *Inter. J. Poultr. Sci.* **2006,** *5* (7), 613–617.
133. Orayaga, K. T., Oluremi, O. I. A., Tuleun, C. D., Carew, S. N. Utilization of Composite Mango (Mangifera indica) Fruit Reject Meal in Starter Broiler Chicks Feeding. *J. Exp. Agricul. Int.* **2017,** *17* (2), 1–9.
134. Padam, B. S.; Tin, H. S.; Chye, F. Y.; Abdullah, M. I. Banana By-products: An Under-Utilized Renewable Food Biomass with Great Potential. *J. Food Sci. Technol.* **2014,** *51* (12), 3527–3545.
135. Palmeira, S. M. V.; Gois, L. M.; Souza, L. D. Extraction of Phenolic Compounds from Mango Peels. *Lat. Am. Appl. Res.* **2012,** *42*, 77–81.
136. Pazmiño-Durán, E. A.; Giusti, M. M.; Wrolstad, R. E.; Glória, M. B. A. Anthocyanins from Banana Bracts (Musa X Paradisiaca) as Potential Food Colorants. *Food Chem.* **2001,** *73* (3), 327–332.
137. Percival, S. S.; Talcott, S. T.; Chin, S. T.; Mallak, A. C.; Lounds-Singleton, A.; Pettit-Moore, J. Neoplastic Transformation of Balb/3t3 Cells and Cell Cycle of Hl-60 Cells Are Inhibited By Mango (*Mangifera indica*) Juice And Mango Juice Ex-*tracts. J. Nutr.* **2006,** *136*, 1300–1304.
138. Pietta, P. G. Favonoids as Antioxidant. *J. Nat. Prod.* **2000,** *63* (7), 1035–1042.
139. Platel, K.; Srinivasan, K. Digestive Stimulant Action of Spices: A Myth or Reality? *Ind. J. Med. Res.* **2004,** *119*, 167–179.
140. Pourhossein, Z.; Qotbi, A. A. A.; Seidavi, A.; Laudadio, V.; Centoducati, G.; Tufarelli, V. Effect of Different Levels of Dietary Sweet Orange (*Citrus sinensis*) Peel Extract on Humoral Immune System Responses in Broiler Chickens. *Anim. Sci. J.* **2015,** *86* (1), 105–110.
141. Rabayaa, E.; Abo Omar, J. M.; Othman, R. A. Utilization of Olive Pulp in Broiler Rations. *An-Najah Univ. J. Res.* **2001,** *15*, 133–144.
142. Ragab, M. S.; Hassan, H. A. Effects of Using Dried Egyptian Clover and Orange Peels as Natural Feed Additives on Egg Production Egg Quality and Immune Response of Laying Hens. **2007,** *21,* 188–205.
143. Rehman, Z. U.; Salariya, A. M.; Habib, F.; Shah, W. H. Utilization of Mango Peels as a Source of Pectin. *J. Chem. Soc. Pak.* **2004,** *26*, 73–76.
144. Ribeiro, R. S. M.; Queiroz, J. H.; Ribeiro, L.; De Queiroz, M. E.; Campos, F. M.; Sant'ana, P. H. M. Antioxidant in Mango (*Mangifera indica* L.) Pulp. *Plant Foods Hum. Nutr.* **2007,** *62*, 13–17.
145. Rodríguez, J.; Di Pierro, D.; Gioia, M.; Monaco, S.; Delgado, R.; Coletta, M.; Marini, S. Effects of a Natural Extract From *Mangifera indica* L, and Its Active Compound,

Mangiferin, on Energy State and Lipid Peroxidation of Red Blood Cells. *Biochim. et Biophys. Acta* **2006,** *1760* (9), 1333–1342.

146. Ronquillo, M. G.; Hernandez, J. C. A. Antibiotic and Synthetic Growth Promoters in Animal Diets: Review of Impact and Analytical Methods. *Food Contr.* **2017,** *72*, 255–267.
147. Ross, S. A.; Ziska, D. S.; Zhao, K.; ElSohly, M. A. Variance of Common Flavonoids by Brand of Grapefruit Juice. *Fitoterapia* **2000,** *71* (2), 154–161.
148. RUMA. Responsible Use of Medicines in Agriculture Alliance (Ruma) Information on Antibiotic Resistance. http://Ruma.org.Uk/About/Position-Papers/Ruma-Information-Note-Antibiotics-Responsible-Use-Antibiotics-Farm-Animals, **2016** (accessed June 20, 2018).
149. Sabutan, M. G. I. Banana Peelings Help Broilers Grow. *Misset World Poult.* **1996,** *12*, 59–61.
150. Saito, M.; Hosoyama, H.; Ariga, T.; Kataoka, S.; Yamaji, N. Antiulcer Activity of Grape Seed Extract and Procyanidins. *J. Agricul. Food Chem.* **1998,** *46* (4), 1460–1464.
151. Sandhu, K. S.; Lim, S. T. Structural Characteristics And In Vitro Digestibility of Mango Kernel Starches (*Mangifera indica* L). *J. Food Chem.* **2007,** *107*, 92–97.
152. Santos-Buelga, C.; Francia-Aricha, E.; Escribano-Bailón, M. Comparative Flavan-3-ol Composition of Seeds from Different Grape Varieties. *Food Chem.* **1995,** *53* (2), 197–201.
153. Saravanan, P.; Muthuvelayudham, R.; Viruthagiri, T. Enhanced Production of Cellulase from Pineapple Waste by Response Surface Methodology. *J. Eng.* **2013,** 1–8.
154. Sarmah, A. K.; Meyer, M. T.; Boxall, A. B. A Global Perspective on the Use, Sales, Exposure Pathways, Occurrence, Fate and Effects of Veterinary Antibiotics (VAs) in the Environment. *Chemosph.* **2006,** *65* (5), 725–759.
155. Schieber, A., Hilt, P. Conrad, J., Beifuss, U., Carle, R. Elution Order of Quercetin Glycosides From Apple Pomace Extracts on a New HPLC Stationary Phase With Hydrophilic End Capping. *J. Separ. Sci.* **2002,** *25*, 361–364.
156. Schieber, A.; Stintzing, F. C.; Carle, R. By-products of Plant Food Processing as a Source of Functional Compounds—Recent Developments. *Trend. Food Sci. Technol.* **2001,** *12* (11), 401–413.
157. Schieber, A.; Ullrich, W.; Carle, R. Characterization of Polyphenols in Mango Puree Concentrate by HPLC with Diode Array and Mass Spectrometric Detection. *Innov. Food Sci. Emerg. Technol.* **2000,** *1* (2), 161–166.
158. Schwaiger, K.; Huther, S.; Hölzel, C.; Kämpf, P.; Bauer, J. Prevalence of Antibiotic-Resistant Enterobacteriaceae Isolated from Chicken and Pork Meat Purchased at the Slaughterhouse and At Retail in Bavaria, Germany. *Int. J. Food Microbiol.* **2012,** *154* (3), 206–211.
159. Singh, P.; Karimi, A.; Devendra, K.; Waldroup, P. W.; Cho, K. K.; Kwon, Y. M. Influence of Penicillin on Microbial Diversity of the Cecal Microbiota in Broiler Chickens. *Poultr. Sci.* **2013,** *92* (1), 272–276.
160. Siyal, F. A.; Wagan, R.; Bhutto, Z. A.; Tareen, M. H.; Arain, M. A.; Saeed, M.; Soomro, R. N. Effect of Orange and Banana Peels on the Growth Performance of Broilers. *Adv. Anim. Vet. Sci.* **2016,** *4* (7), 376–380.

161. Somda, M. K.; Savadogo, A.; Quattara, C. A. T.; Quattara, A. S.; Traore, A. S. Thermotolerant And Alcohol-Tolerant Yeasts Targeted To Optimize Hydrolyzation From Mango Peel For High Bioethanol Production. *Asian J. Biotechnol.* **2011,** *3*, 77–83.
162. Soomro, H.; Rind, M. I.; Sanjrani, S. N.; Magsi, A. S.; Barham, G. S.; Pirzada, S. A.; Sahito, H. A. Effect of Partial Mango Pulp Mixing in Ration on Behaviour and Production of Broiler. *Inter. J. Plant Anim. Sci.* **2013,** *1* (1), 030–036.
163. Soong, Y. Y.; Barlow, P. J. Antioxidant Activity and Phenolic Content of Selected Fruit Seeds. *Food Chem.* **2004,** *88* (3), 411–417.
164. Soong, Y. Y.; Barlow, P. J. Quantification of Gallic Acid and Ellagic (*Mangifera indica L) Kernel and Their Effects on Antioxidant Activity. Food Chem.* **2006,** *97*, 524–530.
165. Souquet, J. M.; Cheynier, V.; Brossaud, F.; Moutounet, M. Polymeric Proanthocyanidins From Grape Skins. *Phytochem.* **1996,** *43* (2), 509–512.
166. Stimpel, M.; Proksch, A.; Wagner, H.; Lohmann-Matthes, M. L. Macrophage Activation and Induction of Macrophage Cytotoxicity by Purified Polysaccharide Fractions from the Plant *Echinacea purpurea. Infect. Immun.* **1984,** *46* (3), 845–849.
167. Subagio, A.; Morita, N.; Sawada, S. Carotenoids and Their Fatty-acid Esters in Banana Peel. *J. Nutri. Sci. Vitami.* **1996,** *42* (6), 553–566.
168. Subba, M. S.; Soumithri, T. C.; Rao, R. S. Antimicrobial Action of Citrus Oils. *J. Food Sci.* **1967,** *32* (2), 225–227.
169. Sun, J., Chu, Y.; Wu, X.; Liu, R. H. Antioxidant and Anti-proliferative Activities of Common Fruits. *J. Agricul. Food Chem.* **2002,** *50*, 7449–7454.
170. Sundaram, S.; Anjum, S.; Dwivedi, P.; Rai, G. K. Antioxidant Activity and Protective Effect of Banana Peel Against Oxidative Hemolysis of Human Erythrocyte at Different Stages of Ripening. *Appl. Biochem. Biotechnol.* **2011,** *164* (7), 1192–1206.
171. Teleszko, M.; Wojdyło, A. Comparison of Phenolic Compounds and Antioxidant Potential Between Selected Edible Fruits and Their Leaves. *J. Funct. Foods*, **2015,** *14*, 736–746.
172. Tewe, O. O. Replacing Maize with Plantain Peels in Broiler Diets. *Nutri. Rep. Int.* **1983,** *28*, 23–29.
173. Toghyani, M.; Toghyani, M.; Gheisari, A.; Ghalamkari, G.; Eghbalsaied, S. Evaluation of Cinnamon and Garlic as Antibiotic Growth Promoter Substitutions on Performance, Immune Responses, Serum Biochemical and Haematological Parameters in Broiler Chicks. *Livest. Sci.* **2011,** *138* (1), 167–173.
174. Tollba, A. A. H. Using Some Natural Additives to Improve Physiological and Productive Performance of Broiler Chicks Under High Temperature Condition. 2. Black Cumin (*Nigella sativa*) or Garlic (*Allium sativum*). *Egypt. Poult. Sci.* **2003,** *23*, 327–340.
175. Torok, V. A.; Allison, G. E.; Percy, N. J.; Ophel-Keller, K.; Hughes, R. J. Influence of Antimicrobial Feed Additives on Broiler Commensal Post Hatch Gut Microbiota Development and Performance. *Appl. Environ. Microbiol.* **2011,** *77* (10), 3380–3390.
176. Torres, J. L.; Bobet, R. New Flavanol Derivatives from Grape (*Vitis vinifera*) Byproducts. Antioxidant Aminoethylthio–Flavan-3-ol Conjugates from a Polymeric Waste Fraction Used as a Source of Flavanols. *J. Agricult. Food Chem.* **2001,** *49* (10), 4627–4634.

177. Torres, J. L.; Varela, B.; García, M. T.; Carilla, J.; Matito, C.; Centelles, J. J.; Bobet, R. Valorization of Grape (*Vitis vinifera*) Byproducts. Antioxidant and Biological Properties of Polyphenolic Fractions Differing in Procyanidin Composition and Flavonol Content. *J. Agricult. Food Chem.* **2002,** *50* (26), 7548–7555.
178. Wadhwa, M.; Bakshi, M. P. S. Utilization of Fruit and Vegetable Wastes as Livestock Feed and as Substrates for Generation of Other Value-added Products. *Rap Publ.* **2013,** *4*, 30.
179. Wallenfels, K. Symbiose und Antibiose. *Angewandte Chemie* **1945,** *58* (1–4), 1–16.
180. Wang, S.; Chen, F.; Wu, J.; Wang, Z.; Liao, X.; Hu, X. Optimization of Pectin Extraction Assisted by Microwave from Apple Pomace Using Response Surface Methodology. *J. Food Eng.* **2007,** *78* (2), 693–700.
181. Waterhouse, A. L.; Walzem, R. L. Nutrition of Grape Phenolics. **1998**.
182. Wheeler, G. E.; Fields, R. Use of a Herbal Supplement to Reduce the Effects of Stress in Intensively Housed Chickens. *Int. Symp. Med. Arom. Plants* **1993,** *344*, 496–511.
183. Widjastuti, T.; Hernawan, E. Utilizing of Banana Peel (*Musa sapientum*) in the Ration and Its Influence on Final Body Weight, Percentage of Carcass and Abdominal Fat on Broilers Under Heat Stress Condition. *Seria Zootech* **2012,** *57*, 104–109.
184. Wijngaard, H. H.; Rößle, C.; Brunton, N. A Survey of Irish Fruit and Vegetable Waste and By-products as a Source of Polyphenolic Antioxidants. *Food Chem.* **2009,** *116* (1), 202–207.
185. Windisch, W.; Schedle, K.; Plitzner, C.; Kroismayr, A. Use of Phytogenic Products as Feed Additives for Swine and Poultry. *J. Anim. Sci.* **2008,** *86* (14), 140–148.
186. World Health Organization. *Antimicrobial Resistance: Global Report on Surveillance*. World Health Organization, **2014**.
187. Yamakoshi, J.; Kataoka, S.; Koga, T.; Ariga, T. Proanthocyanidin-rich Extract from Grape Seeds Attenuates the Development of Aortic Atherosclerosis in Cholesterol-fed Rabbits. *Atheroscl.* **1999,** *142* (1), 139–149.
188. Yang, S. J; Chung, C. C. Feeding Value of Dried Citrus By-products Fed to Layer. *Kor. J. Anim. Sci.* **1985,** *27*, 239–245.
189. Yi, Z.-B.; Yu, Y.; Liang, Y.-Z.; Zeng, B. Evaluation of the Antimicrobial Mode of Berberine by LC/ESI-MS Combined with Principal Component Analysis. *J. Pharma. Biomed. Anal.* **2007,** *44* (1), 301–304.
190. Yulistiani, R.; Praseptiangga, D.; Raharjo, D.; Shirakawa, T. Prevalence of Antibiotic-resistance Enterobacteriaceae Strains Isolated from Chicken Meat at Traditional Markets in Surabaya, Indonesia. *Mat. Sci. Eng.* **2017,** *193* (1), 012007.
191. Zafar, F.; Idrees, M.; Ahmed, Z. Use of Apple By-products in Poultry Rations of Broiler Chicks in Karachi. *Pak. J. Physiol.* **2005,** *1*, 1–2.
192. Zhang, Y. N.; Wang, J.; Qi, B.; Wu, S. G.; Chen, H. R.; Luo, H. Y.; Qi, G. H. Evaluation of Mango Saponin in Broilers: Effects on Growth Performance, Carcass Characteristics, Meat Quality and Plasma Biochemical Indices. *Asian-Australasian J. Anim. Sci.* **2017,** *30* (8), 1143.
193. Zhao, C.; Ge, B.; De Villena, J.; Sudler, R.; Yeh, E.; Zhao, S.; Meng, J. Prevalence of Campylobacter spp., *Escherichia coli*, and *Salmonella serovars* in Retail Chicken, Turkey, Pork, and Beef from the Greater Washington, DC, Area. *Appl. Environ. Microbiol.* **2001,** *67* (12), 5431–5436.

194. Zhou, H.; Gong, J.; Brisbin, J. T.; Yu, H.; Sanei, B.; Sabour, P.; Sharif, S. Appropriate Chicken Sample Size for Identifying the Composition of Broiler Intestinal Microbiota Affected by Dietary Antibiotics, Using the Polymerase Chain Reaction-denaturing Gradient Gel Electrophoresis Technique. *Poult. Sci.* **2007,** *86* (12), 2541–2549.
195. Ziauddin, M.; Phansalkar, N.; Patki, P.; Diwanay, S.; Patwardhan, B. Studies on the Immunomodulatory Effects of Ashwagandha. *J. Ethnopharmacol.* **1996,** *50* (2), 69–76.
196. Zupan, A.; Mikulic-Petkovsek, M.; Cunja, V.; Stampar, F.; Veberic, R. Comparison of Phenolic Composition of Healthy Apple Tissues and Tissues Affected by Bitter Pit. *J. Agricult. Food Chem.* **2013,** *61* (49), 12066–12071.

PART II
Bioactive Compounds and Health Claims

CHAPTER 5

CROCIN: A MECHANISTIC TREATISE

MUHAMMAD IMRAN[1], ALI IMRAN[2], MUHAMMAD UMAIR ARSHAD[2], IAHTISHAM-UL-HAQ[3], and HAFIZ ANSAR RASUL SULERIA[4,*]

[1]Faculty of Allied Health Sciences, University Institute of Diet & Nutritional Sciences, The University of Lahore, Pakistan

[2]Institute of Home and Food Sciences, Government College University, Faisalabad, Pakistan

[3]Department of Diet and Nutritional Sciences, Faculty of Health and Allied Sciences, Imperial College of Business Studies, Lahore, Pakistan

[4]School of Life and Environmental Sciences, Faculty of Science, Engineering and Built Environment, Centre for Chemistry and Biotechnology, Deakin University, 75 Pigdons Road, Geelong, VIC 3216, Australia

**Corresponding author. E-mail: hafiz.suleria@uqconnect.edu.au*

ABSTRACT

Crocin is a carotenoid pigment that is a vigorous and water-soluble constituent derived from saffron (*Crocus sativus*) and gardenia (*Gardenia jasminoides*). Saffron is conventionally used as a flavoring and coloring agent in different recipes. Crocin alongside imparting eye-catching color to the food items has possessed significant beneficial effects on human health attributed to its antioxidant and disease-preventing properties. Its role in prophylaxis of cardiovascular disorders, diabetes, kidney malfunctions, inflammatory conditions, cancer, and neural problems has been identified. In this respect, this chapter summarizes sufficient scientific evidence with possible mechanisms to better understand the functionality of crocin for the use in therapeutic clinical settings.

5.1 INTRODUCTION

Since centuries, saffron (*Crocus sativus* L.) has gained fame as a medicinal and cuisine herb due to its unique culinary characteristics, like taste and flavor. Its usage dates back to 3000 years in various continents, cultures, and civilizations.[24] It is the member of Iridaceae family that was originated from the Middle East. Its price ranged from $40 to $50 per gram globally, presenting it as the most expensive spices of the world.[55] Crocetin and its glucoside derivatives such as crocin, safranal, and picrocrocin are the main components of saffron mentioned in Figure 5.1. Quercetin and kaempferol[19–21] are the main flavonoids present in it.

Functional ingredients having health-promoting and disease-preventing properties are of vital consideration these days. Various plant-based bioactives have been endorsed for their therapeutic effects (Fig. 5.2) against different ailments.[67] In this context, crocin is a well-known carotenoid consisting of ester linkage, that is, [digentiobiosyl all-*tarns*-crocetin](8,8′-di-apocarotene-8,8′-dioic acid) and the color of saffron is mainly due to the presence of crocin. Its main extraction site is the dark red stigmas of *C. sativus* and *Gardenia jasminoides* fruits.[33] Several studies have shown that absorbance of crocin in gastrointestinal tract takes a series of steps for its absorption followed by metabolism. After oral administration, crocin is initially hydrolyzed to crocetin followed by absorption in the intestine. Then, absorbed crocetin partially metabolized products are the mono and diglucuronide conjugates.[10] Further findings suggested that most of the unabsorbed crocin is excreted via intestinal route. Moreover, for the crocetin, absorption and hydrolysis of saffron intestinal part is considered as the vital part. According to findings, it is cleared that single or frequent doses of crocin demonstrated no rise in the plasma concentration of crocetin.[88] As evident from Ames assay, crocin from saffron has shown nonmutagenic nontoxic and nonantimutagenic perspectives.[1] The detailed health perspectives of crocin are discussed herein.

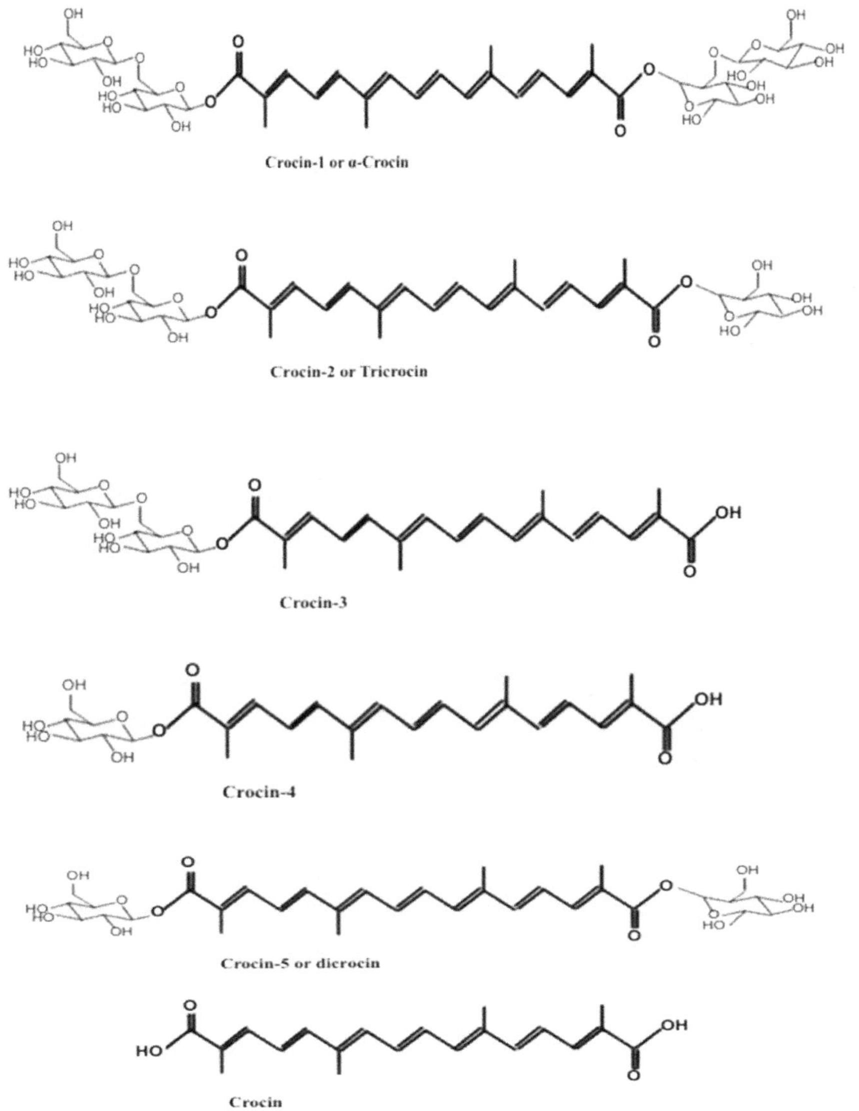

FIGURE 5.1 Types of crocin and their chemical structures.

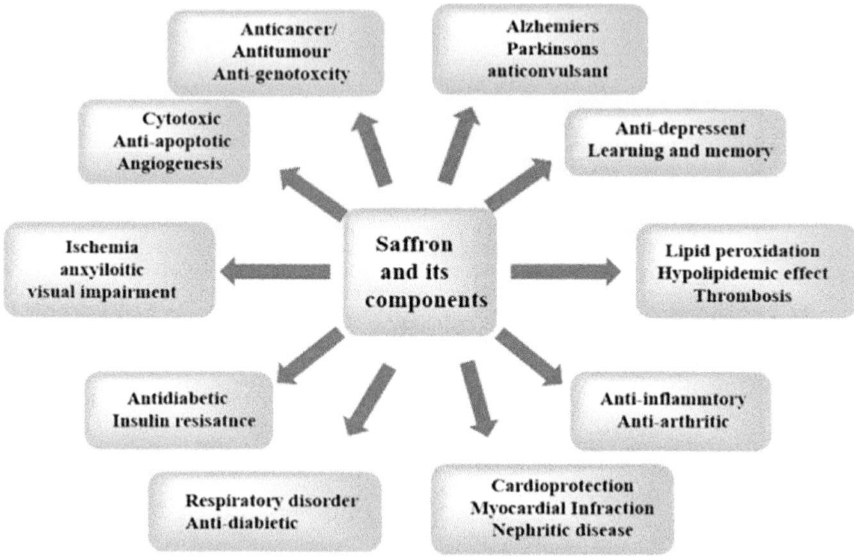

FIGURE 5.2 **(See color insert.)** Pharmacological aspects of saffron and its components.

5.2 HEALTH PERSPECTIVES

5.2.1 ANTICANCER

Globally, cancer is among the leading killers of mankind. Moreover, high-cost medical treatment having side effects and long-term therapeutic procedures further worsened the scenario to tackle this disease. The situation demands certain cost-effective and safe solutions to overcome the dilemma. Purposely, scientists are looking for plant extracts having anticancer activities as a natural remedy to alleviate cancer insurgence.[91] In this context, the crocin has been found to effectively reduce the proliferation in a dose-dependent manner of cells in A2780 that is human ovarian cancer cell line. However, the meager in cisplatin-resistant derivatives of A2780/registered cardiovascular invasive specialist (RCIS) cells compared to A2780 have been found to show a reduction in cell proliferation. Crocin also reduces multidrug resistance-associated protein 1 (MRP1) and MRP2 that are the gene expression in cancers cells at the mRNA level. Compared to drug-sensitive A2780 cells, doxorubicin (DOX) cytotoxicity has also been observed to increase in resistant A2780/RCIS cells.[50]

In another investigation, crocin cleaved caspase 3 and downregulated the caspase 8 and 9 after 24 h (@1, 1.5 and 2 mg kg^{-1} BW) as well as in breast cancer cells (MCF-7) of Swiss albino mice. Both crocin and coriander seed extract (CSE) considerably provoked DNA damage at each concentration in Michigan cancer foundation (MCF)-7 cells. Furthermore, the toxicity profile evaluated by histological exploration unveiled no perceptible differences in kidney, spleen, liver, heart, and lungs in untreated and CSE-treated groups. It has further been observed that time- and dose-dependent yeast phagocytosis by peritoneal macrophages was significantly provoked by crocin treatment. It was delineated that MCF-7 cell death that was mediated by caspase and proapoptotic effect with DNA damage mechanisms are involved at molecular levels upon treatment with crocin and CSE.[15]

In an investigation, crocin effectively inhibited cancer cell proliferation in dose–time-dependent mode in combination with hyperthermia. A synergism was found when using this combined treatment. Interestingly, the cytotoxic effect of crocin was not evident in normal cells. The combined therapy reduced up to 94% of the cancer cell's colony formation. It was also observed that the combined treatment possesses more apoptotic effect than crocin only as indicated by the increase of lactate dehydrogenase (LDH) and altered nuclear morphology of the cells. Also, a marked Bax/B-cell lymphoma 2 (Bcl-2) increase ratio was noticed in treated cells, while genes that induce heat represented a decrease in expression. Furthermore, treated cell proteins, such as Hsp70 and Hsp90, were also lowered.[57] It is observed that crocin momentously diminishes the growth of MCF-7 cells by inducing apoptosis followed by mitochondrial signaling pathways involving upregulation of Bax, promotion of Caspase-8 activation, cytochrome *c* release, and disruption of MMP that is mitochondrial membrane potential.[48]

A research model involving A549 and SPC-A1 cells revealed that dose-dependent apoptosis induction and cell proliferation diminishing perspective of crocin alongside an increase in arresting G0/G1. mRNA levels are considerably increased in associated B-cell lymphoma 2-X protein (Bax) and p53, whereas mRNA expression is lowered in Bcl-2. Moreover, additive effects related to cell proliferation have been noticed in lung cancer cell lines while using crocin in combination with cisplatin or pemetrexed. Crocin significantly promotes the chemosensitivity of adenocarcinoma cells in human lung and suppressed the cell proliferation of these cells. The possible molecular mechanisms may involve apoptosis Bcl-2 downregulation and upregulation of p53, and upregulation of Bax with cell-cycle detection.[22,84] In hepatocellular carcinoma, crocin effects in HepG2 by apoptosis induction and inflammation due to downregulation and particular cell-cycle arrest at phases of S and G2 as well as M phase. It has also been shown to suppress NF-κB.[5] In another exploration, it has been found that crocin suppressed the cyclin D1 and p21 (Cip1) expression in NMU-induced breast tumors. Additionally, it also suppressed p21 (Cip1) in a p53-dependent way.[11] The crocin effectually inhibited HO-8910 cells growth and raised their proportion alongside increasing their apoptosis rate in G0/G1 phase. It was observed that upregulation of the Fas/APO-1, caspase-3, and p53 expressions carried out by crocin. Resultantly, during G0/G1 phase, HO-8910 cell growth is retarded and cell-cycle arrest takes place by the treatment with crocin. Furthermore, crocin can promote apoptosis in HO-8910 ovarian cancer cells in perspective, also enhancing expression of Fas/APO-1 and p53-activating Caspase-3 regulated apoptotic pathway.[6,89]

Crocin also revealed significant anticancer activity in prostate cancer (PCa) cells by promoting cell-cycle arrest. Oral treatment of crocin in xenografted male nude mice showed antitumor effects as evident from reversion of epithelial–mesenchymal transdifferentiation accompanied by momentous reduction of beta-catenin and N-cadherin expression whereas showed increment in E-cadherin expression. Alongside, PCa cell migration is also inhibited and downmodulation of urokinase via invasion and metalloproteinase activities/expression.[28] It has also been noticed that crocin also reduced the LDH-A protein expression and nuclear factor erythroid 2-related factor 2 (Nrf2) is activated in human cancer cells.[44] In another investigation, the preventive role of crocin has been found in T-cell leukemia cell line in human as depicted through reduced cell viability and reactive oxygen production alongside elevated DNA fragmentation and MOLT-4-induced apoptosis and necrosis.[70] The crocin has depicted an effective time- and concentration-dependent role against PCa cell lines both in malignant and nonmalignant manner through G0/G1 phase arrest, cell proliferation is also reduced along with induction of apoptotic cell death, upregulation of Bax, and downregulation of Bcl-2 expression and activation of caspase-9 activity.[23,66]

5.2.2 ANTIDIABETIC

Diabetes mellitus (DM) is a serious health condition that compromises absorption and subsequent metabolism of glucose at cellular levels. Alongside, lipid metabolism is also affected seriously in this metabolic syndrome. The situation arises due to malfunctioning in insulin hormone which could be related to its secretion from pancreatic cells or improper functioning.[37,67] According to the World Health Organization, 336 million individuals are estimated to be a victim of DM by the year 2030 that demands serious actions to be taken to control the prevalence of this metabolic dysfunction.[87]

The intraperitoneal administration of 10, 20, and 30 mg kg^{-1} day^{-1} dose rate of crocin to diabetic rats by streptozotocin (STZ) induction momentously lowered blood glucose, nitric oxide, malondialdehyde (MDA), triglycerides, total lipids, and cholesterol; enhanced glutathione (GSH) and catalase (CAT), superoxide dismutase (SOD); and activated in a dose-dependent manner. Paradoxically, it repressed inflammatory cytokines expression in the abdominal aorta.[27,29] Likewise, another study demonstrated protective perspectives of crocin against STZ-induced diabetes in rats receiving 30 mg kg^{-1} crocin, intraperitoneally. The results depicted that there was a marked decrease in sciatic nerve MDA and blood-glucose levels. Crocin also prevented from edema and degenerative alterations of the sciatic nerve, cold allodynia, sciatic nerve MDA, and hyperglycemia arising from STZ administration. Furthermore, improvement in STZ-induced histopathological, behavioral, and biochemical changes have been noticed while treatment with crocin in biologic model feed trial.[3,12]

In another investigation, Tamaddonfard et al.[80] showed that alongside the reduction in blood glucose levels, crocin demonstrated a significant lowering effect on blood urea nitrogen, MDA, creatinine, and xanthine oxidase (XO) activities. Furthermore, it also elevated the GSH levels that relate to significant increment in the antioxidant status of an individual. The 15 and 30-mg kg^{-1} administration of crocin to the diabetic rats reduces MDA, transfer latency acquisition and transfer latency retention and improves transfer latency shortening and number of neurons in the hippocampus. Furthermore, it prophylaxes loss of neurons from the hippocampus of diabetic rats as well.[63] A research study by Shirali et al.[77] endorsed marked lowering behavior of blood glucose and thiobarbituraic acid-reactive species levels and improvement in hepatic and renal total thiol concentrations of diabetic animals upon treatment with 15, 30, and 60 mg kg^{-1} crocin doses as body weight (BW).

Alongside controlling altered metabolism of glucose in diabetic rats, crocin has also demonstrated an ameliorating effect on compromised lipid metabolism. The study conducted by Azimi et al.[13] showed that crocin considerably lowers adiponectin, total cholesterol, low-density lipoprotein concentration (LDL-C) and high-density lipoprotein concentration (HDL-C), triglycerides, and serum glucose levels. Furthermore, the momentous reduction has been noticed in insulin resistance, microalbuminuria, and HbA1c in homeostatic insulin resistance model assessment.[14,31] Antihyperlipidemic effect of crocin from *G. jasminoides* has also been reported extensively.[45] In a nutshell, crocin has significant ability to control hyperglycemic conditions as sufficient evidence have proven its biologic worth in diabetic conditions with a different type of extracts mentioned in Table 5.1.

TABLE 5.1 Antidiabetic Property of Saffron and Its Components.

Components	Results	References
Methanolic extract of saffron	80 and 240 mg kg^{-1} significantly reduce the blood glucose	[46]
Crocin and safranal	Observed positive impact on glycated hemoglobin and insulin production without altering the SGPT and SGOPT and renal functions in drug-induced glycemic rats	[46]
Saffron extract	Provision of saffron@ of 50 mg kg^{-1} for a period of 14 days caused a significant decline in glucose, insulin, and lipid biomarkers	[56]
Crocetin	40 mg kg^{-1} dose prevented dexamethasone-induced insulin resistance	[40]

SGOPT, serum glutamic oxalacetic-pyruvic transaminase; SGPT, serum glutamic-pyruvic transaminase.

5.2.3 CARDIOVASCULAR HEALTH

Unhealthy lifestyle and poor dietary practices have led to the prevalence of cardiovascular diseases around the globe, which can possibly be managed through dietary modification and adopting an active lifestyle. In this context, crocin has been found to provide cardioprotective effects.[69]

In a research trial, Elsherbiny et al.[26] showed a protective role of crocin against induced myocardial toxicity induced by DOX in male Sprague–Dawley rats. In their experiment, they administered crocin 10 and 20 mg kg^{-1} orally to the DOX-induced myocardial stressed rats for 3 weeks. The biochemical analyses showed a significant elevation in cardiac-specific LDH, creatine kinase [creatine phosphokinase MB (CK-MB)], aspartate aminotransferase, and cardiac Troponin T with impaired electrocardiogram (ECG) profile because of DOX toxicity in rats. Furthermore, examination of cardiac specimen unveiled multifocal myocardial degeneration/necrosis and myocardial inflammatory infiltration due to DOX treatment. It momentously increased the positivity of anti-Cd 68 stained cells number and sufficiently induced myocardial apoptosis. Resultantly, increased cardiac Tumor necrosis factor α (TNF-α), interleukin (IL)-1β, and caspase-3 expression were seen alongside notably decreased IL-10. In contrast, administration of crocin significantly attenuated DOX-induced cardiac toxicity in a dose-dependent manner. The crocin treatment was found to be effective in restoring normal cardiac architecture and improved ECG profile. Moreover, it also reduced oxidative stress, apoptosis, and improved host antioxidant defenses. In addition, crocin also reinstate balance between pro- and anti-inflammatory cytokines as well.[26]

Ischemia and reperfusion (IR) injury mainly affect remote organs and lower extremities. It is a significant problem arising from abdominal aortic surgery. In a research trial comprising crocin-treated IR-mediated adult male rats showed a momentous reduction in tissue TNF-α, interferon gamma, total oxidant status, IL-6, IL-1β, IL-1, oxidative stress, blood urea nitrogen, urea, and creatinine. Furthermore, it also indicated lowered histopathologic injuries, Bax, Tunel, and caspase-3 expression levels alongside improving Bcl2 expression.[2,69]

The crocin administration has also been seen to reduce systolic blood pressure and provide relaxant effects because of β2-adrenoceptors activation and inhibition of calcium channels, muscarinic receptors, and histamine H1, while modulation nitric oxide also occurs.[47,56] In a research trial, it has been indicated that vitamin E and crocin attenuate lipid peroxidation, restore decline in aortal ACh relaxation, and improve contractile responses by KCl and phenylephrine (PE). Crocin reduced toxic effects caused by diazinon (DZN)-induced vascular toxicity, both of them, causing oxidative damage. Moreover, crocin prevent lipid peroxidation cascade and restore altered relaxant and contractive aortal responses.[68]

In another investigation, Thushara et al.[83] showed that crocin ameliorates A23187- and collagen-induced endogenous production of various reactive oxygen species (ROS) and H_2O_2 as a function of concentration. Further, H_2O_2-mediated injuries were also abolished via intrinsic apoptosis pathways and platelet adhesion and aggregation due to collagen-induced toxicity were also hindered by crocin treatment. Similarly, it has also been observed in DZN-induced cardiotoxicity model that (25 and 50 mg kg^{-1}) crocin reduced CK-MB and GSH as well as MDA levels in results improved histopathological characters and Bax/Bcl2 ratio enhances the release of cytochrome cytosol and activation of caspase 3. Furthermore, it also reduced lipid peroxidation and alleviated apoptosis in DZN-induced cardiotoxicity.[68]

5.2.4 OXIDATIVE STRESS

ROS are normally produced during oxidative phosphorylation that acquires significant biological functions in cells.[30,58] However, hypergeneration of free radicals and oxidative stress either from intrinsic or extrinsic factors is responsible, which is involved in pathologies of numerous degenerative disorders.[36] Hence, the control of hypergenerated free and reduction of chances of oxidative stress-mediated malfunctions for this free radical is necessary. In this regard, plant bioactives representing antioxidant properties are pivotal to a load of ROS reduce resulting in a lowered incidence of metabolic disorders.[20,36]

Saffron has been implicated in various medicinal treatments since centuries. Crocin and apocarotenoids from *C. sativus* have been extensively studied for their potential pharmacological benefits.[20] Oxidative damage is often associated with oxidants being absorbed from diet or environment. For instance, in human and animals, sodium nitrite ($NaNO_2$) induces oxidative stress. It is a common contaminant of food, feed, and drinking water chain. In $NaNO_2$-induced oxidative stress, reported a protective effect of crocin unveiled in human erythrocytes. It appreciably attenuated lipids and protein damage oxidation and plasma membrane. Crocin has also been witnessed to ameliorate morphological changes in erythrocytes by improving the antioxidant capacity of the cells and reduced methemoglobin contents in nitrite intoxication.[8]

In an investigation, XO enzymes and MDA levels were controlled by crocin treatment. According to an observation, crocin significantly reduced the levels of XO and MDA and improved GSH levels in subjects in exercising groups compared to untreated subjects. Exercise-induced oxidative tissue damage was satisfactorily controlled by crocin administration via improved antioxidant activities.[4] Similarly, crocin treatment in HCT116 cells significantly improved survival of the cells and inhibited free radical generation and SOD modulation alongside CAT activities reducing MDA levels. Furthermore, reduction in mitochondrial membrane potential, DNA fragmentation, and activation of caspases were also hindered by crocin treatment.[72]

Recently, Ohba et al.[60] evaluated the effect of crocin in the prophylaxis of fibroblast cells derived from human skin (NB1-RGB) which may be resulted due to exposure to UV-A irradiation. It was witnessed that crocin shielded the cells against the production of ROS and cell death. The crocin treatment (1000 mg kg^{-1} p.o.) ameliorated oxidative damage caused by oxidative stress regulation in oxidative stress detector (OKD) mice skin upon exposure to UV-A irradiation. Further, induction of UV-A mediates the caspase-3 activation, and hence, crocin suppressed this expression.

In a rat model study, the intraperitoneal crocin (10, 20, and 30 mg kg^{-1} day^{-1}) administration has been noticed to be associated with improved antioxidant enzyme activities and GSH content. Further, the extent of lipid peroxidation was also reduced in kidney tissues of aged rats. Additionally, a significant reduction in elevated serum renal functional biomarkers was noticed by crocin treatment. Moreover, oxidative stress and proinflammatory cytokines were also momentously reduced by crocin administration. Alongside, a significant reduction in expression of inflammatory genes was exhibited by the administration of crocin in rats.[52,75] In another investigation by Lv et al.[49] showed that crocin protects retinal ganglion cells. Glaucoma results in irreversible blindness owing to degenerative disorder in nerve cells. Apoptosis of retinal ganglion cells has been regarded as a hallmark of glaucoma. The crocin treatment (0.1–1 μM) significantly enhanced cell viability, retinal ganglion cell (RGC)-5 cells protection from apoptosis and LDH release is lowered. It also reduced ROS levels and protein expression of Bax downregulation and cytochrome *c*, increased $\Delta\Psi m$, activated NF-κB, and promoted Bcl-2 protein expression.[49]

Patulin (PAT) production by numerous kinds of fungal species of the genera including *Penicillium*, *Aspergillus*, and *Byssochlamys* as a secondary metabolite found in rotting fruits especially apple and its products. This secondary metabolite causes severe toxic effects on various organs of the body. PAT-induced toxicity in embryonic kidney cells (HEK293) can be prevented by crocin and also in human colon carcinoma (HCT116). Crocin prevents lipid peroxidation and ER stress activation via the reduction in GADD34 and glucose-regulated protein (GRP78) expression and MDA production. Moreover, it also demonstrated the ability to inhibit DNA fragmentation and caspase 3 activation by reestablishing loss of mitochondrial membrane potential.[17] Another mycotoxin namely zearalenone (ZEN) produced by *Fusarium* species can cause reproductive disorders. The crocin treatment has been shown to prevent hepatic and renal tissues toxicity of Balb/c by the toxicity induction of ZEN. It has been indicated that crocin depicted significant reduction at the dose rate of 50, 100, and 250 mg kg^{-1} body weight in oxidative stress markers altered by ZEN-induced toxicity.[71,73]

In another investigation, Boussabbeh et al.[19] explored the ability of crocin to offset the toxic effects of PAT. They employed crocin-based pretreatment to the mice to avert PAT-induced hepatic and renal oxidative stress. The results indicated that crocin reduced the protein oxidation, lipid peroxidation, and restored redox status via regulation of endogenous antioxidant enzyme systems. Alongside hepatotoxicity and nephrotoxicity, the findings suggested a promising role of crocin in prevention of neurotoxicity.[18] In this context, intraperitoneal crocin administration at 12.5, 25, and 50 mg kg^{-1} markedly improved histopathological indications and behavioral index in acrylamide (ACR)-induced neurotoxic rats. It was observed that lipid peroxidation was reduced alongside increment in GSH levels in cerebellum and cerebral cortex. Also, significant improvement in behavioral and histopathological indicators was observed with crocin treatment in Wistar rats exposed to ACR.[54]

5.2.5 NEUROPROTECTIVE ROLE

Oxidative stress and other degenerative factors can severely affect the neural health of an individual that may further influence overall life quality and behavioral attitude of the suffering population. Furthermore, learning and mental performance are also influenced by poor nutrition and other lifestyle factors. In this context, prevention from neural damage may improve the overall performance of the individuals with better life quality. Neuroprotective properties of saffron and its functional ingredients crocin have shown to improve mental health and reduce neural damage caused by various cardinal factors.[78,80]

Wister rats-based malathion-induced depressive-like behavior research model has been used to check the neuroprotective effects of crocin. Malathion is chemically organophosphate insecticide that causes depressive-like behavior in rat. In the research study, brain-derived neurotrophic factor decrementation in rats such as cerebral cortex and hippocampus of the rats was evaluated. Dosage with crocin at 10, 20, and 40 mg kg^{-1} day^{-1} was estimated intraperitoneally in male Wister rats or in malathion-treated rats. It was observed that in the hippocampus and cerebral cortex of the rats' malathion is lowered while glutathione (GSH) increased the MDA contents. However, crocin treatment causes a significant reduction in MDA levels, while increasing GSH depicting the protective effect of this bioactive component. Furthermore, crocin also decreased immobility time in forced swimming test and increased plasma acetylcholinesterase activity in malathion-treated rats.[61,64,78,82]

Asadi et al.[9] proposed a study in which he revealed that crocin improves spatial memory indicators. Intrahippocampal as well as intraperitoneal administration of crocin was significantly correlated to the improvement in escape latency, time, and traveled distance covered in the target quadrant in comparison to beta-amyloid injection was noticed. Additionally, bax/Bcl-2 ratio is decreased in crocin treatment and caspase-3 level cleavage. A report by Mehri et al.[54] is about the intraperitoneal administration of crocin significantly improves behavioral index and morphological characters of ACR-induced neurotoxicity in Wistar rats. They deduced that crocin treatment markedly increased GSH levels in cerebellum and cerebral cortex alongside the reduction in MDA contents, hence improving behavioral and histopathological responses in ACR-induced neurotoxicity.

In a rat model study, Chen et al.[22] investigated the neuroprotective effects of crocin in prevention of retinal damage due to oxidative stressed induced IR. They deduced that reduction in retinal thickness by crocin treatment at a dose rate of 50 mg kg^{-1} via HE-staining results in protection from the reduction in RGCs. Furthermore, crocin also lowered extracellular-signal-regulated kinase (p-ERK) protein and caspase-3 expression as evident from Western blot analysis and immunohistochemistry. The study revealed that crocin enhanced the activity of GSH, T-SOD, and lowered ROS and MDA contents resulting due to IR-induced injury in retinal cells. The immunohistochemical staining showed the presence of positive cells in the inner-nuclear layer and RGC layer of retinal tissue sections which indicates that crocin has a protective ability in IR-induced retinal damage.[22] Similarly, lipid peroxidation generates acrolein which is involved in neurodegenerative disorders pathologies of like Alzheimer's disease. Acrolein causes brain aging and oxidative stress-mediated neurological dysfunctions. It has been seen that crocin prevents its countereffects via enhancing the concentration of GSH and reducing the MDA, phosphor-tau, and amyloid-beta in the brain. Furthermore, signaling pathway modulation by crocin involves mitogen-activated protein kinases (MAPKs) as reported by Rashedinia et al.[65]

In a research trial, Sarshoori et al.[76] described that crocin treatment significantly lowered the cortical infarct volume corresponding to 48% and 60% against a dose rate of 50 and 80 mg kg^{-1}, respectively. Moreover, striatal infarct volume was also reduced by 45% and 75% via crocin treatment. Additionally, NDS of ischemic rats was significantly improved by both doses of the crocin, while 80 mg kg^{-1} crocin administration showed better histological markers as indicated by decreased fiber demyelination and axonal damage and reduced number of eosinophilic neurons in ischemic regions.[76] Traumatic brain injury (TBI) that cause brain damage is also maintained by the use of crocin. In this experiment, the scientist employed c57BL/6 mice for induction of TBI in a controlled cortical impact (CCI) and found that 20 mg kg^{-1} of crocin improved brain edema and neurological severity score while decreased release of inflammatory cytokines, cell apoptosis, and microglial activation depicting a protective role of the treatment against CCI-induced TBI. The Notch activation behavior is improved by the correct dose of crocin, however, a pretreated gamma-secretase inhibitor, that is, DAPT (100 mg kg^{-1}) significantly blocked crocin-induced Notch signaling activation and that's why crocin reduces the ability to shield mice against apoptosis induced by TBI and inflammation.[86] In a nutshell, numerous studies have endorsed the protective effects of crocin in the maintenance of neurological health.

5.2.6 NEPHROPROTECTION

Various environmental factors and lifestyle practices are involved in the etiology of kidney disorders. Furthermore, high intake of antibiotic drugs can also cause serious damage to the kidney tissues. In this context, using a proactive approach to avoid such illness is necessary. For instance, gentamicin is a potent antibiotic usually employed in the treatment of gram-negative bacterial infections. However, its metabolites are associated with the onset of nephrotoxicity. A research trial designed by Yarijani et al.[92] used 32 male Wister rats induced with gentamicin nephrotoxicity for the evaluation of nephroprotective effects of crocin. The results indicated that gentamicin significantly elevated kidney MDA, plasma urea–nitrogen and creatinine levels while considerably lowered the ferric reducing antioxidant power (FRAP). In contrary, crocin treatment (100 mg kg^{-1}) reduced tissue MDA levels and plasma urea–nitrogen and creatinine concentrations alongside improving tissue antioxidant potential as indicated by increased FRAP activity in renal tissues. Moreover, several cellular damages resulted from gentamicin-induced toxicity such as cellular desquamation, tubular fibrosis, glomerular atrophy and necrosis, perivascular congestion, epithelial edema in proximal tubules, intratubular proteinaceous casts, and vascular congestion. Crocin administration recovered all of these possible cellular injuries that depict the protective effect of crocin against gentamicin-induced nephrotoxicity.[92] Crocin also increases tissue antioxidant enzymes involving glutathione reductase, glutathione peroxidase (GPx) and SOD and lowers lipid peroxidation by-products, that is, MDA levels in liver, kidney and brain tissues.[16]

A most potent mycotoxin, that is, PAT is acquired by *Penicillium* production, *Aspergillus*, and *Byssochlamys*. Due to its highly toxic effects, this compound has been studied by various researchers which showed that it is involved in teratogenicity, genotoxicity, and mutagenicity. It also has shown severe damages in various organs of the body in animals. PAT-induced toxicity is protected by the use of crocin. Boussabbeh et al.[19] designed an in vivo research trial. They showed that crocin normalized biochemical indicators in kidney and liver tissues. They exposed that PAT-induced toxicity promotes apoptosis through increased Bax, p53, and cytochrome *c* and decreased Bcl2 expressions. PAT was also found to trigger DNA fragmentation and caspase 3 activation, whereas crocin administration was seen to reduce the induction of apoptosis through regulation of all the biochemical markers proving that crocin is effective in prevention of PAT-induced toxicity.[19] α-Zearalenol (α-ZOL) and β-zearalenol (β-ZOL) are the major metabolites that is produced by ZEN and further involved in induction of various toxic effects. The toxic effects of these metabolites have been studied in human kidney cells (HEK293) based on endoplasmic reticulum (ER) stress. The ER stress produced by the α- and β-ZOL metabolites by 78 kDa glucose-regulated protein (GRP78) upregulation, GADD34, that is, a DNA damage-inducible protein and growth stress. ER stress activation is further associated with the activation of apoptosis via a mitochondrial pathway that involves an increase in production of ROS and lipid peroxidation, caspases activation at the end mitochondrial transmembrane potential is lost ($\Delta\Psi m$). HEK293 cells toxicity is resulted due to effects of α- and β-ZOL induced apoptosis that is recovered by the administration of crocin and then finally reduce the toxic effect in ER stress.[74] Similar protective effects of crocin have also been seen in cardiac tissues.[73]

Cisplatin-induced nephritic stress involves ROS as the major contributor to acute renal failure. There is a significant reduction in blood glucose urea, urinary protein, and creatinine by the use of crocin and glucose concentrations as compared to the cisplatin-treated group. Histopathological evaluation of kidney tissues showed that crocin administration protected the cellular damage caused by cisplatin toxicity and also resulted that MDA levels in lower concentration and elevated total thiol and GPx levels. Crocin also depicted higher mRNA expression of GPx evidencing the promising role of crocin in kidney health promotion.[59]

In the management of acute kidney injury (AKI) that is induced by IR inflammation is a major challenge to tackle. In this context, crocin significantly lowered plasma MDA, TNF-α, creatinine concentration, and ICAM-1 expression and leukocyte infiltration in IR-induced AKI.[51] Furthermore, crocin may also help to protect calcium oxalate nephrolithiasis that is induced by ethylene glycol. The possible mechanism involved including the antioxidant abilities and stone-forming constituents of the urine are also affected by these concentration.[7]

5.2.7 ANTI-INFLAMMATORY ACTIVITIES

Homeostasis mechanism involved in ER stress often used by the cell for adaptation purposes in accordance with intercellular and intracellular changes. Crocin suppressed XBP-1/s, syncytin-1-expressing plasmid, NOS2, and BiP in primary human fetal astrocytes. It also protected from the cytotoxic effects of NO and syncytin-1-expressing astrocytes mediated oligodendrocytotoxicity. Crocin treatment on daily basis post-experimental autoimmune encephalomyelitis (EAE) induction suppressed inflammatory gene expression and ER stress in spinal cords accompanied by preserved axonal density and myelination alongside, lowered macrophage activation and T-cell infiltration. Crocin also ameliorated EAE-associated neurobehavioral deficits.[25]

In hepatotoxicity, crocin has been found to be an effective remedy to alleviate rats' liver toxicity due to cyclophosphamide (CP). Six days of treatment with an oral dose of crocin (10 mg kg^{-1}) to CP-induced toxic rats showed an ameliorative effect of crocin against the hepatotoxicity. Oral administration of crocin significantly improves the tissue antioxidant status of rats as evident from improved SOD, CAT endogenous antioxidant enzymes, and GSH and reduced oxidative stress markers, that is, oxidation of proteins and lipids. Furthermore, serum function enzymes also depicted lowering trend upon treatment with crocin that signifies the role of this compound in normalizing hepatic health biomarkers.[38]

In an investigation, doses of (25, 50, and 100 mg kg^{-1} crocin and (0.5, 1, and 2 mg kg^{-1}) of safranal have been found to reduce the carrageenan produced edema in paw tissues. Further, these compounds decreased neutrophils and restricted inflammatory pain responses in rats.[79] Heme-oxygenase-1 (HO-1) expression that is involved in anti-inflammatory responses are also resulted from the administration of crocin. Nitric oxide production inhibition by the crocin stopped nitric as well as nitric oxide synthase (iNOS) expression induction through downregulating NF-κB activity in lipopolysaccharide (LPS)-stimulated RAW 264.7 macrophages attribute to blocking of HO-1 expression or activity that abrogated these effects. Ca(2+) mobilization from intracellular pools and phosphorylation of calmodulin-dependent protein kinase 4 (CAMK4) is also induced by crocin. Apart from this, HO-1 expression induction, crocin-mediated suppression of iNOS blocked by CAMK4 inhibition by inducing via Ca(2+)/calmodulin–CAMK4–phosphoinositide 3-kinases (PI3K)/protein kinase (Akt)–Nrf2 signaling cascades.[43]

Crocin expressively reserved catabolic enzymes and Toll-like receptor (TLR)-2 and the LPS-induced overexpression in a dose-dependent manner during in vitro study. Considerably, crocin partly prohibited the type II collagen (collagen-II) downregulation and aggrecan. Additionally, LPS-induced activation of MAPK pathway curbed and inhibition of phosphorylation of c-Jun N-terminal kinase (JNK) curbed by crocin. Trials in ex vivo exhibited intervertebral disc protection by crocin against LPS-induced depletion of extracellular matrix components involving collagen-II and proteoglycan.[46] Overall, crocin showed various biological activities that signify its potential use in pharmacological applications as summarized in Table 5.2.

TABLE 5.2 Health Perspectives of Crocin.

Disorders	Mechanisms	References
Anticancer	Decreased gene expressions of MRP1 and MRP2 at mRNA level	[50]
	Downregulated the caspase 8 and 9 induced MCF-7 cell death	[15]
	Upregulated Bax, activated caspase 8, and released cytochrome *c*	[48]
	Reduced cancer cell's colony formation, increased Bax/Bcl-2 ratio, and lowered Hsp70 and Hsp90 proteins	[57]
	Induced apoptosis and caused G0/G1 arrest	[22,84]
	Decreased Bcl-2 expressions	
	Upregulated p53 and Bax	
Antidiabetic	Lowered blood glucose, nitric oxide, MDA, triglycerides, total lipids, and cholesterol	[27,29]
	Enhanced superoxide dismutase, glutathione, and catalase activities	
	Repressed inflammatory cytokines expression	
	Decreased behavior of blood glucose and TBARS levels	[77]
	Improved hepatic and renal total thiol concentrations	
	Lowered blood urea nitrogen, creatinine, and xanthine oxidase activities	[80]
	Reduced TL1 and TL2	[63]
	Improved TLs and number of neurons	
	Prevented from edema and degenerative alterations of the sciatic nerve, cold allodynia, sciatic nerve MDA	[3,12]

TABLE 5.2 *(Continued)*

Disorders	Mechanisms	References
Cardiovascular role	Attenuated DOX-induced cardiac toxicity	[26]
	Restored normal cardiac architecture and host anti-oxidant defenses	
	Decreased tissue tumor necrosis factor α, interferon gamma, total oxidant status	[2,69]
	Lowered IL-6, IL-1β, IL-1, oxidative stress, blood urea nitrogen, urea, and creatinine	
	Lowered histopathologic injuries, Bax, Tunel, and caspase-3 expression levels	
	Improved Bcl2 expression	
Oxidative stress	Decreased XO enzymes and MDA levels	[4]
	Enhanced cell viability, protected RGC-5 cells from apoptosis and lowered LDH release	[49]
	Reduced ROS levels and downregulated the protein expression of Bax and cytochrome *c*, increased ΔΨm, activated NF-κB	
	Promoted Bcl-2 protein expression	
	Improved antioxidant enzyme activities and GSH content	[52,75]
	Reduced elevated serum renal functional biomarkers and proinflammatory cytokines	
	Suppressed the induction of UV-A-mediated caspase-3 activation	[60]
	Modulated SOD and CAT activities	[72]
	Hindered reduction in mitochondrial membrane potential, DNA fragmentation, and activation of caspases	
Neuroprotective	Improved spatial memory indicators	[9]
	Decreased bax/Bcl-2 ratio and cleaved caspase-3 level	[54]
	Improved behavioral index and morphological characters of ACR-induced neurotoxicity	
	Lowered p-ERK protein and caspase-3 expression	[22]

TABLE 5.2 *(Continued)*

Disorders	Mechanisms	References
Nephroprotection	Prevented from cellular damages including cellular desquamation, glomerular atrophy, tubular fibrosis, necrosis, and perivascular congestion	[92]
	Decreased Bax, p53 and cytochrome *c*, and increased Bcl2 expressions	[19]
	Lowered lipid peroxidation, that is, MDA levels	[16]
Anti-inflammatory activities	Suppressed XBP-1/s, syncytin-1-expressing plasmid, NOS2, and BiP inhibited inflammatory gene expression and ER stress	[25]
	Ameliorated EAE-associated neurobehavioral deficits	
	Improved the endogenous antioxidant enzymes like SOD, CAT, and glutathione	[38]
	Reduced the carrageenan produced edema	[79]
	Lowered neutrophils and restricted inflammatory pain responses	
	Induced HO-1 expression	[43]
	Stopped nitric oxide production and iNOS expression	
	Induced Ca(2+) mobilization	

ACR, acrylamide; CAT, catalase; HO-1, hemeoxygenase-1; iNOS, inducible nitric oxide synthase; IL, interleukin; MDA, malondialdehyde; SOD, superoxide dismutase; TBARS, thiobarbituraic acid-reactive species TL1, acquisition transfer latency; TL2, retention transfer latency; TLs, transfer latency shortening; XO, xanthine oxidase;.

5.2.8 MISCELLANEOUS PROPERTIES

The crocin administration has been found to decrease nitric oxide production in bronchial epithelial cells. Further, it exhibited a lowering effect on peroxynitrite ion generation and induced iNOS and release of cytochrome *c* prevention. Moreover, safranal has been identified to lower the oxidative stress via iNOS reduction as well as in brochial cell prevention by apoptosis. In an asthmatic model of murine characterized by crocin anti-inflammatory effect, by enhanced by airway cellular infiltration, airway hyperresponsiveness, and epithelial cell injury . The crocin reduced the overall biomarkers and lowered iNOS production, Th2-type cytokines generation, and bronchial epithelial cell apoptosis in the lungs.[21] In another trial of BALB/c mice, allergic airway inflammation was induced by sensitization challenged by ovalbumin (OVA). The administration of crocin significantly reduced the hyperreactivity and airway inflammation and lowered the concentrations of tryptase, BALF IL-4, IL-5, and IL-13, serum OVA-specific IgE and lung eosinophil peroxidase and inhibited lung eotaxin, p-JNK, p-ERK, and p-p38 expressions in OVA-challenged mice.[90]

The administration of crocin to the rats has shown an increased latency for the first transition to the dark side and time spent in a light/dark box in the illuminated chamber. It relates that crocin reduces the anxiety levels without influencing the motor activity of rodents in the procedure of light/dark exposure. This shows the involvement of the anxiolytic effect of crocin as a possible pharmacological mechanism.[62] In some traditional medicinal therapies, crocin has exhibited the properties of an aphrodisiac agent through its capacity to promote the mounting frequency and ejaculation and intromission latency behaviors.[35] Crocetin significant controls endothelium-dependent relaxation of thoracic aorta possibly due to increased endothelial nitric oxide synthase activity that enhances nitric oxide production in hypercholesterolemic rabbits. It is supposed that crocin behaves in a similar way as PDE-5 inhibitors such as sildenafil as crocin converts to crocetin that increases serum nitric oxide levels.[81]

Previously, the pharmacological worth of crocin was explored through numerous studies. Accordingly, Hosseinzadeh and Ghenaati[32] observed the therapeutic worth of crocin. On the other hand, indomethacin-induced gastric mucosal damage was studied by Kianbakht and Mozafari[42] and found that crocin is quite effective in reducing the damage caused by indomethacin. This possible reason behind the phenomenon is the lowered lipid peroxidation status and increased GSH based antioxidant levels in diabetic and nondiabetic rats resemble with the effect of omeprazole.[42] Crocin has also been found to possess significant immunomodulatory functions by effectively stimulating cellular as well as hormonal immune responses.[40] Also, crocin is a prominent agent to counteract the negative effects of alloxan-induced hyperglycemia. It has been reported that crocin improves insulin levels and reduce hemoglobin A1c and fasting glucose concentrations.[41] Furthermore, crocin has also been endorsed to improve sleep as evident from the findings of Masaki et al.[53] who claimed that crocin accelerates nonrapid eye movement sleep in rats as indicated by locomotor activity.

5.2.9 CROCIN TOXICITY

The toxicity studies of crocin are of immense importance to establish its potential toxicity, if any, as it is the main compound from *C. sativus* L. and *G. jasminodes* which is used as a promising coloring pigment in a number of culinary items to impart color and also used as a spice. For the purpose, rats were treated intraperitoneally with aqueous extracts of stigma at a dose rate of 0.16, 0.32, and 0.48 g kg^{-1} and petal (1.2, 2.4, and 3.6 g kg^{-1}) and the trial was continued for 2 weeks. The biochemical indicators showed extracts of both exhibited anemia; however, extract from petal caused toxic effects on lungs and liver.[39] Previously, it has also been documented that crocin-based dyes exhibit hepatic protective role. The study unveiled that 50 mg kg^{-1} dose for 8 days did not alter hepatic functions but a higher dose, that is, 100 mg kg^{-1} administered for 2 weeks was associated with black pigmentation and hepatic damage. Reversible pigmentation occurred when the rats were fed with 1% crocin dyes for 4 months that depicts its safer nature for use as a coloring agent.[85] Furthermore, acute and subacute toxicity of crocin was evaluated by Hosseinzadeh et al.[34] They deduced that crocin did not exhibit marked damage to the organs at pharmacological doses. In the last decades, the scope of constituents of saffron has gained a paramount position in order to treat various ailments as therapeutic medicines. However, for the future development and assessment of toxicity appraise the suitable perspective of saffron and its ingredients.[21] A dose of a substance 1–5 g kg^{-1} with LD50 are particular low toxic according to the toxicity classification, as well as substances with greater LD50 concentration from 5.0 g kg^{-1}, are classified as practically nontoxic.[73] The safe ranges include the intake of saffron 1.5 g day^{-1} of saffron but 5 g kg^{-1} considered as toxic and 20 g kg^{-1} is the lethal dose. The water-soluble components of saffron and its components donate toward its nontoxicity.[86] Moreover, the toxicity of saffron for pregnant women in case of daily consumption of 10 g can be responsible for abortion.[87] So, the saffron is prohibited during pregnancy and the doctor does not prescribe it.[88] Higher doses of saffron are responsible for types of allergies such as anaphylactic reaction.[89] Such kinds of allergies and sensitivities are very rare in nature. A lot of interesting information about crocin pharmacological activities showed its acute and subacute toxicity. The acute and subacute toxicities were investigated in mice and rats. Long-term consumption of crocin caused changes in the metabolism of rats. So crocin at pharmaceutical level is considered to be safe and nontoxic.[90] Several studies about the nonmutagenicity, subacute, and acute toxicity has been reported.[91] Intraperitoneally administration of safranal is considered to be low or nontoxic when orally administrated.[92] The present literature assured the low dose or pharmaceutical doses as safe. However, a higher concentration is considered as toxic and lethal.

5.3 SUMMARY

Crocin and its metabolic derivatives like crocetin are promising agents in alleviating several maladies. These compounds possess sufficient pharmacological worth to effectively protect from numerous disorders arising from oxidative stress and negative effects of various toxic compounds. A number of evidence have endorsed the preventive and therapeutic worth of crocin and unveiling of its possible positive effects are still being explored further. However, the use of crocin and its derivatives in routine meals and specialty products should be investigated to enhance its consumption to translate its benefits in physical terms. Furthermore, precisely designed research models should be devised in clinical setups to strengthen the evidence received from in vivo animal and ex vivo cell line trials. Nevertheless, scientific evidence have clearly shown its worth in alleviating metabolic malfunctions.

KEYWORDS

- **crocin**
- **neuroprotection**
- **oxidative stress**
- **physiological function**
- **phytochemical**
- **spice**
- **traditional medicine and physiological disorders**

REFERENCES

1. Abdullaev, F. I. Cancer Chemopreventive and Tumoricidal Properties of Saffron (*Crocus sativus* L.). *Exp. Biol. Med.* **2002,** *227* (1), 20–25.
2. Adali, F.; Gonul, Y.; Aldemir, M.; Hazman, O.; Ahsen, A.; Bozkurt, M. F.; Sen, O. G.; Keles, I.; Keles, H. Investigation of the Effect of Crocin Pretreatment on Renal Injury Induced by Infrarenal Aortic Occlusion. *J. Surg. Res.* **2016,** *203* (1), 145–153.
3. Altinoz, E.; Oner, Z.; Elbe, H.; Cigremis, Y.; Turkoz, Y. Protective Effects of Saffron (Its Active Constituent, Crocin) on Nephropathy in Streptozotocin-Induced Diabetic Rats. *Hum. Exp. Toxicol.* **2015,** *34* (2), 127–134.
4. Altinoz, E.; Ozmen, T.; Oner, Z.; Elbe, H.; Erdemli, M. E.; Bag, H. G. Effect of Crocin on Oxidative Stress in Recovery from Single Bout of Swimming Exercise in Rats. *Gen. Physiol. Biophys.* **2016,** *35* (1), 87–94.
5. Amin, A.; A Hamza, A.; Daoud, S.; Khazanehdari, K.; Al Hrout, A.; Baig, B.; Chaiboonchoe, A.; E Adrian, T.; Zaki, N.; Salehi-Ashtiani, K. Saffron-Based Crocin Prevents Early Lesions of Liver Cancer: In Vivo, In Vitro and Network Analyses. *Rec. Pat. Anti-Canc. Drug Discov.* **2016,** *11* (1), 121–133.
6. Amin, A.; Bajbouj, K.; Koch, A.; Gandesiri, M.; Schneider-Stock, R. Defective Autophagosome Formation in p53-Null Colorectal Cancer Reinforces Crocin-Induced Apoptosis. *Int. J. Mol. Sci.* **2015,** *16* (1), 1544–1561.
7. Amin, B.; Feriz, H. M.; Hariri, A. T.; Meybodi, N. T.; Hosseinzadeh, H. Protective Effects of the Aqueous Extract of *Crocus sativus* Against Ethylene Glycol Induced Nephrolithiasis in Rats. *EXCLI J.* **2015,** *14*, 411.
8. Ansari, F. A.; Ali, S. N.; Mahmood, R. Crocin Protects Human Erythrocytes from Nitrite-Induced Methemoglobin Formation and Oxidative Damage. *Cell Biol. Int.* **2016,** *40* (12), 1320–1331.
9. Asadi, F.; Jamshidi, A. H.; Khodagholi, F.; Yans, A.; Azimi, L.; Faizi, M.; Vali, L.; Abdollahi, M.; Ghahremani, M. H.; Sharifzadeh, M. Reversal Effects of Crocin on Amyloid β-Induced Memory Deficit: Modification of Autophagy or Apoptosis Markers. *Pharmacol. Biochem. Behav.* **2015,** *139*, 47–58.
10. Asai, A.; Nakano, T.; Takahashi, M.; Nagao, A. Orally Administered Crocetin and Crocins Are Absorbed into Blood Plasma as Crocetin and Its Glucuronide Conjugates in Mice. *J. Agric. Food Chem.* **2005,** *53* (18), 7302–7306.
11. Ashrafi, M.; Bathaie, S. Z.; Abroun, S.; Azizian, M. Effect of Crocin on Cell Cycle Regulators in *N*-Nitroso-*N*-Methylurea-Induced Breast Cancer in Rats. *DNA Cell Biol.* **2015,** *34* (11), 684–691.
12. Asri-Rezaei, S.; Tamaddonfard, E.; Ghasemsoltani-Momtaz, B.; Erfanparast, A.; Gholamalipour, S. Effects of Crocin and Zinc Chloride on Blood Levels of Zinc and Metabolic and Oxidative Parameters in Streptozotocin-Induced Diabetic Rats. *Avicenna J. Phytomed.* **2015,** *5* (5), 403.
13. Azimi, P.; Ghiasvand, R.; Feizi, A.; Hariri, M.; Abbasi, B. Effects of Cinnamon, Cardamom, Saffron, and Ginger Consumption on Markers of Glycemic Control, Lipid Profile, Oxidative Stress, and Inflammation in Type 2 Diabetes Patients. *Rev. Diab. Stud.* **2014,** *11* (3), 258.

14. Azimi, P.; Ghiasvand, R.; Feizi, A.; Hosseinzadeh, J.; Bahreynian, M.; Hariri, M.; Khosravi-Boroujeni, H. Effect of Cinnamon, Cardamom, Saffron and Ginger Consumption on Blood Pressure and a Marker of Endothelial Function in Patients with Type 2 Diabetes Mellitus: A Randomized Controlled Clinical Trial. *Blood Press.* **2016,** *25* (3), 133–140.
15. Bakshi, H. A.; Hakkim, F. L.; Sam, S. Molecular Mechanism of Crocin Induced Caspase Mediated MCF-7 Cell Death: In Vivo Toxicity Profiling and Ex Vivo Macrophage Activation. *Asian Pac. J. Cancer Prev.* **2016,** *17* (3), 1499–1506.
16. Bandegi, A. R.; Rashidy-Pour, A.; Vafaei, A. A.; Ghadrdoost, B. Protective Effects of *Crocus sativus* L. Extract and Crocin Against Chronic-Stress Induced Oxidative Damage of Brain, Liver and Kidneys in Rats. *Adv. Pharm. Bull.* **2014,** *4* (Suppl. 2), 493.
17. Boussabbeh, M.; Prola, A.; Ben Salem, I.; Guilbert, A.; Bacha, H.; Lemaire, C.; Abis-Essefi, S. Crocin and Quercetin Prevent PAT-Induced Apoptosis in Mammalian Cells: Involvement of ROS-Mediated ER Stress Pathway. *Environ. Toxicol.* **2016,** *31* (12), 1851–1858.
18. Boussabbeh, M.; Salem, I. B.; Belguesmi, F.; Bacha, H.; Abid-Essefi, S. Tissue Oxidative Stress Induced by Patulin and Protective Effect of Crocin. *Neurotoxicology* **2016,** *53*, 343–349.
19. Boussabbeh, M.; Salem, I. B.; Belguesmi, F.; Neffati, F.; Najjar, M. F.; Abid-Essefi, S.; Bacha, H. Crocin Protects the Liver and Kidney from Patulin-Induced Apoptosis In Vivo. *Environ. Sci. Pollut. Res.* **2016,** *23* (10), 9799–9808.
20. Bukhari, S. I.; Manzoor, M.; Dhar, M. A Comprehensive Review of the Pharmacological Potential of *Crocus sativus* and Its Bioactive Apocarotenoids. *Biomed. Pharmacother.* **2018,** *98*, 733–745.
21. Bukhari, S. I.; Pattnaik, B.; Rayees, S.; Kaul, S.; Dhar, M. K. Safranal of *Crocus sativus* L. Inhibits Inducible Nitric Oxide Synthase and Attenuates Asthma in a Mouse Model of Asthma. *Phytother. Res.* **2015,** *29* (4), 617–627.
22. Chen, L.; Qi, Y.; Yang, X. Neuroprotective Effects of Crocin against Oxidative Stress Induced by Ischemia/Reperfusion Injury in Rat Retina. *Ophthalm. Res.* **2015,** *54* (3), 157–168.
23. D'alessandro, A. M.; Mancini, A.; Lizzi, A. R.; De Simone, A.; Marroccella, C. E.; Gravina, G. L.; Tatone, C.; Festuccia, C. *Crocus sativus* Stigma Extract and Its Major Constituent Crocin Possess Significant Antiproliferative Properties Against Human Prostate Cancer. *Nutr. Cancer* **2013,** *65* (6), 930–942.
24. Deo, B. Growing Saffron—The World's Most Expensive Spice. *Crop Food Res.* **2003,** *20* (1), 1–4.
25. Deslauriers, A. M.; Afkhami-Goli, A.; Paul, A. M.; Bhat, R. K.; Acharjee, S.; Ellestad, K. K.; Noorbakhsh, F.; Michalak, M.; Power, C. Neuroinflammation and Endoplasmic Reticulum Stress Are Coregulated by Crocin to Prevent Demyelination and Neurodegeneration. *J. Immunol.* **2011,** *187* (9), 4788–4799.
26. Elsherbiny, N. M.; Salama, M. F.; Said, E.; El-Sherbiny, M.; Al-Gayyar, M. M. Crocin Protects against Doxorubicin-Induced Myocardial Toxicity in Rats through Down-Regulation of Inflammatory and Apoptic Pathways. *Chem.-Biol. Interact.* **2016,** *247*, 39–48.

27. Farshid, A. A.; Tamaddonfard, E. Histopathological and Behavioral Evaluations of the Effects of Crocin, Safranal and Insulin on Diabetic Peripheral Neuropathy in Rats. *Avicenna J. Phytomed.* **2015,** *5* (5), 469.
28. Festuccia, C.; Mancini, A.; Gravina, G. L.; Scarsella, L.; Llorens, S.; Alonso, G. L.; Tatone, C.; Di Cesare, E.; Jannini, E. A.; Lenzi, A. Antitumor Effects of Saffron-Derived Carotenoids in Prostate Cancer Cell Models. *BioMed Res. Int.* **2014,** *2014*, 135048.
29. Ghorbanzadeh, V.; Mohammadi, M.; Dariushnejad, H.; Chodari, L.; Mohaddes, G. Effects of Crocin and Voluntary Exercise, Alone or Combined, on Heart VEGF-A and HOMA-IR of HFD/STZ Induced Type 2 Diabetic Rats. *J. Endocrinol. Invest.* **2016,** *39* (10), 1179–1186.
30. Hamanaka, R. B.; Chandel, N. S. Mitochondrial Reactive Oxygen Species Regulate Cellular Signaling and Dictate Biological Outcomes. *Trends Biochem. Sci.* **2010,** *35* (9), 505–513.
31. Hemmati, M.; Asghari, S.; Zohoori, E.; Karamian, M. Hypoglycemic Effects of Three Iranian Edible Plants; Jujube, Barberry and Saffron: Correlation with Serum Adiponectin Level. *Pak. J. Pharm. Sci.* **2015,** *28* (6), 2095-2099.
32. Hosseinzadeh, H.; Ghenaati, J. Evaluation of the Antitussive Effect of Stigma and Petals of Saffron (*Crocus sativus*) and Its Components, Safranal and Crocin in Guinea Pigs. *Fitoterapia* **2006,** *77* (6), 446–448.
33. Hosseinzadeh, H.; Nassiri-Asl, M. Avicenna's (*Ibn sina*) the Canon of Medicine and Saffron (*Crocus sativus*): A Review. *Phytother. Res.* **2013,** *27* (4), 475–483.
34. Hosseinzadeh, H.; Shariaty, M.; Sameni, A.; Vahabzadeh, M. Acute and Sub-acute Toxicity of Crocin, a Constituent of *Crocus sativus* L. (Saffron), in Mice and Rats. *Pharmacologyonline* **2010,** *2*, 943–951.
35. Hosseinzadeh, H.; Ziaee, T.; Sadeghi, A. The Effect of Saffron, *Crocus sativus* Stigma, Extract and Its Constituents, Safranal and Crocin on Sexual Behaviors in Normal Male Rats. *Phytomedicine* **2008,** *15* (6), 491–495.
36. Iahtisham-Ul-Haq; Butt, M. S.; Randhawa, M. A.; Shahid, M. Hepatoprotective Effects of Red Beetroot Based Beverages against CCl4-Induced Hepatic Stress in Sprague Dawley Rats. *Asian Pac. J. Trop. Med.* **2018,** *11* (4).
37. Ibrahim, S. S. Protective Effect of Hesperidin, a Citrus Bioflavonoid, on Diabetes-Induced Brain Damage in Rats. *J. Appl. Sci. Res.* **2008,** *4* (1), 84.
38. Jnaneshwari, S.; Hemshekhar, M.; Santhosh, M. S.; Sunitha, K.; Thushara, R.; Thirunavukkarasu, C.; Kemparaju, K.; Girish, K. S. Crocin, a Dietary Colorant Mitigates Cyclophosphamide-Induced Organ Toxicity by Modulating Antioxidant Status and Inflammatory Cytokines. *J. Pharmacy Pharmacol.* **2013,** *65* (4), 604–614.
39. Karimi, G.; Taiebi, N.; Hosseinzadeh, H.; Shirzad, F. Evaluation of Subacute Toxicity of Aqueous Extract of *Crocus sativus* L. Stigma and Petal in Rats. *J. Med. Plants* **2004,** *4* (12), 29–35.
40. Khajuria, D. K.; Asad, M.; Asdaq, S.; Kumar, P. The Potency of *Crocus sativus* (Saffron) and Its Constituent Crocin as an Immunomodulator in Animals. *Latin Am. J. Pharm.* **2010,** *29*.
41. Kianbakht, S.; Hajiaghaee, R. Anti-Hyperglycemic Effects of Saffron and Its Active Constituents, Crocin and Safranal, in Alloxan-Induced Diabetic Rats. *J. Med. Plants* **2011,** *3* (39), 82–89.

42. Kianbakht, S.; Mozafari, K. Effects of Saffron and Its Active Constituents, Crocin and Safranal, on Prevention of Indomethacin Induced Gastric Ulcers in Diabetic and Nondiabetic Rats. *J. Med. Plants* **2009,** *8*, 30–38.
43. Kim, J.-H.; Park, G.-Y.; Bang, S. Y.; Park, S. Y.; Bae, S.-K.; Kim, Y. Crocin Suppresses LPS-Stimulated Expression of Inducible Nitric Oxide Synthase by Upregulation of Heme Oxygenase-1 via Calcium/Calmodulin-Dependent Protein Kinase 4. *Mediat. Inflamm.* **2014**, *2014*, 728709.
44. Kim, S. H.; Lee, J. M.; Kim, S. C.; Park, C. B.; Lee, P. C. Proposed Cytotoxic Mechanisms of the Saffron Carotenoids Crocin and Crocetin on Cancer Cell Lines. *Biochem. Cell Biol.* **2014,** *92* (2), 105–111.
45. Lee, I.-A.; Lee, J. H.; Baek, N.-I.; Kim, D.-H. Antihyperlipidemic Effect of Crocin Isolated from the Fructus of *Gardenia jasminoides* and Its Metabolite Crocetin. *Biol. Pharm. Bull.* **2005,** *28* (11), 2106–2110.
46. Li, K.; Li, Y.; Ma, Z.; Zhao, J. Crocin Exerts Anti-Inflammatory and Anti-Catabolic Effects on Rat Intervertebral Discs by Suppressing the Activation of JNK. *Int. J. Mol. Med.* **2015,** *36* (5), 1291–1299.
47. Llorens, S.; Mancini, A.; Serrano-Díaz, J.; D'alessandro, A. M.; Nava, E.; Alonso, G. L.; Carmona, M. Effects of Crocetin Esters and Crocetin from *Crocus sativus* L. on Aortic Contractility in Rat Genetic Hypertension. *Molecules* **2015,** *20* (9), 17570–17584.
48. Lu, P.; Lin, H.; Gu, Y.; Li, L.; Guo, H.; Wang, F.; Qiu, X. Antitumor Effects of Crocin on Human Breast Cancer Cells. *Int. J. Clin. Exp. Med.* **2015,** *8* (11), 20316.
49. Lv, B.; Chen, T.; Xu, Z.; Huo, F.; Wei, Y.; Yang, X. Crocin Protects Retinal Ganglion Cells Against H_2O_2-Induced Damage Through the Mitochondrial Pathway and Activation of NF-κB. *Int. J. Mol. Med.* **2016,** *37* (1), 225–232.
50. Mahdizadeh, S.; Karimi, G.; Behravan, J.; Arabzadeh, S.; Lage, H.; Kalalinia, F. Crocin Suppresses Multidrug Resistance in MRP Overexpressing Ovarian Cancer Cell Line. *DARU J. Pharm. Sci.* **2016,** *24* (1), 17.
51. Mahmoudzadeh, L.; Najafi, H.; Ashtiyani, S. C.; Yarijani, Z. M. Anti-Inflammatory and Protective Effects of Saffron Extract in Ischaemia/Reperfusion-Induced Acute Kidney Injury. *Nephrology* **2017,** *22* (10), 748–754.
52. Maleki, E. M.; Eimani, H.; Bigdeli, M. R.; Narenji, A. G.; Abedi, R. Effects of Crocin Supplementation During In Vitro Maturation of Mouse Oocytes on Glutathione Synthesis and Cytoplasmic Maturation. *Int. J. Fertil. Steril.* **2016,** *10* (1), 53.
53. Masaki, M.; Aritake, K.; Tanaka, H.; Shoyama, Y.; Huang, Z. L.; Urade, Y. Crocin Promotes Non-Rapid Eye Movement Sleep in Mice. *Mol. Nutr. Food Res.* **2012,** *56* (2), 304–308.
54. Mehri, S.; Abnous, K.; Khooei, A.; Mousavi, S. H.; Shariaty, V. M.; Hosseinzadeh, H. Crocin Reduced Acrylamide-Induced Neurotoxicity in Wistar Rat through Inhibition of Oxidative Stress. *Iran. J. Basic Med. Sci.* **2015,** *18* (9), 902.
55. Melnyk, J. P.; Wang, S.; Marcone, M. F. Chemical and Biological Properties of the World's Most Expensive Spice: Saffron. *Food Res. Int.* **2010,** *43* (8), 1981–1989.
56. Mokhtari-Zaer, A.; Khazdair, M. R.; Boskabady, M. H. Smooth Muscle Relaxant Activity of *Crocus sativus* (Saffron) and Its Constituents: Possible Mechanisms. *Avicenna J. Phytomed.* **2015,** *5* (5), 365.

57. Mostafavinia, S. E.; Khorashadizadeh, M.; Hoshyar, R. Antiproliferative and Proapoptotic Effects of Crocin Combined with Hyperthermia on Human Breast Cancer Cells. *DNA Cell Biol.* **2016,** *35* (7), 340–347.
58. Murphy, M. P. How Mitochondria Produce Reactive Oxygen Species. *Biochem. J.* **2009,** *417* (1), 1–13.
59. Naghizadeh, B.; Mansouri, S. M. T.; Mashhadian, N. V. Crocin Attenuates Cisplatin-Induced Renal Oxidative Stress in Rats. *Food Chem. Toxicol.* **2010,** *48* (10), 2650–2655.
60. Ohba, T.; Ishisaka, M.; Tsujii, S.; Tsuruma, K.; Shimazawa, M.; Kubo, K.; Umigai, N.; Iwawaki, T.; Hara, H. Crocetin Protects Ultraviolet A-Induced Oxidative Stress and Cell Death in Skin In Vitro and In Vivo. *Eur. J. Pharmacol.* **2016,** *789*, 244–253.
61. Oruc, S.; Gönül, Y.; Tunay, K.; Oruc, O. A.; Bozkurt, M. F.; Karavelioğlu, E.; Bağcıoğlu, E.; Coşkun, K. S.; Celik, S. The Antioxidant and Antiapoptotic Effects of Crocin Pretreatment on Global Cerebral Ischemia Reperfusion Injury Induced by Four Vessels Occlusion in Rats. *Life Sci.* **2016,** *154*, 79–86.
62. Pitsikas, N.; Boultadakis, A.; Georgiadou, G.; Tarantilis, P.; Sakellaridis, N. Effects of the Active Constituents of *Crocus sativus* L., Crocins, in an Animal Model of Anxiety. *Phytomedicine* **2008,** *15* (12), 1135–1139.
63. Rajaei, Z.; Hadjzadeh, M.-A.-R.; Nemati, H.; Hosseini, M.; Ahmadi, M.; Shafiee, S. Antihyperglycemic and Antioxidant Activity of Crocin in Streptozotocin-Induced Diabetic Rats. *J. Med. Food* **2013,** *16* (3), 206–210.
64. Rao, S. V.; Yenisetti, S. C.; Rajini, P. S. Evidence of Neuroprotective Effects of Saffron and Crocin in a *Drosophila* Model of Parkinsonism. *Neurotoxicology* **2016,** *52*, 230–242.
65. Rashedinia, M.; Lari, P.; Abnous, K.; Hosseinzadeh, H. Protective Effect of Crocin on Acrolein-Induced Tau Phosphorylation in the Rat Brain. *Acta Neurobiol. Exp.* **2015,** *75*, 208–219.
66. Rastgoo, M.; Hosseinzadeh, H.; Alavizadeh, H.; Abbasi, A.; Ayati, Z.; Jaafari, M. R. Antitumor Activity of PEGylated Nanoliposomes Containing Crocin in Mice Bearing C26 Colon Carcinoma. *Planta Med.* **2013,** *79* (06), 447–451.
67. Raza, A.; Butt, M. S.; Iahtisham-Ul-Haq; Suleria, H. a. R. Jamun (*Syzygium cumini*) Seed and Fruit Extract Attenuate Hyperglycemia in Diabetic Rats. *Asian Pac. J. Trop. Biomed.* **2017,** *7* (8), 750–754.
68. Razavi, B. M.; Hosseinzadeh, H.; Movassaghi, A. R.; Imenshahidi, M.; Abnous, K. Protective Effect of Crocin on Diazinon Induced Cardiotoxicity in Rats in Subchronic Exposure. *Chem. Biol. Interact.* **2013,** *203* (3), 547–555.
69. Razmaraii, N.; Babaei, H.; Nayebi, A. M.; Assadnassab, G.; Helan, J. A.; Azarmi, Y. Cardioprotective Effect of Grape Seed Extract on Chronic Doxorubicin-Induced Cardiac Toxicity in Wistar Rats. *Adv. Pharm. Bull.* **2016,** *6* (3), 423.
70. Rezaee, R.; Mahmoudi, M.; Abnous, K.; Zamani Taghizadeh Rabe, S.; Tabasi, N.; Hashemzaei, M.; Karimi, G. Cytotoxic Effects of Crocin on MOLT-4 Human Leukemia Cells. *J. Complement. Integr. Med.* **2013,** *10* (1), 105–112.
71. Salem, I. B.; Boussabbeh, M.; Helali, S.; Abid-Essefi, S.; Bacha, H. Protective Effect of Crocin Against Zearalenone-Induced Oxidative Stress in Liver and Kidney of Balb/c Mice. *Environ. Sci. Pollut. Res.* **2015,** *22* (23), 19069–19076.

72. Salem, I. B.; Boussabbeh, M.; Kantaoui, H.; Bacha, H.; Abid-Essefi, S. Crocin, the Main Active Saffron Constituent, Mitigates Dichlorvos-Induced Oxidative Stress and Apoptosis in HCT-116 Cells. *Biomed. Pharmacother.* **2016,** *82*, 65–71.
73. Salem, I. B.; Boussabbeh, M.; Neffati, F.; Najjar, M.; Abid-Essefi, S.; Bacha, H. Zearalenone-Induced Changes in Biochemical Parameters, Oxidative Stress and Apoptosis in Cardiac Tissue: Protective Role of Crocin. *Hum. Exp. Toxicol.* **2016,** *35* (6), 623–634.
74. Salem, I. B.; Boussabbeh, M.; Prola, A.; Guilbert, A.; Bacha, H.; Lemaire, C.; Abid-Essefi, S. Crocin Protects Human Embryonic Kidney Cells (HEK293) from α- and β-Zearalenol-Induced ER Stress and Apoptosis. *Environ. Sci. Pollut. Res.* **2016,** *23* (15), 15504–15514.
75. Samarghandian, S.; Azimi-Nezhad, M.; Borji, A.; Farkhondeh, T. Effect of Crocin on Aged Rat Kidney Through Inhibition of Oxidative Stress and Proinflammatory State. *Phytother. Res.* **2016,** *30* (8), 1345–1353.
76. Sarshoori, J. R.; Asadi, M. H.; Mohammadi, M. T. Neuroprotective Effects of Crocin on the Histopathological Alterations Following Brain Ischemia–Reperfusion Injury in Rat. Iran. *J. Basic Med. Sci.* **2014,** *17* (11), 895.
77. Shirali, S.; Zahra Bathaie, S.; Nakhjavani, M. Effect of Crocin on the Insulin Resistance and Lipid Profile of Streptozotocin-Induced Diabetic Rats. *Phytother. Res.* **2013,** *27* (7), 1042–1047.
78. Soeda, S.; Aritake, K.; Urade, Y.; Sato, H.; Shoyama, Y. Neuroprotective Activities of Saffron and Crocin. *The Benefits of Natural Products for Neurodegenerative Diseases*; Springer: Berlin-Heidelberg, 2016.
79. Tamaddonfard, E.; Farshid, A.-A.; Eghdami, K.; Samadi, F.; Erfanparast, A. Comparison of the Effects of Crocin, Safranal and Diclofenac on Local Inflammation and Inflammatory Pain Responses Induced by Carrageenan in Rats. *Pharmacol. Rep.* **2013,** *65* (5), 1272–1280.
80. Tamaddonfard, E.; Farshid, A. A.; Asri-Rezaee, S.; Javadi, S.; Khosravi, V.; Rahman, B.; Mirfakhraee, Z. Crocin Improved Learning and Memory Impairments in Streptozotocin-Induced Diabetic Rats. *Iran. J. Basic Med. Sci.* **2013,** *16* (1), 91.
81. Tang, F.; Qian, Z.; Liu, P.; Zheng, S.; He, S.; Bao, L.; Huang, H. Crocetin Improves Endothelium-Dependent Relaxation of Thoracic Aorta in Hypercholesterolemic Rabbit by Increasing eNOS Activity. *Biochem. Pharmacol.* **2006,** *72* (5), 558–565.
82. Terraf, P.; Kouhsari, S. M.; Ai, J.; Babaloo, H. Tissue-Engineered Regeneration of Hemisected Spinal Cord Using Human Endometrial Stem Cells, Poly-ε-caprolactone Scaffolds, and Crocin as a Neuroprotective Agent. *Mol. Neurobiol.* **2017,** *54* (7), 5657–5667.
83. Thushara, R.; Hemshekhar, M.; Santhosh, M. S.; Jnaneshwari, S.; Nayaka, S.; Naveen, S.; Kemparaju, K.; Girish, K. Crocin, a Dietary Additive Protects Platelets from Oxidative Stress-Induced Apoptosis and Inhibits Platelet Aggregation. *Mol. Cell. Biochem.* **2013,** *373* (1–2), 73–83.
84. Vali, F.; Changizi, V.; Safa, M. Synergistic Apoptotic Effect of Crocin and Paclitaxel or Crocin and Radiation on MCF-7 cells, a Type of Breast Cancer Cell Line. *Int. J. Breast Cancer* **2015,** *2015*, 7.

85. Wang, C.; Hwang, L. S.; Lin, J. Reversible Hepatic Black Pigmentation and Enzyme Alteration Induced by Prolonged Feeding of High Dose of Crocin Dyes in Rats. *Proc. Natl. Sci. Council, Repub. China, B: Life Sci.* **1984,** *8* (3), 246–253.
86. Wang, K.; Zhang, L.; Rao, W.; Su, N.; Hui, H.; Wang, L.; Peng, C.; Tu, Y.; Zhang, S.; Fei, Z. Neuroprotective Effects of Crocin against Traumatic Brain Injury in Mice: Involvement of Notch Signaling Pathway. *Neurosci. Lett.* **2015,** *591*, 53–58.
87. WHO. Prevention of Blindness from Diabetes Mellitus: Report of a WHO Consultation in Geneva, Switzerland, November 9–11, 2005; World Health Organization: Geneva, 2006.
88. Xi, L.; Qian, Z.; Du, P.; Fu, J. Pharmacokinetic Properties of Crocin (Crocetin Digentiobiose Ester) Following Oral Administration in Rats. *Phytomedicine* **2007,** *14* (9), 633–636.
89. Xia, D. Ovarian Cancer HO-8910 Cell Apoptosis Induced by Crocin In Vitro. *Nat. Prod. Commun.* **2015,** *10* (2), 249–252.
90. Xiong, Y.; Wang, J.; Yu, H.; Zhang, X.; Miao, C. Anti-Asthma Potential of Crocin and Its Effect on MAPK Signaling Pathway in a Murine Model of Allergic Airway Disease. *Immunopharmacol. Immunotoxicol.* **2015,** *37* (3), 236–243.
91. Yang, C. S.; Landau, J. M.; Huang, M.-T.; Newmark, H. L. Inhibition of Carcinogenesis by Dietary Polyphenolic Compounds. *Annu. Rev. Nutr.* **2001,** *21* (1), 381–406.
92. Yarijani, Z. M.; Najafi, H.; Madani, S. H. Protective Effect of Crocin on Gentamicin-Induced Nephrotoxicity in Rats. *Iran. J. Basic Med. Sci.* **2016,** *19* (3), 337.

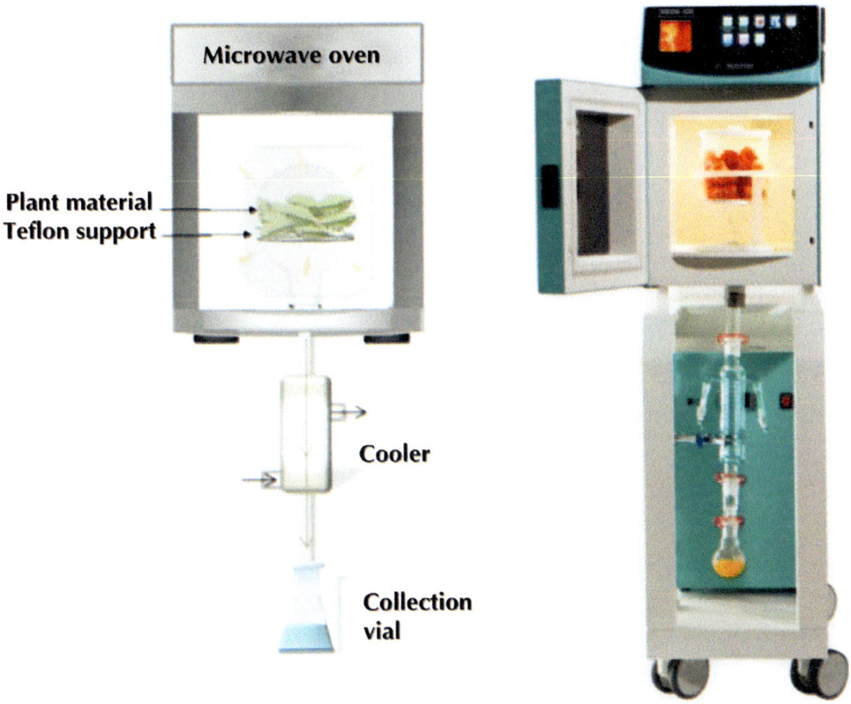

FIGURE 1.5 Microwave hydrodiffusion and gravity (MHG) extraction system.

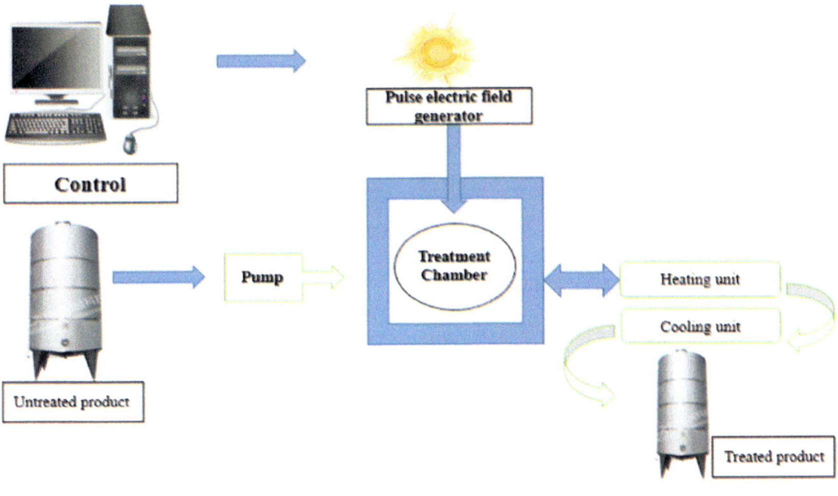

FIGURE 1.8 Schematic diagram of pulsed electric field-assisted extraction.

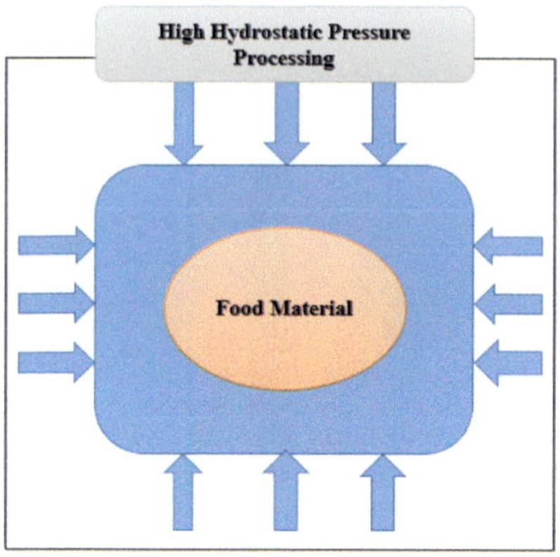

FIGURE 1.10 Schematic diagram showing mechanism of high-pressure system.

FIGURE 4.1 Fruit-processing waste.

FIGURE 4.2 Antibacterial modes of action of phytochemicals derived from fruit waste.

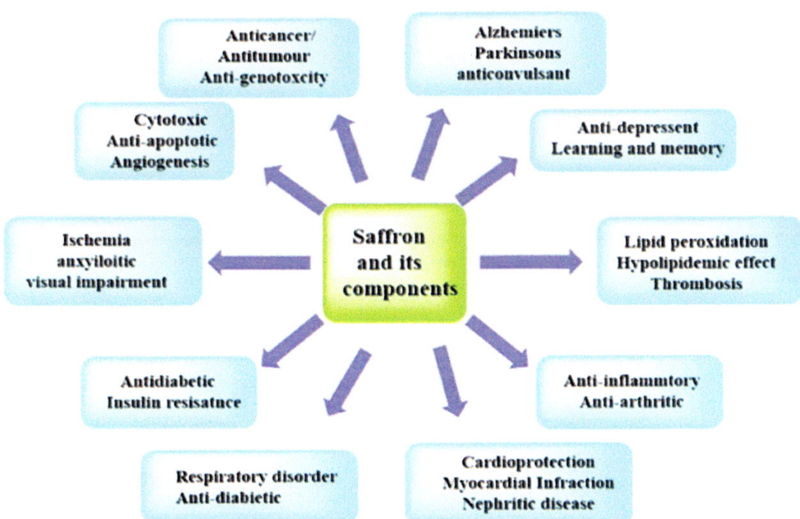

FIGURE 5.2 Pharmacological aspects of saffron and its components.

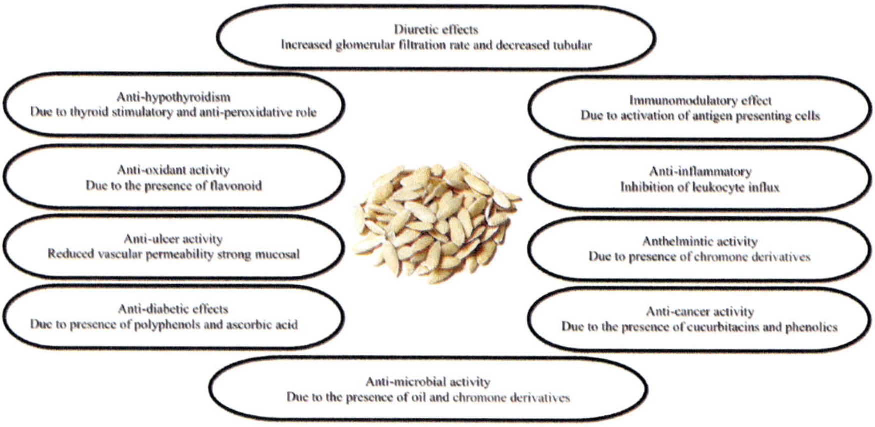

FIGURE 7.2 Potential therapeutic benefits and mechanisms of bioactive compounds present in the melon seeds.

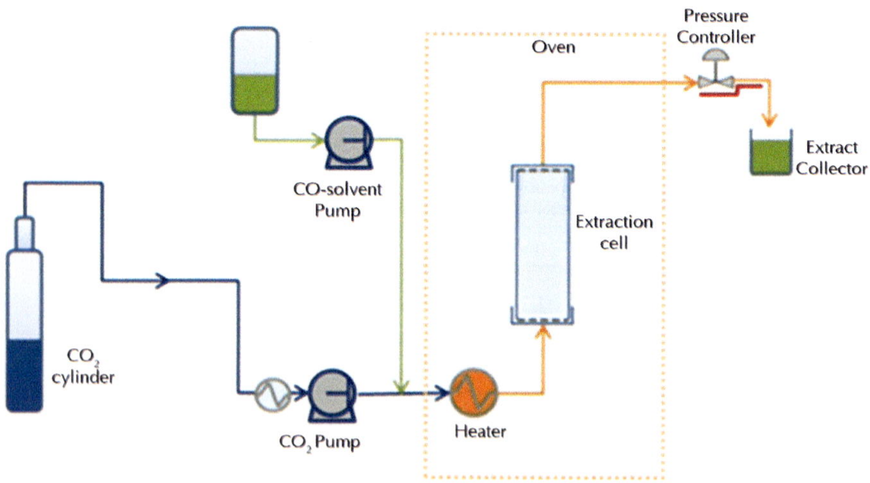

FIGURE 9.2 A typical supercritical fluid extraction system.

CHAPTER 6

GRAPES: A NUTRITIONAL KEY TO COMBAT VARIOUS MALADIES

TABUSSAM TUFAIL[1], FARHAN SAEED[1], HUMA BADER UL AIN[1], AHMAD DIN[2], and HAFIZ ANSAR RASUL SULERIA[3,*]

[1]*Institute of Home and Food Sciences, Government College University, Faisalabad, Pakistan*

[2]*Ayub Agricultural Research Institute, Faisalabad, Pakistan*

[3]*School of Life and Environmental Sciences, Faculty of Science, Engineering and Built Environment, Centre for Chemistry and Biotechnology, Deakin University, 75 Pigdons Road, Geelong, VIC 3216, Australia*

**Corresponding author. E-mail: hafiz.suleria@uqconnect.edu.au*

ABSTRACT

In the recent times, functional foods have gained greater attention for their role in the maintenance of health of human beings by lowering the risk of many lifestyle diseases like diabetes, obesity, inflammation, cardiovascular diseases, and cancer. Dietary modification by addition of functional food as one of the lifestyle modifications can help a consumer to provide protection against this variety of diseases. This nutritional and therapeutic potential is owing to the presence of various bioactive and functional ingredients in functional foods. In this milieu, grapes hold a firm position amongst fruits due to the presence of a broad spectrum of bioactive compounds mainly polyphenols, resveratrol, melatonin, and anthocyanins. These all compounds are principal components amongst grapes' bioactives. These functional ingredients are of considerable significance in the medicinal and therapeutic use of this commodity in various traditional and modern dietary therapies. Anthocyanins, resveratrol, polyphenols, and melatonin have a diverse range of biological activities and shown favorable effects in alleviating oxidative stress, diabetes mellitus, cardiovascular complications, obesity, inflammation conditions, and cancer. In this proposed chapter, we discuss the importance of grapes and their bioactive compounds in different lifestyle diseases management.

6.1 INTRODUCTION

Good health is based on the nutrients and vitamins that are abundantly provided by the fruits as they are suitable for approximately all the individuals.[2] An incredible amount and different varieties of tasty and full of flavor fruits are produced whole year in the tropical zones around the globe.[27,82] Fruits are ripened and eatable parts of the plants having seeds and pulp along with surrounding tissue, sweet as well as sour taste. Fruits are considered as a major source of juices. They comprehend a vast range of constructive and advantageous nutrients mainly micronutrients, polysaccharides, fiber, volatile components, and a vast variety of bioactive compounds.[67,82]

Among fruits, grapes are considered as the most significant and fundamental crops all over the world. Nearly 60 species of grapes are integrated with the genus *Vitis*. *Vitis vinifera*, also known as European grapes, is the most cultivated fruits crops globally.[91] They are called as European grapes because their cultivation rate is much higher in this region. Grapes are nonclimacteric drupes and cultivated as the deciduous and perennial woody creeping plants owing to their important and elemental constituents. It was investigated by the archeological statistics that approximately 80,000 years before, the production of *V. vinifera* subsp. *vinifera*, the farm grapes begun from their wild ancestors in the Near East of the globe.

The primary production of these climbing plants was started from the Middle East and was first preserved in jars in nearly about 5000 BC.[57] In the production of grapes, the 20th position is secured by Brazil around the world. Investigation reveals that the cultivation of grapes annually ranges between 1.3 and 1.4 million tons in 2012.[38] The production of grapes in the whole world was approximately 27.44 GL in 2015 and the utilization rate is about 24 GL in the same year. Among all fruits, the most utilized fruit all over the world is grapes. This is attributed to the intrinsic sensory properties such as aroma, taste, and freshness along with commercial availability, reasonable price, and in recent times the functional properties explain the enormous in vitro and in vivo factors, and epidemiological or clinical research.[54,88]

According to the evidence, products of grapes including candies, jams, sauces, wines, and juices are utilized globally due to their significant production. The most utilized grape product all over the world is the wine that has gained much more attention than that of ethyl alcohol. In France, Italy, the United States, Spain, Chile, and Turkey, grape juice is produced at large scale. Many countries are emerging as a good producer in the last few years; these include Iran, Argentina, Brazil, and Australia.[31] Dried grapes or raisins are produced nowadays globally. The major raisins producing and consuming country is the United States with 28% of production. The second one is Turkey with the production of 26% raisins. The seventh largest production is from Argentina with 3.4% manufacturing.[39] The American grapes are scientifically called as *Vitis labrusca* and commonly known as table grapes. This variety is consumed generally as raw and utilized as table wines, juices, and cooking.[17]

Grapes contain 23 species and 20 accepted taxa. Currently, grapes are considered to be one of the most valuable, conventional as well as cherished crops of the globe and mainly deal in nonalcoholic as well as alcoholic beverages. The configuration of grapes chemically fluctuates due to variety, soil, cultivar, and climate. The possessions of their derivatives can be verified by the chemical as well as the physical properties of the grapes. The involvement of organic acids such as malic, citric, and tartaric specifies the presence of acidity that is capable of giving the product a lower pH and establishing equilibrium between the sour and sweet flavor[22–24] to give the product a taste that mesmerizes the taste buds.

The occurrence, as well as the dissemination in the world, is not because of the only unique properties of the grapes and wine, but mostly it is dependent on its extraordinary nutritional importance and health-promoting assets. Grapes become the most important constituent of the customary Mediterranean diet as they played a vital role in promoting healthy effects in living organisms. Bioactive products of grapes that are consumed by humans on a daily basis possess anti-inflammatory, antioxidant, as well as cardioprotective properties.[1,3]

6.2 CHEMICAL COMPOSITION

Grapes' chemical composition is reliant on various factors such as quality, composition, and the characteristics of soil, extent at which the grapes mature, nature of the farming system, hydrological stress, preprocessing, agronomical techniques (conventional, biodynamic, and organic), disclosure to sunlight, pathogenesis, and way of pressing.[33,37,62] Grapes are a good source of carbohydrates (12–18%), a terrace amount of fat is also observed (0.3–0.4%), a significant amount of proteins is also present (0.5–0.6%), and a very good amount of water is there about 82%. In addition to these, vitamin A and vitamin C are also present, 0.001–0.0015% and 0.01–0.02%, respectively. A substantial amount of calcium (0.01–0.02%), phosphorus (0.08–0.01%), and potassium (0.1–0.2%) is also observed. Additionally, boron, which is good for bone health, is also present in grapes. Grape seeds are rich in polyphenols especially oligomeric proanthocyanidins and monomeric flavan-3-ols.[63] The dimers and trimers are the grape-seed proanthocyanidins and they are extremely polymerized oligomers of monomeric catechins. Moreover, in grapes, sugar covers a large part of soluble solids, that is, 150–250 g/L in ripe grape juice. Among sugars, mainly the fructose and the glucose are available in the grape juice. Furthermore, glucose is the only main sugar in unripe berries. In ripening berries, both fructose and glucose are available in equivalent proportions (1:1). Afterward, after ripening, fructose becomes more predominant than glucose. This is how fructose and glucose vary in grape varieties.

Along with the sugars, some of the organic acids are also predominant constituents of solids in juice and wine. They give tart taste to the wine and affect its pH, stability, and color. Tartaric acid and malic acid are the main organic acids in the grapes. However, tartaric acid and malic acid constitutes 90% of the total acids present; citric acid and amino acids are also present in grapes. Their concentration increases in initial periods of the growth of berry. When the fruit has ripened, with the increase in sugar content in fruit, acid concentration decreases. Normally, the malic acid decreases in a larger extent than tartaric acid in fruits. Tartaric acid is rarely present in fruits; grape is one of them to have it. Tartaric acid could be present in free form or as a salt; potassium bitartrate is a good example of salt. Potassium bitartrate affects the pH and cold stability of wine.

Additionally, various nitrogenous compounds are also present in grapes. Organic nitrogenous compounds and ammonium cations are the two main nitrogenous compounds and protein, amino acids, and peptides are good examples. The nitrogen content is affected by a couple of factors; these include the soil, variety, cultural practices, fertilization, and climate. During the maturation period, total nitrogen contents of fruit are increased. The minerals are also absorbed from the soil. The weight of the fruit, about 0.2–0.6%, is made up of them. Chloride, sulfate, iron, phosphate, potassium, iron, and sodium are important mineral compounds. Potassium is the most important among all of these. About 50–70% of cations are accounted by potassium in the juice. The concentration of potassium is increased during the ripening season. Formation of potassium bitartrate is also observed. Potassium bitartrate has some important function like reducing the acidity and increasing the pH. One important thing to be renowned is that there is a tartaric acid that is involved in the wine instability problems due to the presence of potassium salt.

6.3 BIOACTIVE COMPOUNDS

6.3.1 *RESVERATROL*

A natural phytoalexin, existing in red wine and grape, which shows a good antioxidant activity, is resveratrol (3,5,40-*trans*-trihydroxystilbene). Under the stress condition (UV radiation), resveratrol is formed by 70 or more than that plant species as defense from the pathogens. It is a good resistant to fungal diseases. It is a fat-soluble compound. It is present in both of the isomeric forms, that is, the *cis* form (structure is given in Fig. 6.1) and the *trans* form (its structure is given in Fig. 6.2). These forms can bind to the glucose; hence, it naturally occurs as glucoside called piceid. It occurs naturally in the grapevine and is produced in the skin of berry; it is also found in the seeds in muscadine grapes.[42–50]

Grape skin is composed of 50–100 μg of resveratrol per gram wet weight.[7] The geographic origin, exposure to fungal disease, some stress condition, and grape cultivar influence the concentrations of resveratrol in grapes. During the time of the winemaking, resveratrol increases its contents that depend on the time of fermentation during which wine spends in contact with the skin of a grape. So, its concentrations in the white wines are significantly lower as these wines are made after the removal of the skin of a grape. The red wines can be fermented with the skin of grape hence allow wines to absorb greater the quantity of the resveratrol. In the duration of winemaking, piceids (resveratrol glucosides) converts itself to the resveratrol; so, a higher concentration of resveratrol in the wine compared to fresh grape juice is explained.[50]

The biological activity of the phytoalexins and their effect on the health in human is a subject of great interest of scientists these days. As per the data collected by WHO in 1980–1990s that concerns incidences of the coronary heart disease, French people have reduced suffering from coronary heart disease though they have a diet that is quite rich in the saturated fats. French paradox is the name given to this fact.[50] This fact has been investigated by a couple of research groups and the French people were compared with the Americans too to answer why there is less possibility of the presence of coronary heart diseases in French people when they consume more saturated fats (15.6%) than American people (11.3%).[28]

French people consumed red wine in their diets. The consumption of wine reduces coronary heart diseases as compared to other alcoholic beverages and this fact is considered to be a factor for the French paradox. This increased the interest of researchers toward resveratrol. Resveratrol is considered to be the strong antiglomerulonephritic food factor that is capable of suppressing hyperlipidemia, proteinuria, and hypoalbuminemia.[64] Across the intestinal epithelium, it diffuses rapidly.[53]

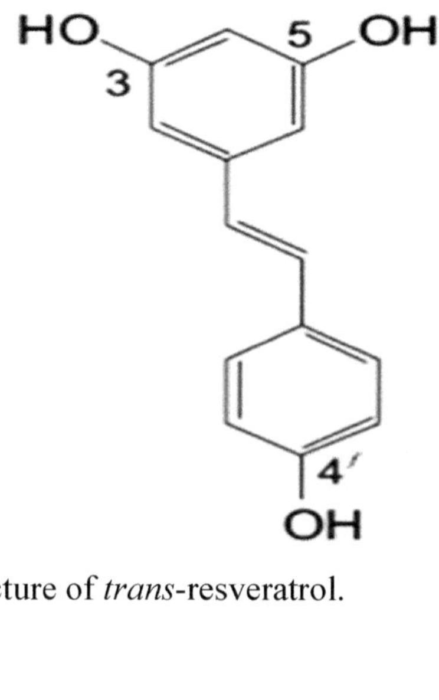

FIGURE 6.1 Structure of *trans*-resveratrol.

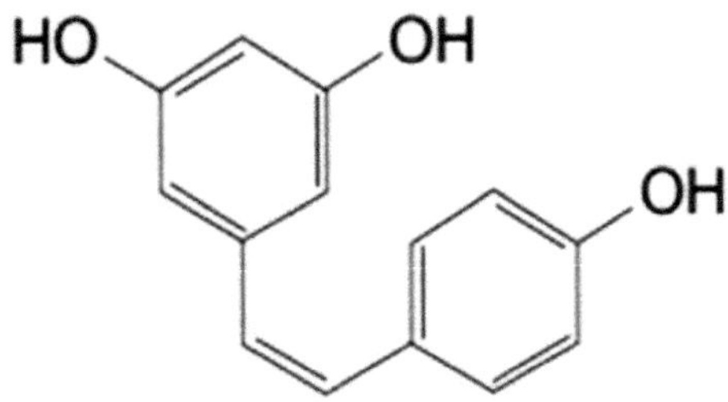

FIGURE 6.2 Structure of *cis*-resveratrol.

6.3.2 *MELATONIN*

Naturally occurring in microbes, plants, and animals, melatonin (*N*-acetyl-5-methoxytryptamine) is an indoleamine as shown in Figure 6.3. For induction of the immune system, regulation of sleep–wake cycles, vascular tone, and sexual development melatonin plays an important role in the human body.[43] In the past, melatonin was considered to be a neurohormone that synthesizes in the vertebrates; present studies have declared that including the invertebrates, it is also found in bacteria, animals, plants, algae, and fungi.[41] Iriti et al.[43] exposed melatonin presence in grape cultivars, this boosted up the researchers to conduct studies that concerned presence and biosynthesis of melatonin in the grapes and in the wines.[35,58,59]

Melatonin is found in grapes (5–96 pg/g), in Spanish wines (50–80 pg/mL) and Italian wines (400–500 pg/mL).[42] As by a study, it was revealed that after intake of 100 mL of red wine, human serum concentration of the melatonin was considerably increased after 1 h,[35] This also resulted in the increased plasma antioxidant capacity. So by consuming this wine, the effect of neurohormone in humans can get modulated. In recent years, studies are being conducted on melatonin for treatment of cancer, cardiovascular diseases (CVDs), circadian, sexual dysfunction, sleep disorders, depression, and immune disorders. So, the hypothesis is that consuming moderate wine is beneficial for health is supported by the presence of melatonin in grapes and wines.

Iriti et al.[43] discoverers one more important substance in the grapes—melatonin, researchers consider melatonin as a very powerful antioxidant that involves the physiological functioning in the human body.[83]

Aside from the regulation of circadian rhythm it possesses a wide range of functions. The presence of secondary metabolites in grapes is the health-promoting effects of them. The antioxidant activity of melatonin is focused on the protection of the nuclear and the mitochondrial DNA by scavenging OH, O_2^-, H_2O_2, and NO and by inhibiting the lipid peroxidation. With the change of the phenological stage, melatonin contents in the grapes also change. The maximum melatonin content is observed from the skin during before veraison, it then decreases by 47% during veraison, and modification from the preveraison to the veraison maximizes its quantity by 63% from the grape seed and by 95% from grape flesh. Furthermore, the concentration of melatonin in the seed is considerably more than the concentration that is measured in grape skin and the grape flesh. Quantity of melatonin is observed to be maximum in whole berries in grapes being harvested in an early stage of veraison at the time of maturation of seed.[61]

In a study by Reiter et al.,[74] it has been stated that the melatonin is a good scavenger of the free radicals in the mitochondria. Furthermore, he also reported that melatonin overpowers apoptotic signals that originate in the mitochondria and, in this way, diminish the disorders related to mitochondrial dysfunction, that is, age-related disease, brain damages. The melatonin is very efficient against the damages of the brain that are mostly caused by the discharge of the free radicals that contained the atom oxygen;[90] it also reduces Parkinson's disease which causes brain damages.

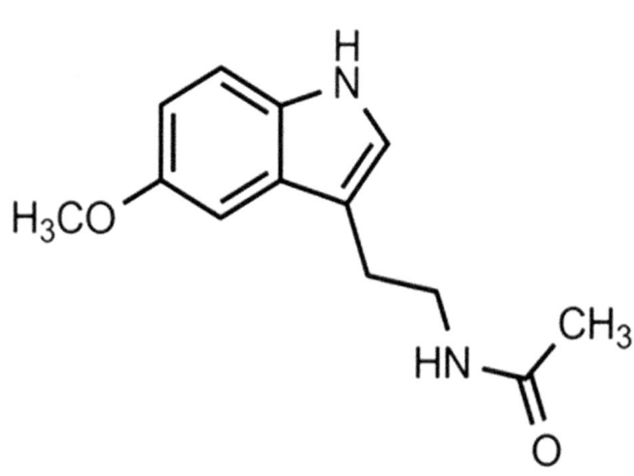

FIGURE 6.3 Structure of melatonin.

6.3.3 POLYPHENOLS

Kushnerova et al.[47] suggested that diprim, a natural polyphenol, found in crest of Amur grape has the ability to scavenge the free radical which might have glutathione-saving effect as well as inhibit lipid peroxidation. The consumption of grapes is a part of therapeutic regimens that could overpower threats that are associated with oxidative stress. Polyphenols along with the bioactive potential resist the humanity from antibacterial, antioxidant, anticarcinogenic and anti-inflammatory diseases.[15,51] The polyphenols are considered as the xenobiotics according to humans. It was investigated that they have the ability to metabolize extensively in living organisms. Polyphenols occur by way of glycosides such as stilbenes and flavonoids and oligomers such as proanthocyanidins (Fig. 6.4). In vitro studies showed that flavonoids, abundantly present in Concord grape, have more radical scavenger properties than that of *α*-tocopherol. Similarly, serum antioxidant ability has been increased due to the consumption of juice of Concord grapes, thus, has more protective properties against LDL (low-density lipoprotein) as compared to *α*-tocopherol in healthy subjects.

Oxidative stress, as well as free radical which otherwise results in the onset of chronic disease, have been reduced by the flavonoids (Fig. 6.5) of Concord grape juice in the hosts.[65] In animals, regeneration of B cells and improvement of chemically induced diabetes has been stimulated by the flavonoids isolated from grapes.[20] After alcohol distillation of wine, color extraction, and even exposure to a high temperature of grape seeds still dried seeds contained substantial flavone (Fig. 6.6) content which showed radical scavenger properties.[34]

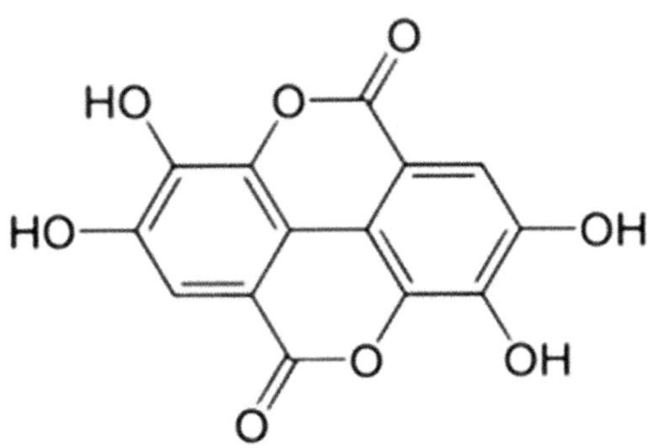

FIGURE 6.4 Structure of polyphenol.

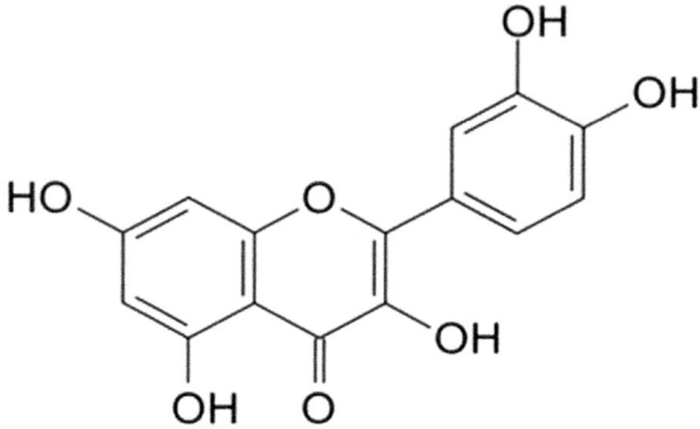

FIGURE 6.5 Structure of flavonoid.

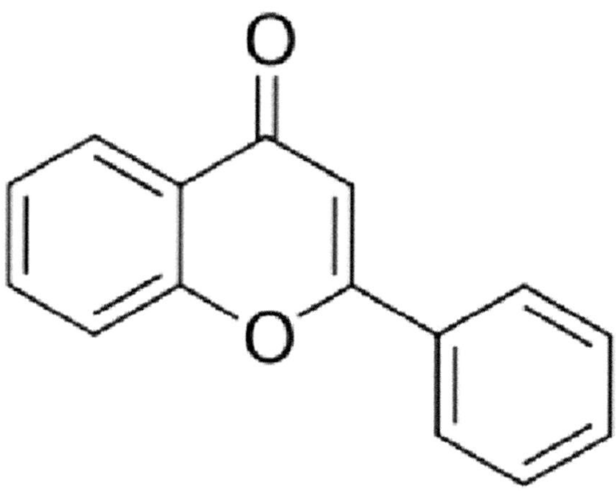

FIGURE 6.6 Structure of flavone.

6.3.4 *ANTHOCYANINS*

Anthocyanins, belong to flavonoid family, are another important moiety present in grapes. Different biological activities like antimicrobial, anti-cancinogenic, anti-inflammatory, antioxidant activity, etc. have been demonstrated by the anthocyanins in several recent research articles (Fig. 6.7).[56] Wang and his fellow[92] suggested about the antioxidant activity of anthocyanins, which is extensively reviewed and studied elsewhere. Anthocyanin exhibits the red color of exocarp derived from grapes. Other phytoalexins are stilbenes and flavonoids. Including stilbenes, resveratrol save tissues of grapes by dangerous emission of UV rays. Phytophagous, proanthocyanidins, or reduced tannins are mostly involve in the conversion of premature fruits into mature ones.[40] Anthocyanins are accrued from cytoplasmic vesicles as well as vacuoles and are known as anthocyanoplasts.[85]

Due to rich secondary metabolites of the grapes like isoprenoids and phenylpropanoids, it made the grapes a popular fruit crop in scientific society. Phenylalanine derivatives are ingenuous phenolic acids including hydroxycinnamates, hydroxybenzoates, and polyphenols, for example, stilbenes, flavonoids, and proanthocyanidins are known as phenylpropanoids. Isoprenoids consist of different constituents that are derived after acetyl CoA, polyterpenes, diterpenes, triterpenes, sesquiterpenes, hemiterpenes, tetraterpenes, and monoterpenes. These compounds are fundamental and exhibit a significant role in chemo-ecology in regard to grapes. Polyphenols like catechins, quercetin, and resveratrol reflect a rich profile of grape and wine during the study of grape metabolomics by different research groups. Resveratrol, a key moiety of grape, has responsible for therapeutic abilities after examination of the composition of grapes and second their effect on the health of human in recent years.

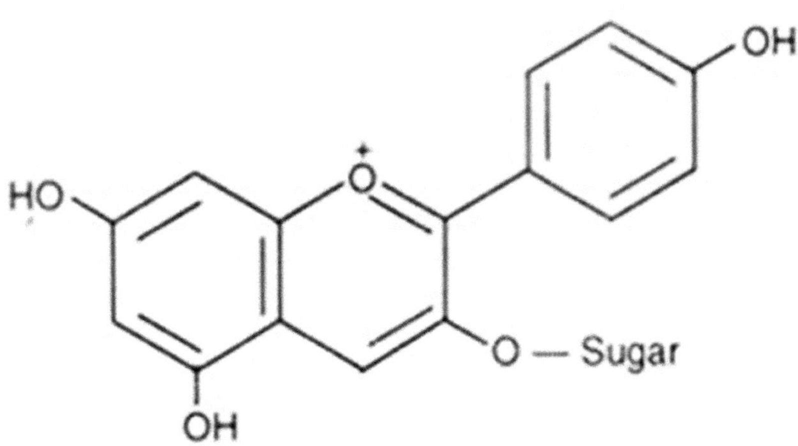

FIGURE 6.7 Structure of anthocyanin.

6.4 HEALTH CLAIMS

Nutritional phytochemicals are beneficial to health and are dependent on the bioavailability. The bioavailability is dependent on the properties and composition of the compounds, the individual biotransformation systems, and the intricacy of the food medium that comprises seeds, skin, tissues, skin as well as the skin of the plant. Different moieties, extracted from the grapes have few medicinal properties which are immunomodulatory, antidiabetes, neuroprotective, antiobesity, antiinfection, antiaging, anticarcinogenic, and antioxidant. The functional properties of grapes are responsible for the beneficial effects on the human bodies by the action on the physiology and catabolism as well as anabolism of the body. The bioactive components extracted from the grapes add value to the food products and also subsidizes in the development and amendment in the diet plan of humans and help in the prevention of Alzheimer's disease, cancer, CVDs, deteriorating diseases without economic as well as environmental forfeitures.[26] The phytochemical of grapes are preclinically studied both in vitro and in vivo methods in modern era that impute the surfeit of biological events to the main components. During the last few decades, considerable improvement has been made about different grape-derived moieties toward biological activates, apart from the grape compound chemistry. Different metabolites like resveratrol, caffeic acid plays an important role in the radical scavenging produce during the oxidative stress. Likewise, resveratrol also showed some therapeutic and preventive abilities. Regarding clot formation as well as bone health, vitamin K imparts a substantial role that is abundantly present in grapes. Different varieties of colors are a rich source of antioxidants as well as polyphenols.[73]

6.4.1 ANTI-INFLAMMATORY EFFECT

Panico and his fellows in 2006 suggested that human and animal models showed that phenolic compound like flavonols, procyanidins, and flavanols mainly present in the seed of grape possess a noteworthy anti-inflammatory effect.[19,81,86] Grape seed and skin extract suppressed ear inflammation as well as polymorphonuclear leukocyte infiltration tempted by 12-*O*-tetradecanoylphorbol 13-acetate for 30 min in rats.[11] Similarly, findings were observed in another study where combined extracts of seed and skins of grapes used against degenerative diseases of joint have a similar effect as that of drug known as indomethacin. Two different pathways followed by the grape phenolics are antioxidative action and immunomodulation. Li and his colleagues[52] studied the inhibition of cytokine gene by a basic pathway which was then studied by Panico and his fellow in 2006 to prove the human chondrocytes assay. Likewise, another study related to inhibition of *N*-acetyl-β-D-glucosaminidase and nitric oxide synthase activity were also successfully done by the proanthocyanidins which were introduced in the rat by injection of 10 mg/kg.[52]

Human adipocytes and macrophage-like cell lines formed less MCP-1 and IL-6 which results in the enhancement of anti-inflammatory adiponectin and adipokine after the induction of inflammatory stimulus. Before the induction of inflammatory stimuli, they are pretreated with extracts of grape seed procyanidin which results in the modulation of cytokine and adipokine gene expression related to anti-inflammation.[19] Additionally, the same trend was seen in case if C-reaction protein (CRP), TNF-α, and IL-6 in the mesenteric WAT (white adipose tissues), when the rats were fed with high-fat feed.[86] Adiponectin mRNA expression increases in the mesenteric WAT, whereas opposite results were observed in the liver regarding CRP mRNA. These studies proved that procyanidins extract by grapes suppressed inflammation at mRNA level and similar results were also seen to decrease the risk related to high-fat diet and obesity which leads toward metabolic and cardiovascular disorders.

6.4.2 ANTIMICROBIAL EFFECTS

Majority of the plant polyphenols have antibacterial, antifungal as well as antiviral activities.[8,9,13,44] Distinct part of grape showed distinct antimicrobial activities due to the presence of phenolic compounds. Flesh region of grapes showed maximum followed by an extract of whole fruit, skin, fermented pomace seed, and leaves. But Thtmothe et al.[87] pointed out that fermented pomace is effective as compared to the whole fruit extract against the microbial activity. Likewise, in 2009, Brown and his fellows suggested that grape skin is more effective against *Helicobacter pylori* in comparison with the synergistic effect of both between the skin and seed.[12]

Resveratrol, a phenolic compound in grape, at 10–20 μL found to be effective against *Candida albicans*, human pathogenic fungi, whereas nonpigmented polymeric phenolic obtained from the seed, independent of pH, showed antilisterial activity as compared to pigmented polymeric phenolic, which were pH dependent, obtained from skin and juice. The result of the investigation between antimicrobial activity and compound reveals that castalagin, prodelphinidin, and epigallocatechin with 3,4,5-trihydroxyphenyl groups might be effective against antibacterial activities. Both degrees of polymerization and the number of hydroxyls may be crucial for antimicrobial activity of polyphenolic compounds of grapes.[84]

6.4.3 ANTIOXIDANT EFFECT

A major hallmark of the number of health-related problems is oxidative stress. Antioxidants which scavenge free radicals and help to avoid the damage of cell are the natural compounds present in grapes (Table 6.1). Free radicals cause a condition which is termed as oxidative stress. Reiter and his fellows in 2003 reported about the scavenging property of melatonin which scavenges different free radicals like H_2O_2, NO, O_2^-, and OH, thus protect mitochondrial and nuclear DNA and also inhibit peroxidation of lipid.[74] Likewise, another study showed that melatonin may be effective against the signals which were generated during apoptosis to support mitochondrial dysfunction—age-related diseases, brain damage, etc. Thus, melatonin is quite good against brain damage by scavenge-free radicals, which contained the atom oxygen.[90] Another study showed the contribution of melatonin against free radicals, thus, protect from mutagenic as well as carcinogenic actions by improving longevity.[90]

Reiter and his fellows in 2003 also discussed the positive effects of melatonin for the effective action on the septic shock that occurs in the newborns. Clinical trials as well as in vitro studies showed the mode of action, both direct and indirect, which has promising effects in the treatment and prevention of diseases that are caused by the free radical generations. Direct scavenging of free radicals, stimulation of mitochondrial function, stimulation of antioxidant enzymes, and synergy with classic antioxidant are the multiple pathways of melatonin to reduce the oxidative stress, which were obtained from plenty of studies some of which are illustrated above. Another natural phytoalexin resveratrol (3,5,40-*trans*-trihydroxystilbene), abundantly found in the red wine as well as in grapes, have potent antioxidant property.

Cai and his colleagues[14] stated that different synthesized form of resveratrol, that is,. 3,5-DHS, 3,4-dihydroxy-*trans*-stilbene (3,4-DHS), and 4,40-DHS, 4-hydroxy-*trans*-stilbene have reduced the free radicals during the induction of peroxidation of rat liver microsomes. Iron-induced peroxidation and 2,20-azobis (2-amidinopropane hydrochloride) were reduced by the *trans*-stilbene derivatives which are potent antioxidants. Estrogen and catechol are synthesized by 17b-estradiol which is oxidized by the expression of cytochrome P450 1A1 and 1B1 as well as 2,3,7,8-tetrachlorodibenzo-*p*-dioxin-induced aryl hydrocarbon receptor DNA binding activity which were inhibited by the potent antioxidant activity of resveratrol.[21] Hence, both catechol estrogen, as well as oxidative stress induced damage in tissues, can be retarded by resveratrol.

Both in vivo and in vitro studies showed that proanthocyanidin which can be isolated from grape seed demonstrated antioxidant protection due to the alternation in p53 and Bcl-2 expression induced in smokeless tobacco-induced cellular injury.[6] Furthermore, antioxidant dietary supplements can be produced from the phytochemicals like epicatechin, catechin, and gallic acid which are present in the seed and skin of the grapes.[94] Likewise, thioconjugates are present in the white grape pomace.[78] Another important group component of polyphenols, that is, anthocyanins are also present in the grape which along others contributes to suppressing the threats of oxidative stress.

TABLE 6.1 Antioxidant Activities of the Extracts from Grapes and Its Products.

Resource	Antioxidant activity	References
Grape seed	Decreasing the oxidated LDL in plasma	[78]
Juice	Reducing oxidative stress in serum	[32]
Red wine	Shield of *Saccharomyces cerevisiae* induced by H_2O_2 against membrane oxidation	[26]
Fruit beverage (grapes)	Protection against antioxidant and microbial systems against oxidative stress induced by H_2O_2	[20]
Grape wine	Protection from the accumulation of hyper-cholesterolemic hamsters against the aortic fatty streak	[5]

TABLE 6.1 *(Continued)*

Resource	Antioxidant activity	References
Defatted milled grape seed	It elevates the levels of GSH and ATP by oxidative stress induced by chemical anti-cancer adriamycin	[91]
Grape seed extract	An excellent preservative for meat flesh and oils	[55]
White grape dietary fiber concentrate	It is an antioxidant for fatty acids especially polyunsaturated fatty acid	[79]

ATP, adenosine triphosphate; GSH, glutathione.

6.4.4 ANTIDIABETIC EFFECT

At present, the prevalence of diabetes increased day by day as a result of which a large population of adult suffered from this disease. As a result of its increasing prevalence, various therapeutic approaches have been applied to treat this disease, unless no particular treatment yet discovered. Among different approaches, the use of natural component especially the grape is very useful. Bolton and his fellows in 1998 stated about the response of grape against insulin response is significantly high rather than juice alone which led to the speculation that oranges and apples are less insulinogenic as compared to that of grape. Glucose and fiber content of the fruit ascribed the plasma insulin and glucose responses.[10] Zhang and his colleagues in 2004 demonstrated that oleanolic aldehyde and oleanolic acid pound in grape skin have insulin-secreting activities.[95]

Glycosylated protein level can be lower and in the same time insulin sensitivity can be enhanced by the combined effects of zinc monothionine and chromium polynicotinate which were obtained by the grape seed extract.[72] Similarly, B-cell regenerations and ameliorate that induces, chemically, diabetes in the animals can be stimulated by the epicatechin and flavonoids isolated from the grapes.[20] Park[70] reported that approval of glucose by the transforming myelocytic cell of human can be retarded by the resveratrol. Similarly, protection from fructose-induced diabetes can be found as a result of activation of SIRT-1 by resveratrol.[71] Insulin and homeostasis sensitivity can be improved by the resveratrol.[60]

6.4.5 CARDIOPROTECTIVE EFFECT

Grape seed extracts or polyphenols present in the grape juice could be helpful against oxidative stress as well as postprandial hyperlipidemia. Malondialdehyde-modified-LDL, plasma lipid hydroperoxides, and serum lipid peroxidation products are the risk factors of oxidative stress. In vivo activities of catalase and superoxide dismutase remained at a normal level as the rat liver is being protected by the extracts of grape seed against oxidative damage induced by irradiation.[18] Despite pure resveratrol, polyphenols in 5–50 μg/mL grape seed extracts were effective against the generation of superoxide anion as well as reduction of platelet adhesion and aggregation.[66] CVDs can be minimized by vasorelaxation through the inhibition of phosphodiesterase-5 activity with the intake of anthocyanins present in the grape skin as well as in wine.[25] In 2006, Falchi and his colleagues found that consumption of grape's flesh and skin extracts demonstrated equivalent conclusion of cardioprotection after 30 days of intake in rats.[30]

Another study was conducted by the Castilla and his colleagues,[16] where an amount of 100 mL of the juice of grapes was given for 2 weeks on daily basis, after which antioxidant capacity and absorption of cholesterol standardized tocopherol plasma level significantly increased and decreased was observed in the LDL and oxidized LDL level, respectively. Likewise, in another study plasma cholesterol concentration was minimized to 11% after feeding the moderate dose of grape extracts to hamsters.[4] Despite these parameters, apolipoprotein A-I and plasma level of HDL was also evaluated which significantly increased by the consumption of red wine,[89] thus coronary heart disease can be controlled. Furthermore, catechin, quercetin, and resveratrol present in grapes enhanced the quantity of plasma apolipoprotein-I to 26%, 22%, and 19%, respectively.

Another study reported that CVDs risk can be lowered by the reduction of plasma monocyte chemoattractant protein 1 through the consumption of red grape juice.[16] Daily 375 mL of red wine for consecutive two weeks showed that maximum quantity of three-phase brushless asynchronous excitation system (TBAES) and conjugated dienes in Cu-oxidized LDL were reduced.[89] Similar study related to copper-induced oxidation of LDL in human can be completely inhibited at the level of 0.01% by the consumption of grape juice.[47]

6.4.6 ANTICANCER EFFECT

The biggest challenge to medical professionals and researches regarding rapid growing health problem is cancer. In 1997, Craig proposed that despite various fruits and vegetables, grapes can also be very useful in the reduction of risk of cancer. Antineoplastic as well as anticarcinogenic are the well-known properties of resveratrol which is a chief compound of grapes (Table 6.2). Cytochrome *c* release by mitochondria which result in formation of the apoptosome complex which results in activating the caspase is one of the proposed molecular mechanism of resveratrol, despite many others, against cancer.[68] Kubota and his fellows in 2003 reported that by interfering the countenance of the heat shock protein (70.9), resveratrol is reported to be useful against the prostate cancer cell.[46]

Resveratrol controls the proliferation and apoptosis into the cancer cells of the breast by accumulating ceramide and the phenolic moiety of stilbenoids that is very important for inducing ceramide-associated growth inhibition.[76] Hakimuddin et al.[36] observed that red wine showed anticancer properties in human cancer cells of breast owing to the presence of polyphenolics. This involved inhibiting the cell proliferation by flavonoids and the activity associated with calcium calmodulin and phosphodiesterase. It is being indicated, flavonoids do interfere with the functioning of second messenger "calcium." Therefore, the certain types of the ingredients of grape wine had anticancer functioning and this is useful in the development of the functional neutraceuticals by the anticancer properties.

In the recent era, Roy et al.[75] in his study report the activity of resveratrol; he elaborates that it inhibits the tumors of mouse skin as it regulates the metabolism in mitochondria and the phosphoinositide 3-kinase Akt pathway. The tissues of breast cancer possess a considerable expression of aromatase compared to the normal tissue in the body. Moreover, in the study of Eng,[29] the inhibitory action of red wine extract fraction on aromatase activity owing to the presence of procyanidin B dimers was observed and it is concluded that the dimmers play a vital role in the treatment of the breast cancers tissues. There is another component from the grape known as epigallocatechin 3-gallate, it has good antineoplastic potential because it inducts the programmed death of cells. Anthocyanin is another bioactive component derived from the grapes; it has anticarcinogenic properties as reported by studies.[49,80] In conclusion, it is the grapes that possess alternative complementary therapeutic regimen which saves from the progression of the tumor in certain types of cancer.

TABLE 6.2 Anticancer Activities of Phenolic Compounds from Grapes.

Phenols	Effects	References
Proanthocyanidins	Inhibited breast cancer metastasis	[68]
Anthocyanin	Activated antioxidant response element upstream of genes, induced two to four times increase in DNA fragmentation, repaired and protected genomic DNA integrity, and retard blood vessel growth in some tumors	[46,69]
Procyanidins	Decreased cell viability and proliferation in the age of 30	[76]
Flavones	Reduced cell proliferation, induced differentiation, and apoptosis	[36]
Flavonoid	More effectively induced apoptosis than antitumor agent	[80]
Resveratrol	Induced apoptotic and antiproliferative effects, inhibited cyclooxygenase-2 transcription	[36]

6.4.7 ANTIAGING EFFECTS

It is revealed that polyphenolics present in the foods have a good benefit in drawing back neuronal and behavioral aging. As they have very good antioxidant properties, like hunting the free radicals, these prevent organ and tissue from any kind of damage due to oxidation; they also help the body against the negative mechanism of the redox status. Rats between the age of 19–21 months were tested and these evidence were drawn. Improvements were observed like the release of dopamine from striatal slices when they were given grape juice (10%), while the 50% grape juice improved action capacity.[79–81] Advance researches discovered the supplement with the extracts of grape seed (100 mg/kg body weight) for 30 days and inhibited the accretion of age-related oxidative DNA damages in neural tissue due to phenolic compounds from the extracts.[8,93]

6.5 SUMMARY

The reliability of humanity is on plants because they are capable to fulfill the nutrient need of a living organism. Modern research has invigorated fruits to repossess humans from different disorders by the use of diet-based physiotherapy. Grapes, as a fruit, are one of the potential fruit having the ability to treat different disorders. The narration of grapes according to researchers is unmatchable. They are notorious in accordance with their medical and counteractive potential. Recent investigation has been made to discover the bioactive components as well as their significance in accordance with CVDs, antioxidant activity, lowering of blood pressure, etc. Grapes are recognized by their health-promoting activities and the presence of an important bioactive component, that is, resveratrol having antiproliferative compounds and exhibiting antioxidant activity. Polyphenols are other components that exhibit health benefits by reducing the chances of CVDs. The function of melatonin in the progression and development of grapes' growth and development is investigated that is extracted from the skin of grapes for the defense in contrast to oxidative stress and in the generation of DNA oxidation catabolites. In short, modern investigation about grapes reveals miracle properties against the different diseases.

KEYWORDS

- **grapes**
- **phytonutrients**
- **polyphenols**
- **procyanidins**
- **proliferation**
- **resveratrol**
- **therapeutic foods**

REFERENCES

1. Albers, W. G.; Amarenco, P.; Easton, J. D.; Sacco, R. L.; Teal, P. P. Antithrombotic and Thrombolytic Therapy for Ischemic Stroke: The Seventh ACCP Conference on Antithrombotic and Thrombolytic Therapy. *Chest* **2004,** *126*, 483S–512S.

2. Ali, W.; Mahmood, N. N.; Hasan, A. M.; Abbood, R. S. Effect of Some Natural Preservatives on Some Homemade Fresh Fruits Juices. *Adv. Environ. Biol.* **2016,** *10* (3), 23–27.
3. Ardevol, A.; Blade, C.; Salvado, M. J.; Arola, L. Changes in Lipolysis and Hormone-Sensitive Lipase Expression Caused by Procyanidins in 3T3-L1 Adipocytes. *Int. J. Obes.* **2000,** *24*, 319–324.
4. Auger, C.; Gerain, P.; Laurent-Bichon, F.; Portet, K. et al. *Phenolics from Commercialized Grape Extracts Prevent Early Atherosclerotic Lesions in Hamsters by Mechanisms Other than Antioxidant Effect*; Food and Agriculture Organization of the United Nations, 2004.
5. Auger, C.; Teissedre, P. L.; Gerain, P.; Lequeux, N.; Bornet, A.; Serisier, S.; Besançon, P.; Caporiccio, B.; Cristol, J. P.; Rouanet, J. M. Dietary Wine Phenolics Catechin, Quercetin, and Resveratrol Efficiently Protect Hypercholesterolemic Hamsters Against Aortic Fatty Streak Accumulation. *J. Agric. Food Chem.* **2005,** *53* (6), 2015–2021.
6. Bagchi, D.; Das, D. K.; Tosaki, A.; Bagchi, M.; Kothari, S. C. Benefits of Resveratrol in Women's Health. *Drugs Exp. Clin. Res.* **2001,** *27* (5–6), 233–248.
7. Baliga, M. S.; Meleth, S.; Katiyar, S. K. Growth Inhibitory and Antimetastatic Effect of Green Tea Polyphenols on Metastasis Specific Mouse Mammary Carcinoma 4T1 Cells In Vitro and In Vivo Systems. *Clin. Cancer Res.* **2005,** *11*, 1918–1927.
8. Balu, M.; Sangeetha, P.; Haripriya, D.; Panneerselvam, C. Rejuvenation of Antioxidant System in Central Nervous System of Aged Rats by Grape Seed Extract. *Neurosci. Lett.* **2005,** *383*, 295–300.
9. Baydar, N. G.; Sagdic, O.; Ozkan, G.; Cetin, S. Determination of Antibacterial Effects and Total Phenolic Contents of Grape (*Vitis vinifera*) Seed Extracts. *Int. J. Food Sci.* **2006,** *41*, 799–804.
10. Bolton, P. F.; Pickles, A.; Murphy, M.; Rutter, M. Autism, Affective and Other Psychiatric Disorders: Patterns of Familial Aggregation. *Psychol. Med.* **1998,** *28* (2), 385–395.
11. Bralley, E. E.; Hargrove, J. L.; Greenspan, P.; Hartle, D. K. Topical Anti-Inflammatory Activities of *Vitis rotundifolia* (Muscadine Grape) Extracts in the Tetradecanoylphorbol Acetate Model of Ear Inflammation. *J. Med. Food* **2008,** *10*, 636–642.
12. Brown, J. C.; Huang, G.; Haley-Zitlin, V.; Jiang, X. P. Antibacterial Effects of Grape Extracts on *Helicobacter pylori*. *Appl. Environ. Microbiol.* **2009,** *75*, 848–852.
13. Bruno, G.; Sparapano, L. Effects of Three ESCA-Associated Fungi on *Vitis vinifera* L.: V. Changes in the Chemical and Biological Profile of Xylem Sap from Diseased cv. Sangiovese Vines. *Physiol. Mol. Plant Pathol.* **2007,** *71*, 210–229.
14. Cai, Y. J.; Fang, J. G.; Yang, L. *et al.* Inhibition of Free Radical-Induced Peroxidation of Rat Liver Microsomes by Resveratrol and Its Analogues. *Biochim. Biophys. Acta* **2003,** *1637*, 31–38.
15. Casazza, A. A.; Aliakbarian, B.; Lagazzo, A.; Garbarino, G.; Carnasciali, M. M.; Perego, P.; Busca, G. Pyrolysis of Grape Marc Before and After the Recovery of Polyphenol Fraction. *Fuel Process. Technol.* **2016,** *153*, 121–128.
16. Castilla, P.; Echarri, R.; Davalos, A.; Cerrato, F.; Ortega, H.; Teruel, J. L.; Lucas, M. F.; Gomez-Coronado, D.; Ortuno, J.; Lasuncion, M. A. Concentrated Red Grape

Juice Exerts Antioxidant, Hypolipidemic, and Anti-Inflammatory Effects in Both Hemodialysis Patients and Healthy Subjects. *Am. J. Clin. Nutr.* **2006,** *84*, 252–262.

17. Cerqueira, J. C.; Machado, B. A. S. Obtaining and Characterization of Grape-Peel Extract Rich in Bioactive Compounds. *Int. J. Agric. Res., Innov. Technol.* **2016,** *1*, 34–44.
18. Cetin, A.; Kaynar, L.; Kocyigit, I.; Hacioglu, S. K.; Saraymen, R.; Ozturk, A.; Orhan, O.; Sagdic, O. The Effect of Grape Seed Extract on Radiation-Induced Oxidative Stress in the Rat Liver. *Turk. J. Gastroenterol.* **2008,** *19* (2), 92–98.
19. Chacona, M. R.; Ceperuelo-Mallafrea, V.; Maymo-Masipa, E.; Mateo-Sanzb, J. M.; Arolac, L.; Guitierreza, C.; Fernandez-Reald, J. M.; Ardevolc, A.; Simona, I. and Vendrella. Grape-Seed Proanthocyanidins Modulate Inflammation on Human Differentiated Adipocytes In Vitro. *Cytokine* **2009,** *47*, 137–142.
20. Chakravarthy, B. K.; Gupta, S.; Gambhir, S. S.; Gode, K. D. Pancreatic Beta-Cell Regeneration in Rats by (–)-Epicatechin. *Lancet* **1981,** *318*, 759–760.
21. Chen, M. L.; Yi, L.; Jin, X., Liang, X. Y., Zhou, Y., Zhang, T., et al. Resveratrol Attenuates Vascular Endothelial Inflammation by Inducing Autophagy Through the cAMP Signaling Pathway. *Autophagy* **2013,** *9*, 2033–2045.
22. Cirqueira, M. G.; Costa, S. S.; Viana, J. D.; Silva, C. A. B. C.; Guez, M. A. U.; Machado, B. A. S. Phytochemical Importance and Utilization Potential of Grape Residue from Wine Production. *Afr. J. Biotechnol.* **2017,** *16*, 179–192.
23. Craig, M. *The Last Freedom*; Hodder and Stoughton: London, 1997.
24. de Bruyne, T.; Pieters, L.; Witvrouw, M.; de Clercq, E.; Berghe, D. V.; Vlietinck, A. J. Biological Evaluation of Proanthocyanidin Dimers and Related Polyphenols. *J. Nat. Prod.* **1999,** *62*, 954–958.
25. Dell'Agli, M.; Galli, G. V.; Vrhovsek, U.; Mattivi, F.; Bosisio, E. In Vitro Inhibition of Human cGMP-Specific Phosphodiesterase-5 by Polyphenols from Red Grapes. *J. Agric. Food Chem.* **2005,** *53* (6), 1960–1965.
26. Deng, Q.; Penner, M. H.; Zhao, Y. Chemical Composition of Dietary Fiber and Polyphenols of Five Different Varieties of Wine Grape Pomace Skins. *Food Res. Int.* **2011,** *44*, 2712–2720.
27. Derradji, Z.; Djeghri-Hocine, B.; Amrane, A.; Melouka, H.; Djabbar, R. Effect of Protein Hydrolysates Added to Date Juice on the Growth of Lactic Acid Bacteria. *Adv. Environ. Biol.* **2016,** *10* (1), 214–220.
28. Elmadfa, I.; Kornsteiner, M. Fats and Fatty Acid Requirements for Adults. *Ann. Nutr. Metab.* **2009,** *55* (1–3), 56–75.
29. Eng C. *PTEN: One Gene, Many Syndromes. Human Mutat.* **2003,** *22*, 183–198.
30. Falchi, M.; Bertelli, A.; Scalzo, R. L.; Morassut, M.; Morelli, R.; Das, S.; Cui, J. H.; Das, D. K. Comparison of Cardioprotective Abilities Between the Flesh and Skin of Grapes. *J. Agric. Food Chem.* **2006,** *54*, 6613–6622.
31. Food and Agriculture Organization (FAO). *Production of Grapes*; 2011. http://faostat.fao.org/site/339/default.aspx (accessed Mar 12, 2014).
32. Garcia-Alonso, J.; Ros, G.; Vidal-Guevara, M. L.; Periago, M. J. Acute Intake of Phenolic-Rich Juice Improves Antioxidant Status in Healthy Subjects. *Nutr. Res.* **2006,** *26*, 330–339.

33. Genova, G.; Tosetti, R.; Tonutti, P. Berry Ripening, Pre-Processing and Thermal Treatments Affect the Phenolic Composition and Antioxidant Capacity of Grape (*Vitis vinifera* L.) Juice. *J. Sci. Food Agric.* **2016,** *96* (2), 664–671.
34. Gonzalez-Paramas, A. M.; Esteban-Ruano, S.; Santos-Buelga, C.; Pascual-Teresa, S.; Rivas-Gonzalo, J. Flavanol Content and Antioxidant Activity in Winery Byproducts. *J. Agric. Food Chem.* **2004,** *52*, 234–238.
35. Guerrero, J. M.; Martiez-Cruz, F.; Elorza, F. L. Significant Amounts of Melatonin in Red Wine: Its Consumption Increases Blood Melatonin Levels in Humans. *Food Chem.* **2008,** *8*, 162–167.
36. Hakinuddin, F.; Paliyath, G.; Meckling, K. Selective Cytotoxicity of a Red Grape Wine Flavonoid Fraction Against MCF-7 Cells. *Breast Cancer Res. Treat.* **2004,** *85*, 65–79.
37. Hopfer, H.; Nelson, J.; Collins, T. S.; Heymann, H.; Ebeler, S. E. The Combined Impact of Vineyard Origin and Processing Winery on the Elemental Profile of Red Wines. *Food Chem.* **2015,** *172*, 486–496.
38. Instituto Brasileiro de Geografia e Estatística—IBGE. *Produção Agrícola Municipal*; IBGE: Brasília, 2013. www.sidra.ibge.gov.br/bda/tabela.
39. INV (Instituto Nacional de Vitivinicultura). *Las pasas de uva en el mundo y en Argentina. Subgerencia de Estadística y Asuntos Técnicos Internacionales*; Dpto. de Estadística y Estudios de Mercado, 2011. www.inv.gov.ar (accessed June 3, 2015).
40. Iriti, M. 2011. Editorial: Introduction to Polyphenols, Plant Chemicals for Human Health. *Mini Rev. Med. Chem.* **2011,** *11*, 1183–1185.
41. Iriti, M.; Faoro, F. Grape Phytochemicals: A Bouquet of Old and New Nutraceuticals for Human Health. *Med. Hypoth.* **2006,** *67*, 833–838.
42. Iriti, M.; Giulia, C.; Sara, V.; Ilaria, M.; Soave, C.; Fico, G., et al. Chitosan-Induced Ethylene-Independent Resistance Does Not Reduce Crop Yield in Bean. *Biol. Cntrl.* **2010,** *54*, 241–247.
43. Iriti, M.; Rossoni, M.; Faoro, F. Melatonin Content in Grape: Myth or Panacea? *J. Sci. Food Agric.* **2006,** *86*, 1432–1438.
44. Jung, H. J.; Hwang, I. A.; Sung, W. S.; Kang, H.; Kang, B. S.; Seu, Y. B.; Lee, D. G. Fungicidal Effect of Resveratrol on Human Infectious Fungi. *Arch. Pharm. Res.* **2005,** *2*, 557–560.
45. Krasnow, M. N.; Murphy, T. M. Polyphenol Glycosylating Activity in Cell Suspensions of Grape (*Vitis vinifera*). *J. Agric. Food Chem.* **2004,** *52* (11), 3467–3472.
46. Kubota, T.; Uemura, Y.; Kobayashi, M.; Taquchi, H. Combined Effects of Resveratrol and Paclitaxel on Lung Cancer Cells. *Anticancer Res.* **2003,** *23* (5A), 4039–4046.
47. Kushnerova, N. F.; Sprygin, V. G.; Fomenko, S. E.; Gordeichuk, T. N.; Rakhmanin, I. A. Efficiency of Diprim for the Restoration of the Functional Status of the Liver After Its Damage by Ethanol Poisoning. *Gigienal Sanitariia* **2002,** *1*, 56–60.
48. Lagouge, M.; Argmann, C.; Gerhart-Hines, Z.; Meziane, H.; Lerin, C.; Daussin, F.; Messadeq, N.; Milne, J.; Lambert, P.; Elliott, P.; Geny, B.; Laakso, M.; Puiqserver, P.; Auwerx, J. Resveratrol Improves Mitochondrial Function and Protects Against Metabolic Disease by Activating SIRT1 and PGC-1 alpha. *Cell* **2006,** *127* (6), 1109–1122.
49. Lala, G.; Malik, M.; Zhao, C.; He, J.; Kwon, Y.; Giusti, M. M.; Magnuson, B. A. Anthocyanin-Rich Extracts Inhibit Multiple Biomarkers of Colon Cancer in Rats. *Nutr. Cancer* **2006,** *54*, 84–93.

50. LeBlance, M. R. Cultivar, Juice Extraction, Ultra Violet Irradiation and Storage Influence the Stilbene Content of Muscadine Grapes (*Vitis rotundifolia* Michx.). Dissertation, Louisiana State University, 2006.
51. Li, S. H.; Zhao, P.; Tian, H. B.; Chen, L. H.; Cui, L. Q. Effect of Grape Polyphenols on Blood Pressure: A Meta-Analysis of Randomized Controlled Trials. *PLoS One* **2015,** *10* (9), e0137665.
52. Li, W. G.; Zhang, X. Y.; Wu, Y. J.; Tian, X. Anti-Inflammatory Effect and Mechanism of Proanthocyanidins from Grape Seeds. *Acta Pharmacol. Sin.* **2001,** *22* (12), 1117–1120.
53. Li, Y.; Shin, Y. G.; Yu, C.; Kosmeder, J. W.; Hirschelman, W. H.; Pezzuto, J. M.; Breemen, R. B. V. Increasing the Throughput and Productivity of Caco-2 Cell Permeability Assays Using Liquid Chromatography–Mass Spectrometry: Application to Resveratrol Absorption and Metabolism. *Combin. Chem. High Throughput Screen* **2003,** *6* (8), 757–767.
54. Lima, M. D. S.; Silani, I. D. S. V.; Toaldo, I. M.; Correa, L. C.; Biasoto, A. C. T.; Pereira, G. E.; Bordignon-Luiz, M. T.; Ninow, L. L. Phenolic Compounds, Organic Acids and Antioxidant Activity of Grape Juices Produced from new Brazilian Varieties Planted in the Northeast Region of Brazil. *Food Chem.* **2014,** *161*, 94–103.
55. Lu, L.; Li, B.; Cheng, M.; Chang, W.; Pan, J.; Zhang, C.; Gao, H. Oral Administration of Grape Seed Proanthocyanidin Extracts Downregulate RAGE Dependant Nuclear Factor-Kappa BP65 Expression in the Hippocampus of Streptozotocin Induced Diabetic Rats. *Exp. Clin. Endocrinol. Diab.* **2008,** *116*, 215–224.
56. Mazza, G. J. Anthocyanins and Heart Health. *Ann. dell'Isit. Super. Sanita* **2007,** *43*, 369–374.
57. McGovern, P. F. *Ancient Wine: The Search for the Origins of Viniculture*; Princeton University Press: Princeton, NJ, 2013.
58. Meng, J. F.; Shi, T. C.; Song, S.; Zhang, Z. W.; Fang, Y. L. Melatonin in Grapes and Grape-Related Foodstuffs: A Review. *Food Chem.* **2017,** *231*, 185–191.
59. Mercolino, L.; Saracino, M. A.; Bugamelli, F.; Ferranti, A.; Malaguti, M.; Hrelia, S.; Raggi, M. A. HPLC-F Analysis of Melatonin and Resveratrol Isomers in Wine Using an SPE Procedure. *J. Separ. Sci.* **2008,** *31*, 1007–1114.
60. Milne, J. C.; Lambert, P. D.; Schenk, S.; Carney, D. P.; Smith, J. J.; Gagne, D. J.; Jin, L.; Boss, O.; Perni, R. B.; Vu, C. B.; Bemis, J. E.; Xie, R.; Disch, J. S.; Ng, P. Y.; Nunes, J. J.; Lynch, A. V.; Yang, H.; Galonek, H.; Israelian, K.; Choy, W.; Iffland, A.; Lavu, S.; Medvedik, O.; Sinclair, D. A.; Olefsky, J. M.; Jirousek, M. R.; Elliott, P. J.; Westphal, C. H. *Nature* **2007,** *450* (7170), 712–716.
61. Murch, S. J.; Hall, B. A.; Le, C. H.; Saxena, P. K. Changes in the Levels of Indoleamine Phytochemicals During Veraison and Ripening of Wine Grapes. *J. Pineal Res.* **2010,** *49* (1), 95–100.
62. Natividade, M. M. P.; Correa, L. C.; Souza, S. V. C.; Pereira, G. E.; Lima, L. C. O. Simultaneous Analysis of 25 Phenolic Compounds in Grape Juice for HPLC: Method Validation and Characterization of Sao Francisco Valley Samples. *Microchem. J.* **2013,** *110*, 665–674.
63. Naziri, E.; Nenadis, N.; Mantzouridou, F. T.; Tsimidou, M. Z. Valorization of the Major Agrifood Industrial By-Products and Waste from Central Macedonia (Greece)

for the Recovery of Compounds for Food Applications. *Food Res. Int.* **2014,** *65* (Part C), 350–358.

64. Nihei, T.; Miura, Y.; Yagasaki, K. Inhibitory Effect of Resveratrol on Proteinuria, Hypoalbuminemia and Hyperlipidemia in Nephritic Rats. *Life Sci.* **2001,** *68*, 2845–2852.
65. O'Byrne, D. J.; Devaraj, S.; Grundy, S. M.; Jialal, I. Comparison of the Antioxidant Effects of Concord Grape Juice Flavonoids Alpha-Tocopherol on Markers of Oxidative Stress in Healthy Adults. *Am. J. Clin. Nutr.* **2002,** *76*, 1367–1374.
66. Olas, B.; Wachowicz, B.; Tomczak, A.; Erler, J.; Stochmal, A.; Oleszek, W. Comparative Anti-Platelet and Antioxidant Properties of Polyphenol-Rich Extracts from Berries of *Aronia melanocarpa*, Seeds of Grape and Bark of *Yucca schidigera* In Vitro. *Platelets* **2008,** *19* (1), 70–77.
67. Omer, H. A. A.; Ahmad, S. M.; Bassuony, N. I.; El-Allawy, H. M. H. Response of Growing Rabbits to Diets Containing Bioactive Mixture Composed of Lemon, Onion and Garlic Juice. *Adv. Environ. Biol.* **2015,** *9* (24), 312–319.
68. Opipari, A. W.; Tan, L.; Boitano, A. E.; et al. Resveratrol-Induced Autophagocytosis in Ovarian Cancer Cells. *J. Cancer Res.* **2004,** *64*, 696–703.
69. Panico, A.; Cardile, V.; Avondo, S.; Garufi, F.; Gentile, B.; Puglia, C.; Ronsisvalle, G. The *In Vitro* Effect of a Lyophilized Extract of Wine Obtained from Jacquez Grapes on Human Chondrocytes. *Phytomedicine* **2006,** *13*, 522–526.
70. Park, J. B. Inhibition of Glucose and Dehydroascorbic Acid Uptakes by Resveratrol in Human Transformed Myelocytic Cells. *J. Nat. Prod.* **2001,** *64* (3), 381–384.
71. Pillai, A.; Warren, G.; Gunathilake, W.; Idris, I. Effects of Sleep Apnea Severity on Glycemic Control in Patients with Type 2 Diabetes Prior to Continuous Positive Airway Pressure Treatment. *Diab. Technol. Therap.* **2011,** *13*, 945–949.
72. Preuss, H. G.; Bagchi, D.; Bagchi, M. Protective Effects of a Novel Niacin-Bound Chromium Complex and a Grape Seed Proanthocyanidin Extract on Advancing Age and Various Aspects of Syndrome X. *Ann. N. Y. Acad. Sci.* **2002,** *957*, 250–259.
73. Preuss, H. G.; Montamarry, S.; Echard, B.; Scheckenbach, R.; Bagchi, D. Long-Term Effects of Chromium, Grape Seed Extract, and Zinc on Various Metabolic Parameters of Rats. *Mol. Cell. Biochem.* **2001,** *223* (1–2), 95–102.
74. Reiter, R. J.; Tan, D. X.; Mayo, J. C.; Sainz, R. M.; Leon, J.; Czarnocki, Z. Melatonin as an Antioxidant: Biochemical Mechanisms and Pathophysiological Implications on Humans. *Acta Biochim. Pol.* **2003,** *50*, 1129–1146.
75. Roy, P.; Madan, E.; Kalra, N.; Nigam, N.; George, J.; Ray, R. S.; Hans, R. K.; Prasad, S.; Shukla, Y. Resveratrol Enhances Ultraviolet B-Induced Cell Death through Nuclear Factor-κB Pathway in Human Epidermoid Carcinoma A431 Cells. *Biochem. Biophys. Res. Commun.* **2009,** *384*, 215–220.
76. Sala, G.; Minutolo, F.; Macchia, M.; Sacchi, N.; Ghidoni, R. Resveratrol Structure and Ceramide-Associated Growth Inhibition in Prostate Cancer Cells. *Drugs Exp. Clin. Res.* **2003,** *29*, 263–269.
77. Sano, A.; Uchida, R.; Saito, M.; Shivo, N.; Komori, Y.; Tho, Y.; Hashizume, N. Beneficial Effects of Grape Seed Extract on Malondialdehyde-Modified LDL. *J. Nutr. Sci. Vitaminol. (Tokyo)* **2007,** *53* (2), 174–182.

78. Selga, A.; Sort, X.; Bobet, R.; Torres, J. L. Efficient One Pot Extraction and Depolymerization of Grape (*Vitis vinifera*) Pomace Procyanidins for the Preparation of Antioxidant Thio-Conjugates. *J. Agric. Food Chem.* **2004,** *52* (3), 467–473.
79. Shanmuganayagam, D.; Warner, T. F.; Krueger, C. G.; Reed, J. D.; Folts, J. D. Concord Grape Juice Attenuates Platelet Aggregation, Serum Cholesterol and Development of Atheroma in Hypercholesterolemic Rabbits. *Atherosclerosis* **2007,** *190* (1), 135–142.
80. Singletary, K. W.; Jung, K. J.; Giusti, M. M. Anthocyanin-Rich Grape Extract Blocks Breast Cell DNA Damage. *J. Med. Food* **2007,** *10*, 244–251.
81. Milne, J. C.; Lambert, P. D.; Schenk, S.; Carney, D. P.; Smith, J. J., Gagne, D. J.; Jin, L.; Boss, O.; Perni, R. B.; Vu, C. B.; Bemis, J. E.; Xie, R.; Disch, J. S.; Ng, P. Y.; Nunes, J. J.; Lynch, A. V.; Yang, H.; Galonek, H.; Israelian, K.; Choy, W.; Iffland, A.; Lavu, S.; Medvedik, O.; Sinclair, D. A.; Olefsky, J. M.; Jirousek, M. R.; Elliott, P. J.; Westphal, C. H. Small Molecule Activators of SIRT1 as Therapeutics for the Treatment of Type 2 Diabetes. *Nature* **2007,** *450* (7170), 712–716.
82. Soofyani, Z.; Hashemiravan, M.; Pourahmad, R. Production of Beverage Based on Probiotic Fermented Mixture of Malt Extract and Red Fruit Juices. *Adv. Environ. Biol.* **2015,** *9* (2), 762–769.
83. Srinivasan, V.; Pandi-Perumal, S. R.; Cardinali, D. P.; Poeggeler, B.; Hardeland, R. Melatonin in Alzheimer's Disease and Other Neurodegenerative Disorders. *Behav. Brain Funct.* **2006,** *2*, 15.
84. Tagurt, T.; Tanaka, T.; Kouno, I. Antimicrobial Activity of 10 Different Plant Polyphenols Against Bacteria Causing Food-Borne Disease. *Biol. Pharm. Bull.* **2004,** *27*, 1965–1969.
85. Teixeira, A.; Eiras-Dias, J.; Castellarin, S. D.; Geros, H. Berry Phenolics of Grapevine Under Challenging Environments. *Int. J. Mol. Sci.* **2013,** *14*, 18711–18739.
86. Terra, X.; Montagut, G.; Bustos, M.; Llopiz, N.; Ardevol, A.; Blade, C.; Arola, L. Grape-Seed Procyanidins Prevent Low-Grade Inflammation by Modulating Cytokine Expression in Rats Fed a High-Fat Diet. *J. Nutr. Biochem.* **2009,** *20*, 210–218.
87. Thtmothe, J.; Bonsi, I. A.; Padilla-Zakour, O. I. Chemical Characterization of Red Wine Grape (*Vitis vinifera* and *Vitis* Interspecific Hybrids) and Pomace Phenolic Extracts and Their Biological Activity against *Streptococcus mutans. J. Agric. Food Chem.* **2007,** *55*, 10200–10207.
88. Toledo, J.; Manzaneque, T.; Ruiz-Diez, V.; Kucera, M.; Pfusterschmied, G.; Wistrela, E. et al. Piezoelectric MEMS Resonators for Density and Viscosity Sensing in Engine Oil with Diesel Fuel. *Transducers* **2015,** *2591*, 436–439.
89. Tsanga, C.; Higginsa, S.; Duthiea, G. G.; Duthiea, S. J.; Howiea, M.; Mullena, W.; Leana, M. E. J.; Crozier, A. The influence of Moderate Red Wine Consumption on Antioxidant Status and Indices of Oxidative Stress Associated with CHD in Healthy Volunteers. *Br. J. Nutr.* **2005,** *93*, 233–240.
90. Tutunculer, F.; Eskiocak, S.; Basaran, U. N.; Ekuklu, G.; Ayvaz, S.; Vatansever, U. The Protective Role of Melatonin in Experimental Hypoxic Brain Damage. *Pediatr. Int.* **2005,** *47*, 434–439.
91. Urcan, D. E.; Giacosa, S.; Torchio, F.; Segade, S. R.; Raimondi, S.; Bertolino, M.; Gerbi, V.; Pop, N.; Roll, L. 'Fortified' Wines Volatile Composition: Effect of Different Postharvest Dehydration Conditions of Wine Grapes cv. *Malvasia moscata* (*Vitis vinifera* L.). *Food Chem.* **2017,** *219*, 346–356.

92. Wang, H.; Cao, G.; Prior, P. L. Oxygen Radical Absorbing Capacity of Anthocyanins. *J. Agric. Food Chem.* **1997,** *45*, 304–309.
93. Xu, L.; Li, B.; Cheng, M.; Zhang, W.; Pan, J.; Zhang, C.; Gao, H. Oral Administration of Grape Seed Proanthocyanidin Extracts Downregulate RAGE Dependant Nuclear Factor-Kappa BP65 Expression in the Hippocampus of Streptozotocin Induced Diabetic Rats. *Exp. Clin. Endocrinol. Diab.* **2008,** *116*, 215–224.
94. Yilmaz, Y.; Toledo, R. T. Major Flavonoids in Grape Seeds and Skins: Antioxidant Capacity of Catechin, Epicatechin, and Gallic Acid. *J. Agric. Food Chem.* **2004,** *52* (2), 255–260.
95. Zhang, L. Y.; Peng, Y. B.; Pelleschi-Travier, S., Fan, Y., Lu, Y. F., Lu, Y. M., et al. Evidence for Apoplasmic Phloem Unloading in Developing Apple Fruit. *Plant Physiol.* **2004,** *135*, 574–586.

CHAPTER 7

PHYTOCHEMISTRY AND POTENTIAL THERAPEUTIC PERSPECTIVE OF MELON SEEDS

AFSHAN SHAFI[1], UMAR FAROOQ[1], KASHIF AKRAM[2,*], and ZAFAR HAYAT[3]

[1]*Department of Food Science & Technology, MNS-University of Agriculture Multan, Pakistan*

[2]*Department of Food Sciences, Cholistan University of Veterinary and Animal Sciences Bahawalpur, Pakistan*

[3]*Department of Animal Sciences, CVAS-Jhang campus, University of Veterinary & Animal Sciences, Lahore, Pakistan*

**Corresponding author. E-mail: kashifakramft@gmail.com*

ABSTRACT

Agroindustrial waste (produced during processing and consumption: like seed, peel, fibrous material, bran, and stalk) is becoming a global issue. Fruit and vegetable waste especially seeds and peels are a rich source of many biologically active components (phytochemicals) and other valuable nutrients like amino acids, fatty acids, vitamins, and minerals. Seeds of sweet melon (*Cucumis melo*) are rich in nutrients as well as biologically active phytochemicals but till now these seeds are underutilized by the food experts to produce valuable functional food products. So by keeping in view the importance of sweet melon (*C. melo*) seeds, the current chapter is focused on the exploration of valuable nutrients and phytochemicals present in the melon. The practical applications of these seeds with their potential therapeutic role in the food products are discussed.

7.1 INTRODUCTION—SWEET MELON (*Cucumis melo*)

Sweet melon (*Cucumis melo*), the most significant member of Cucurbitaceae family, is considered as a vegetable in most of the region of the globe. Its utilization throughout the world produces a large amount of food waste like seeds and peels.[2] The chemical investigation of these wastes materials showed their nutritional properties with the presence of many biologically active compounds. Therefore, underutilization of these seeds and peels is basically the wastage of many important nutrients.[37]

Melon (*C. melo*) cultivation requires a hot and dry climate and, therefore, is a warm season crop. The species requires an optimum temperature of 27–30°C and its seed requires approximately 23–24°C for germination and will not germinate at temperatures less than 18°C. Although light is not as important as temperature, it can affect plant color and fruit flavor. In addition, excess atmospheric humidity adversely affects sugar formation, texture, and flavor.[4]

7.2 NOMENCLATURE OF *Cucumis melo*

According to the nomenclature (Table 7.1), sweet melon belongs to kingdom Plantae—plants and subkingdom Tracheobionta which have vascular tissues for the transportation of water and food that is why they are also known as vascular plants. Then these are Spermatophyta (which means containing seed) in superdivision and division is Magnoliophyta (flowering plants). Melon has its place in the class Dicotyledons (Magnoliopsida), subclass Dilleniidae, and order Violales. The family of sweet melon is Cucurbitaceae, which is also known as the cucumber family. Genus and species of sweet melon are *Cucumis* and *C. melo*, respectively.[27]

TABLE 7.1 Nomenclature of Sweet Melon.[27]

Kingdom	Plantae
Subkingdom	Tracheobionta
Superdivision	Spermatophyta
Division	Magnoliophyta
Class	Dicotyledons
Order	Violales
Family	Cucurbitaceae
Genus	*Cucumis*
Species	*Cucumis melo*

7.3 GROUPS OF *Cucumis melo*

The species of *C. melo* is further divided into five subgroups named as *cantalupensis* (True cantaloupe), *inodorus*, *reticulatus*, *conomon*, and *flexuosus*, which are further categorized into various cultivars (Table 7.2) according to the physiology and physicochemical characters of the fruit. Yellow-colored canary melon belongs to *inodorus* subgroup of *C. melo* species.[45] Seinat melon has its place as another variety of *C. melo* (*tibish*), which is grown in warm climatic zones. This type of melon is economically important as the fruits are well known for their therapeutic potentials.[7]

TABLE 7.2 Subgroups and Cultivars with Fruit Characteristics of *Cucumis melo*.[36]

Subgroups	Important cultivar	Fruit appearance
Cantalupensis (True cantaloupe)	Prescott melon D' Alger Charentais	Plain surface Good aroma After mature, fruit does not slip from the vine No netting
Inodorus	Canary melon Casaba melon Crenshaw melon Honeydew melon	Green or white flesh Fruit does not slip from the vine after ripening Highly aromatic
Reticulatus	Muskmelons Persian melons Galia melon	Netted surface Good aroma Fruit slips from vine on maturity
Conomon	Makuwa uri Chinese melon Sakata's sweet	Aroma less Crispy white flesh
Flexuosus	Armenian cucumber Snake melon	Elongated fruit with no aroma

Similarly, muskmelon, the member of *reticulatus* group of *C. melo* of Cucurbitaceae family, has the characteristics such as netted structure and strong aromatic flavor of the fruit. This member of the *C. melo* is mostly found in the Asian region. This fruit has the consumer preference because of its immense sweetness, texture, and strong flavor or aroma. It is also a rich source of phytonutrients.[25] These are found with high concentrations of various noteworthy nutrients including ascorbic acid, carotene, folic acid, and potassium coupled with high-consumer acceptance due to characteristic taste and aroma.[24]

7.4 WHOLE FRUIT COMPOSITION

Melon fruit is cultivated in all the dry (tropical) regions of the globe and belongs to family Cucurbitaceae. Depending upon the variety of melon fruit, its pulp is sweet, refreshing, and has a pleasant aroma. Large quantities of seeds are present in the fruit which are beneficial because these seeds are a rich source of proteins (54%) and oil (37%) and also possess medicinal values, but in spite of all these benefits, these beneficial seeds are considered as waste. Due to the rich nutritive content of seeds, they can be utilized as toppings in cakes, soups, and other foods.

According to a study, the *inodorus* seeds and *C. melo* contain crude protein (25%), moisture (4.5%), crude fat (25%), fibrous material (23.3%), minerals (2.4%), and carbohydrates (19.8%), while −5.12°C and −59.01°C is melting and crystallization temperature, respectively. Major triacylglycerols are POL, LLL, OLL, and PLL. Here O represents oleic, P represents palmitic, L represents linoleic, and S represents stearic acids. More abundant triglycerols are LLL (24.9%) while in POL are found in lower quantity (12.4%) as compared to others. The oil contains a variety of unsaturated fatty acids (UFAs) (86.1%). Linoleic acid with 69.0% is predominated, while in palmitic acid, it is 8.4% and oleic acid is 16.8%. Existence of different volatile compounds is also indicated by the analysis of electronic nose as compared to refined sunflower oil.

7.4.1 PHYTOCHEMISTRY OF SEED

The seeds consist of mainly two parts: seed kernel and seed coat. The composition of sweet melon (*C. melo*) seed shares 65–72% seed kernel and 27–35% seed coat.

7.4.1.1 NUTRITIONAL COMPOSITION OF *Cucumis melo* SEED

Various research studies indicated that the nutritional composition of *C. melo* seed kernel varies depending upon different factors. These variations might be due to the difference in climacteric conditions, agricultural practices, irrigation water quality, soil structure, and so on. The nutritional value of *C. melo* var. *tibish* as observed by Azhari et al.[6] consists of protein 28.58 ± 0.50%, fat 31.13 ± 0.90%, fiber 24.75 ± 0.34%, moisture 4.27 ± 0.12%, total sugars 6.94 ± 0.55%, and mineral contents 4.33 ± 0.14% (Table 7.3).

Siddeeg et al.[39] performed an experimental trial on melon seeds to evaluate physicochemical as well as functional characteristics. For this purpose, they used whole seed powder, defatted seed powder, and extracted seed protein. All these specimens of seeds or different forms of seed's constituents were investigated for their proximate composition and functional potential. Results of this study indicated that seed flour contained a good quantity of lipid contents and fiber at the concentration of 31.13% and 24.75%, respectively. These studied samples showed different levels and quality of potential functional compounds.

Petkova and Antova[35] studied the chemical composition of three different varieties of melon (*C. melo* L.) in details. They examined the proximate chemical composition of sweet melon seed (*C. melo*) and observed that the seed contained crude fat 41.6–44.5%, crude fiber 4.5–8.5%, soluble sugars 3.7–4.2%, protein 34.4–39.8%, carbohydrates 8.2–12.7%, and total mineral contents 4.6–5.1%.

TABLE 7.3 Nutritional Composition of *Cucumis melo* Seed.

Nutrient	Values	References
Protein	28.58 ± 0.50%	Azhari et al.[5]
Glutelin	81.80 ± 0.26%[a]	
Albumin	73.09 ± 0.13[b]	
Globulin	50.95 ± 0.15[c]	
Prolamin	5.81 ± 0.14[d]	
Fiber	24.75 ± 0.34	
Lipids	31.13 ± 0.90	
Sugars	6.94 ± 0.55	
Moisture	4.27 ± 0.12	
Minerals	4.33 ± 0.14	
Potassium	1148.75 ± 1.53 (mg/100 g)	Mallek-Ayadi et al.[26]
Magnesium	1062.25 ± 0.72	
Calcium	506.13 ± 1.52	
Sodium	336.5 ± 0.72	
Iron	2.69 ± 0.81	
Zinc	2.34 ± 0.64	
Manganese	1.25 ± 0.15	
Copper	0.53 ± 0.12	

7.4.1.2 FATTY ACIDS AND BIOACTIVE COMPOUNDS PRESENT IN *Cucumis melo* SEED

The gas chromatography mass spectrometry analysis conducted by Azhari et al.[5] while determining the fatty acid profile of seinat seed showed 80.54% UFAs concentration with linoleic acid (61.10 ± 0.08%) as a major component (Fig. 7.1). Bouazzaoui et al.[10] also found that linoleic acid (60.1%) followed by the oleic acid (25.3%) as the major UFAs in *C. melo* L. *inodorus* seeds. The polyunsaturated fatty acids concentration observed in a research study conducted on melon (*C. melo*) seed was higher than the abovementioned seed oils and also from sunflower seed oil (66.2%),[17] pumpkin seed oil (42.6%), and soybean seed oil (53.2%).[31] As far as saturated fatty acids contents were concerned, *C. melo* seeds contained lower SFA content (8.97–9.73%) than that of palm oil (49.9%), soybean oil (15.1%), pumpkin seed oil (20.3%),[31] sunflower seed oil (11.3%),[17] and *C.*

melo L. *inodorus* seed oil (14.6%).[11] The occurrence of a high concentration of essential fatty acids such as linoleic acid and oleic acid in edible seed make it nutritionally valuable. Seeds rich in essential fatty acids are gaining market value and consumer acceptance due to multifaceted functional/physiological health effects.[45]

Therefore, melon (*C. melo*) seeds can be utilized as a functional edible oil with high concentration of essential fatty acids especially linoleic acid by the food industries.[17] Bar-Nun and Mayer[8] reported that the *C. melo* seed oil contains 60–70% of linoleic acid (Ω-6 fatty acid), which is beneficial for human health.

Petkova and Antova[35] regarding the fatty acid profile of melon (*C. melo*) seed reported the presence of phospholipids (0.7–1.7%), sterols (0.6%), and tocopherols (435–828 mg/kg) in the oils of three different varieties of melon seed. The main fatty acids in the lipids of melon seed were linoleic from 51.1% to 58.5%, and oleic acid from 24.8% to 25.6%. They further reported that the presence of trilinolein (31.3–32.2%), oleodilinolein (31.0–34.0%), and palmitoyl dilinolein (14.9–22.3%) in the melon seed lipid profile constituted 80% of the total triglyceride composition. The major phospholipids found in the melon (*C. melo*) seed lipids were phosphatidylethanolamine (8.4–17.1%), phosphatidylinositol (24.4–33.9%), and phosphatidylcholine (23.0–33.1%). Among all the observed fatty acids, palmitic acid was observed in the range of 34.4–61.7% as major fatty acid followed by oleic acid in the range of 8.9–27.2%. In case of sterol esters, linoleic acid (32.7–39.1%) and oleic acid (25.1–30.7%) were found as major components along with γ-tocopherol (71.4–91.5%) as main tocopherol fraction in the melon seed oils.

Other than the presence of fatty acids in the *C. melo* seed, it is an important source of bioactive components as well like tocopherol, phytosterols, flavonoids, and phenolic compounds are presented in Figure 7.1 and Table 7.4.[8,26] Silva and Jorge[40] investigated the bioactive components of *C. melo* seed and found that total amount of tocopherol in the seed was 270.2 mg/kg, where γ-tocopherol was in higher quantity (249.6 mg/kg) and α-tocopherol was in low quantity (20.5 mg/kg) but β-tocopherol was not found in the sweet melon seed oil. Total phytosterols observed in the melon seed oil were 327.3 mg/100 g. They also investigated that phytosterols of sweet melon seed oil consist of 210.2 mg/100 g of β-sitosterol and 117.1 mg/100 g of stigmastanol. Similarly, total phenolic compounds 130.7 mg gallic acid equivalent (GAE)/kg, and total carotenoids 6.3 μg were observed in the melon seed oil. According to Mariod and Matthäus,[1] γ-tocopherol was the predominant tocopherol (80.7%) in the *C. melo* var. *agrestis* seed oil, whereas α-tocopherol was only 21%. Among phytosterols (3785 mg/kg), β-sitosterol was the main component with the highest quantity and total phenolic components were 31.9 mg/kg. According to Azhari et al.,[6] total tocopherol contents were 43.20 mg/100 g with γ-tocopherol in abundant quantity. Total phenolic contents were found to be 28.17 mg/100 g along with 302.70 mg/100 g total phytosterols in the *C. melo* var. *tibish* seed oil.

Yanty et al.[44] conducted a research study to investigate the biologically active components present in the melon seed. The results of the study revealed the occurrence of phenolic glycoside and benzyl *O*-*b*-D-glucopyranoside, 3,29-*O*-dibenzoylmultiflor-8-en-3a,7b,29-triol. The high technology of NMR was used for the investigation of chemical structures of these compounds. De Marino et al.[13] also performed a research trial on melon seeds to determine its bioactive compounds. During their study, they observed the occurrence of multiflorane triterpene esters as novel melon constituents. Although the melon seeds contain a variety of bioactive components, major research studies are focused on the investigation of the fatty acid profile, tocopherols contents, and sterols of seed oil.[1]

De Marino et al.[13] investigated the phytochemistry of *C. melo* seeds and found three new biologically active phytochemicals named as 3,29-*O*-dibenzoylmultiflor-8-en-3,7b,29-triol, phenolic glycoside (*E*)-4-hydroxycinnmyl alcohol 4-*O*-(2-*O*-*b*-dapiofuranosyl) (1,2)-*b*-dglycopyranoside, benzyl *O*-*b*-D-glucopyranoside and 3-*O*-*p*-amino-benzoyl-29-*O*-benzoylmultiflor-8-en-3a,7b,29-triol, whereas another phytochemical investigation of *C. melo* conducted by Trease and Evans[42] also resulted in a number of bioactive phytochemicals present in seed and seed oil. They observed isomultiflorenol, multiflorenol, 24-methylenecycloartenol, teraxerol, euphol, α- and β-amyrin, lupeol, 24-methylene-24-dihydrolanosterol, 24-methyl-25(27)-dehydrocycloartanol, 24-methylene-24-dihydroparkeol, cycloartenol, and tirucallol during their study. *C. melo* seed oil also contains a significant amount of cephalin, lecithin, and cerebroside (potential bioactive phytochemicals).[8]

A number of bioactive compounds were also diagnosed by Prajapati et al.[36] while working on phytochemistry of seeds of *C. melo*. They found a substantial amount of clerosterol, stigmasterol, codisterol, 25(27)-dehydroporiferasterol, avenasterol, campesterol, sitosterol, isofucosterol, 25(27)-dehydrochondrillasterol, spinasterol 24ξ-methyllathosterol,

24β-ethyl-25-(27)-dehydrolathosterol, and 22-dihydrospinasterol. These seeds are potential alternatives of various nuts in a variety of beverages especially milk beverages.[41] The bioavailability of *C. melo* seed protein is also high because of the presence of essential amino acids like lysine, histidine, tryptophane, cysteine, myristic acid, arginine, aspartic, glutamic acids, galactane, and citrulline.[29,36]

Three novel chromone (bioactive phytochemicals) were identified by Ibrahim[22] using the methanolic extract of the seeds of *C. melo* L. var. *reticulatus*. These novel bioactive phytochemicals were 5,7-dihydroxy-2-[2-(3,4-dihydroxyphenyl)ethyl] chromone, 5,7-dihydroxy-2-[2-(4-hydroxyphenyl) ethyl] chromone (cucumin S. derivatives) and 7-glucosyloxy-5-hydroxy-2-[2-(4-hydroxyphenyl) ethyl] chromone. He also observed the presence of three already identified bioactive chemicals in the seed of *C. melo* named as beta-amyrin, beta-sitosterol-3-*O*-beta-glucopyranoside, and beta-sitosterol. De Marino et al.[13] identified a new phenolic glycoside (*E*)-4-hydroxycinnamyl alcohol 4-*O*-(2′-*O*-β-D-apiofuranosyl) (1″→2′)-β-D-glucopyranoside during an investigation on phytochemicals present in various parts of *C. melo*. They examine the presence of benzyl *O*-β-D-glucopyranoside, 3-*O*-*p*-amino-benzoyl-29-*O*-benzoylmultiflor-8-en-3α, 7β, 29-triol 3, and 29-*O*-dibenzoylmultiflor-8-en-3α.

Fahamiya et al.[18] detected a large number of bioactive compound groups in the seeds of *C. melo*. They observed 14 various types of dihydroxy triterpenes and their derivatives and only 1 type of oxo-sterol. Other than these compounds, they also identified isokarounidiol-3-*p*-methoxybenzoate, 7-oxodihydrokarounidiol-3-benzoate, karounidiol, karounidiol-3-benzoate, isokarounidiol, 29-hydroxylupeol, 7-oxodihydrokaroudnidiol, 5-dehydrokarounidiol, 25-diol, bryonolol, loranthol, 3-epibryonolol, betulin, erythrodiol, (23*Z*)-cycloart-23-ene-3beta, and 7-oxositosterol. Along with these health beneficial phytochemicals they also found the existence of some nutrient inhibitors like CMeTI-A and CMeTI-B in *C. melo* seeds, while using the affinity chromatography, gel filtration, ion exchange chromatography, and high-performance liquid chromatographic identification technologies. Petkova and Antova[35] reported the presence of many health beneficial constituents including sterols, phospholipids, phenolics (protocatechuic acid), and natural form of tocopherols in melon seed oil. All these components were present in a good quantity. According to Mehra et al.,[28] 2.415 μg GAE/mg of total phenolics and 401.2 μg catechin equivalents (CE)/mg of total flavonoids were present in musk melon seed oil.

Bouazzaoui et al.[10] investigated the mineral and fatty acid composition of melon (*C. melo* L. *inodorus*) seeds. The results indicated the presence of considerable quantity of oil (30.7%) and total mineral contents (4.08%). The physicochemical analysis demonstrated high physicochemical parameters like iodine (104.52 g (I_2)/100 g) and saponification (193.60 mg (KOH)/g). Fatty acid investigation results showed a healthy lipid profile of melon seeds comprised of polyunsaturated fatty acids (60.1%) as well as monounsaturated fatty acids (25.3%).

Proximate composition of seeds of melon hybrid variety AF-522 was estimated by de Melo et al.[14] Results of the experimental study indicated high quantity and quality of edible oil with a healthy fatty acid profile containing UFAs, whereas the other important constituents include a high percentage of protein (up to 14.9%). Moreover, the high oil yield was obtained by using a solvent extraction method. This high-quality edible oil also showed a significant physicochemical analysis outcome including peroxide value (4.96), iodine number (111.8), and saponification values (210.6). Although the detailed investigation regarding the fatty acid profile of melon seed oil indicated the presence of 24 different types of fatty acids, the concentration of UFAs was up to 64%. Other nutritional composition of the melon seed represents the presence of essential amino acids such as arginine, aspartic, and glutamic acids, whereas methionine and lysine were in limiting quantity.[36]

Achu et al.[3] conducted an investigative study on various species of Cucurbitaceae oilseed crops to estimate their biological activity as well as nutritious significance. For this experiment, they used the species from different areas to evaluate the variation among different areas' crops. After the preparatory operations performed on seeds of all tested varieties, oil was extracted from all the seeds/specimens. After oil extraction, seed meal or cake was evaluated for proximate as well as total phenolic contents present in seeds. Results of all these investigations indicated the occurrence of high-quality proteins or peptides comprising essential amino acids. Although the presence of total phenolic components was low, because of the presence of high-quality amino acids, these seeds were recommended for infant diet formula.

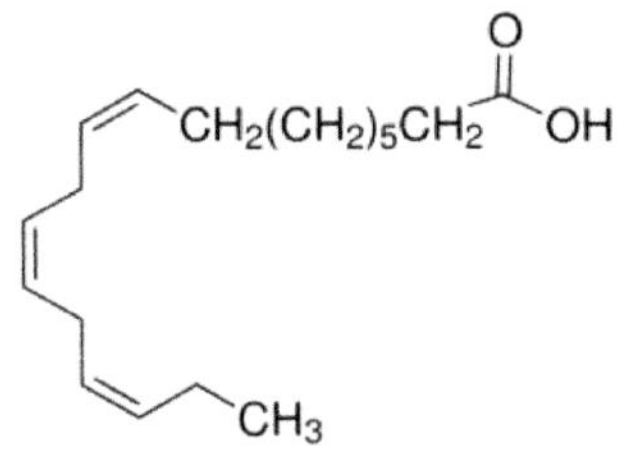	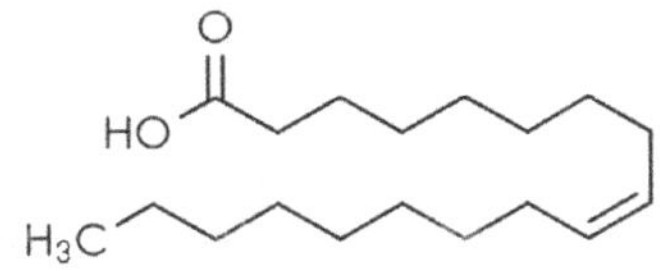
(a) Lenoleic acid	(b) Lenoleic acid
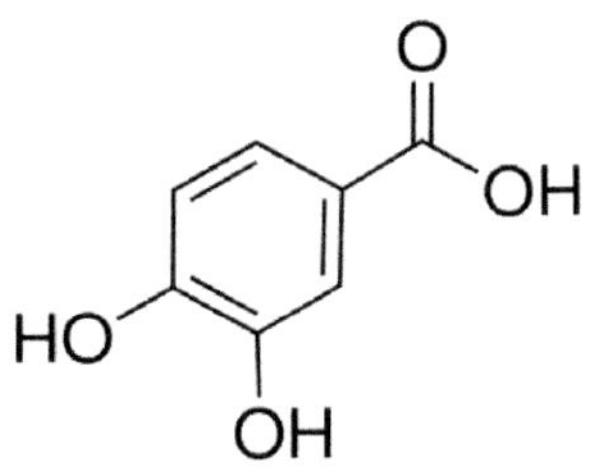	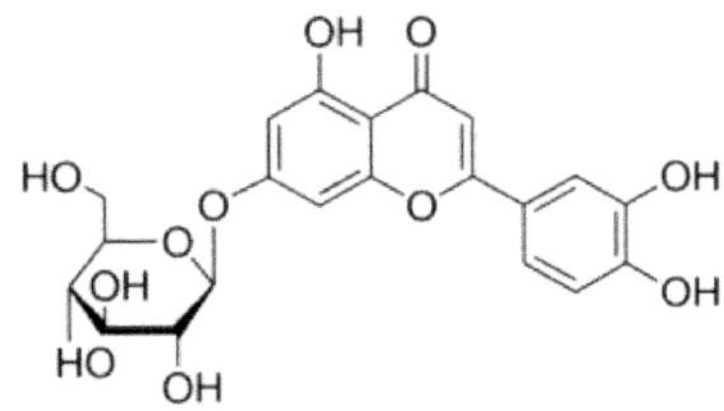
(c) Protocatechuic acid	(d) Luteolin-7-*O*-glycoside
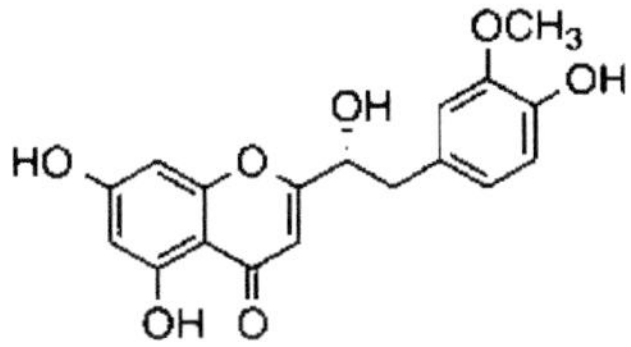	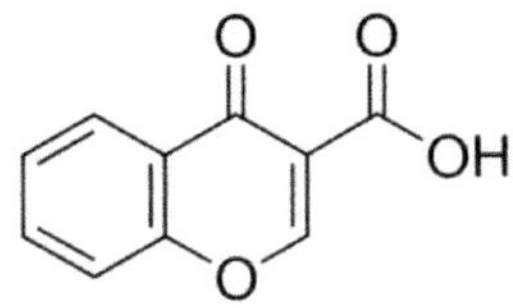
(d) Cucumins. S	(e) Chromones

FIGURE 7.1 *(Continued)*

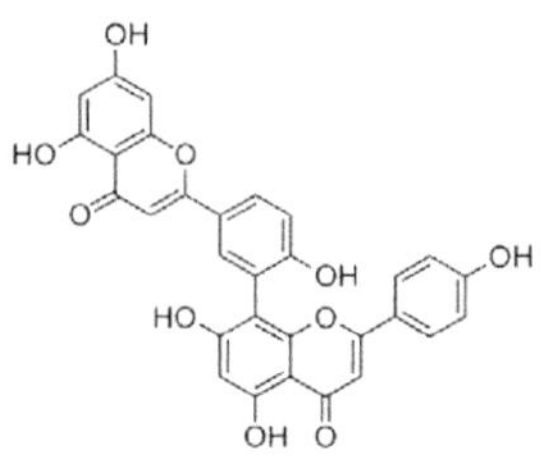

(f) Amentoflavone

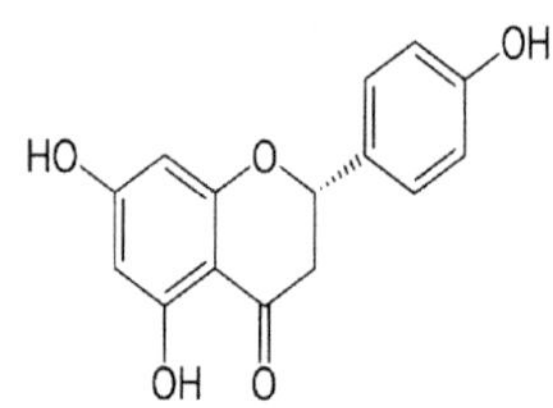

(g) Naringenin

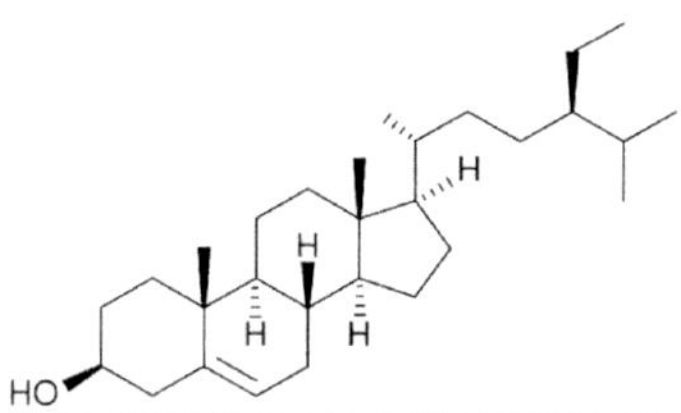

(h) β- sitosterol

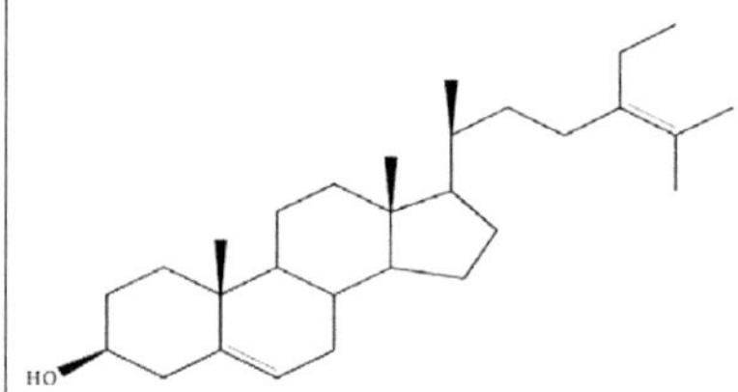

(i) Δ^5,24-stigmastadienol

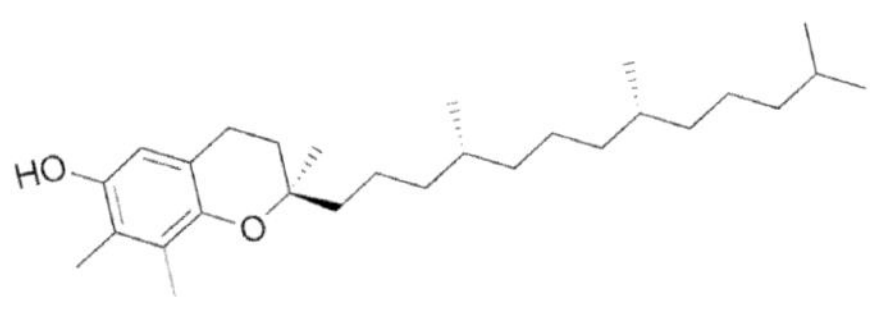

(j) γ-tocopherol

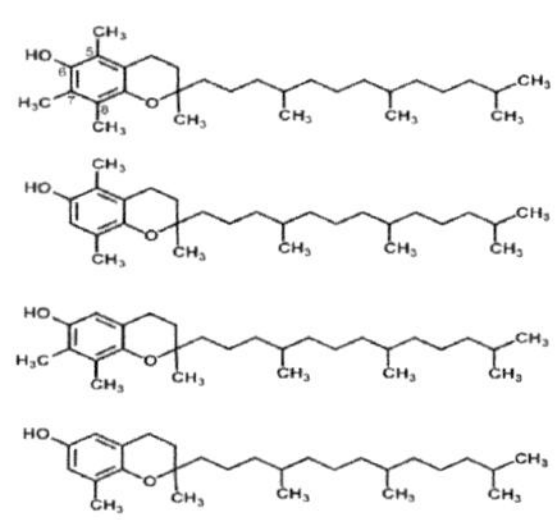

(k) β+γ Tocopherols

FIGURE 7.1 Predominant phytochemicals present in *Cucumis melo* seeds.

TABLE 7.4 Phytochemical Composition of *Cucumis melo* Seed.[26]

		Nomenclature	Values (%)
Fatty acids	C14:0	Myristic acid	0.04 ± 0.01
	C15:0	Pentadecanoic acid	0.03 ± 0.01
	C16:0	Palmitic acid	8.71 ± 0.07
	C17:0	Margaric acid	0.07 ± 0.00
	C18:0	Stearic acid	5.54 ± 0.06
	C20:0	Arachidic acid	0.16 ± 0.01
	C24:0	Lignoceric acid	0.06 ± 0.01
	C16:1 ω-7	Palmitoleic acid	0.08 ± 0.02
	C17:1	Margaroleic acid	0.03 ± 0.01
	C18:1 ω-9	Oleic acid	15.84 ± 0.03
	C20:1	Eicosenoic acid	0.13 ± 0.01
	C20:1 ω-9	Gadoleic acid	0.12 ± 0.01
	C22:1 ω-9	Erucic acid	0.02 ±0.01
	C18:2 ω-6	Linoleic acid	68.98 ±0.05
	C18:3 ω-3	Linolenic acid	0.2 ± 0.00
		Saturated fatty acids	14.61
		Monounsaturated fatty acids	16.22
		Polyunsaturated fatty acids	69.18
Phenolics	Phenolic acids	Gallic acid	7.26 ± 0.02
		Protocatechuic acid	0.89 ± 0.01
		Caffeic acid	3.13 ± 0.00
		Rosmarinic acid	2.91 ± 0.04
	Flavonoids	Luteolin-7-*O*-glycoside	9.60 ± 0.01
		Naringenin	4.72 ± 0.01
		Apigenin	3.88 ± 0.03
		Flavone	1.94 ± 0.02
		Amentoflavone	32.80 ± 0.21
	Secoiridoids	Oleuropein	1.65 ± 0.07
	Lignans	Lignans	3.95 ± 0.03
	Pinoresino	Pinoresinol	3.95 ± 0.03
	Unknown	Unknown	3.48 ± 0.06

TABLE 7.4 *(Continued)*

	Nomenclature	Values (%)
Sterol	Campesterol	0.92 ± 0.05
	Campestanol	0.59 ± 0.17
	24-Methylenecholesterol	0.22 ± 0.02
	Stigmasterol	1.29 ± 0.04
	Chlerosterol	3.25 ± 0.11
	β-Sitosterol	206.42 ± 0.09
	Sitostanol	4.84 ± 0.21
	Δ^7-Campesterol	14.55 ± 0.35
	Δ^7-Stigmasterol	14.18 ± 0.07
	Δ^7-Avenasterol	2.11 ± 0.14
	Δ^5-Avenasterol	2.18 ± 0.02
	Δ^5,23-Stigmastadienol	0.037 ± 0.01
	Δ^5,24-Stigmastadienol	117.91 ± 0.06
	Cholesterol	0.88 ± 0.03
	Brassicasterol	Tr
Tocopherol	-α-Tocopherol	2.85 ± 0.17
	β + γ-Tocopherols	18.13 ± 0.41
	γ-Tocopherol	6.09 ± 0.53

7.5 THERAPEUTIC BENEFITS

A number of researchers have claimed various therapeutic potentials of melon seeds due to the presence of nutritive as well as bioactive compounds.

7.5.1 ANTI-ALZHEIMER ACTIVITY

It was claimed that *C. melo* seeds are therapeutic in nature. This claimed was checked by Parle and Kulwant[32] by specifically targeting the Alzheimer disease. In the study, *C. melo* seed powder and seed extract were used to treat Alzheimer effect in mice. Blood cholesterol, brain acetylcholinesterase activity, and blood glucose levels were also analyzed. After the completion of study duration, the results indicated that *C. melo* seed powder, as well as seed extracts, possessed significant potential to protect the animals from developing symptoms like memory deficits. This study also indicated that *C. melo* seed improves the memory through reduction of transfer latency and escape latency time and increment in time spent in target quadrant (TSTQ), discrimination index, and step-down latency. The *C. melo* seed extract also has the potential to lower the activity of acetylcholinesterase. This enzyme activity is basically the indicator of blood glucose and cholesterol levels. It is, therefore, concluded from the outcomes of the study that *C. melo* seed has a strong potential to cure and manage Alzheimer's disease and can be used as a natural remedy of this disease. Alzheimer's disease management mechanism of *C. melo* seed consists of basically sevenfold mechanism including (1) significant antioxidant potential, (2) presence of polyunsaturated fatty acids like linoleic and arachidonic acid. These polyunsaturated fatty acids are responsible for growth and regeneration of cholinergic neurons (3) presence of phosphatidylethanolamine and phosphatidylcholine serve as the precursors for the synthesis of acetylcholine; (4) α-linoleic acid abundantly present in the seed kernel stimulates the release of neuroprotectin D1, which performs a neuroprotective role; (5) the inhibition of acetylcholinesterase enzyme; (6) lowering of blood cholesterol; and 7) antihyperglycemic effect of *C. melo* seeds help in the prevention of brain damage due to excessive glucose.

7.5.2 ANTIULCER ACTIVITY

Antiulcer activity of methanol extract of seed of *C. melo* was evaluated by Parmer and Kar[33] and found that these seeds exhibit antiulcerogenic activity. The possible reason of this activity might be due to the presence of triterpenoids and sterols, whereas the gastroprotective effect may be recognized to decrease in vascular permeability, diminished lipid peroxidation, and scavenging of free radicals (reactive oxygen species, ROS) along with the strength of mucosal barrier (Fig. 7.2). A peptic ulcer is an abraded area of the stomach caused principally by the digestive action of gastric juice and small intestinal secretions. It is mainly an inflamed split in the skin or lining of the alimentary tract in mucous membrane. The presence of triterpenoids and sterols is responsible for the management of these conditions.[14]

7.5.3 ANTHELMINTIC PROPERTY

Anthelmintic activity of *C. melo* seeds was assessed by Fahamiya et al.[18] while working on the ethnobotany of various parts of *C. melo* and their significance in Unani medicine. The methanolic and *n*-hexane extracts of the seeds of *C. melo* L. have exhibited excellent anthelmintic and antimicrobial activity. Whole seed flour of *C. melo* is also found useful as vermifuge.[15]

7.5.4 ANALGESIC AND ANTI-INFLAMMATORY ACTIVITY

Methanolic extract of seeds of *C. melo* possesses the analgesic activity and also inhibited the leukocytes influx and diminished LTB4 levels.[19] *C. melo* seeds showed effective analgesic activity in a research trial. The possible mechanism of analgesic activity was carrageenan induces gathering of leukocytes in the pleural cavity, as well as the development of LTB4 levels in pleural exudates after stimulation of inflammation. Movement of neutrophils to the exaggerated area constitutes a significant pro-inflammatory factor ROS, as they release ROS in the extracellular medium. *C. melo* abolished the leukocyte influx and reduced LTB4 levels and produced anti-inflammatory effect.[9]

7.5.5 ANTIOXIDANT ACTIVITY

The seed extract was found to have significant scavenging activity 75.59% at 300 μg/ml by 1, 1-dipheny1-2-picrylhydrazy1 picrylhydrazyl method and 69.86% at 400 μg/mL by hydrogen peroxide method as compared to standard (ascorbic acid).[5] *C. melo* methanol extract showed 1, 1-diphenyl-2-picrylhydrazyl and hydroxyl radicals scavenging activity particularly due to the presence of phenolic compounds especially flavonoids.[22]

7.5.6 ANTIDIABETIC

Chen et al.[12] assessed the potential of melon (*C. melo* L. var. *makuwa* Makino) seeds for the management of hyperglycemia linked to type 2 diabetes by extracting useful phytochemicals and utilizing them in foods.

The inhibitory effects of the hexane extracts of roasted melon seed on α-glucosidase and α-amylase were examined under different roasting temperatures and compared with those of the unroasted seeds. The highest hypoglycemic activity was exhibited by the seed roasted at 250°C for 90 min with an inhibition value of 87.7% and 52.3% against α-amylase and α-glucosidase, respectively.

7.5.7 DIURETIC ACTIVITY

The diuretic effect of ethanolic seed extracts of *C. melo* in Albino rats was evaluated by the scientists. Furosemide (5 mg/kg) was used as a drug. The diuretic effect was evaluated by measuring the urine volume, sodium, potassium, chloride, and bicarbonate contents. A significant diuretic effect was observed treated with extracts of *C. melo* individually in experimental animals compared to the control. The results, therefore, suggested the use of *C. melo* seeds as a cure for renal diseases in traditional medical practice.[38] Further, extract of *C. melo* (400 mg/kg) showed more diuretic effect than standard.[47] Moreover, pharmacodynamic investigations into the diuretic activity of *C. melo* seed (ether extract) has also been done by Singh and Sisodia in 1970.[18] The diuretic effects of *C. melo* L. were experienced in anesthetized dogs. Results showed that ether extract of the seeds extensively increased the urinary volume and chloride substance. The mechanism for the rise of chloride substance may be recognized to decrease tubular reabsorption increased and glomerular filtration rate.[43]

It has been reported that the melon seed extracts possess anti-inflammatory activities by inhibition of reactive oxygen species,[19,21,34] antioxidant activity due to flavonoids,[23] antiulcer activity (gastroprotective effect) through reduction in muscular permeability and lipid peroxidation due to the presence of triterpenoids and sterols, anticancer,[10,46] and antihypothyroidism through increment in thyroid hormones (T3 and T4) with a related diminish in tissue lipid peroxidation and antiperoxidative role,[30] excellent anthelmintic activity and antimicrobial activity[15], and antidiabetic activity due to healthy role of ascorbic acid and polyphenols.[16] It has also reported that the melon seed (*C. melo*) or its extracts have hepatoprotective,[33] diuretic,[43] cardiovascular[22] and immuno-modulatory effects[8] due to the presence of phytochemicals. Some important therapeutic effects of melon seed along with their mechanism of action and bioactive components are also shown in Figure 7.2.

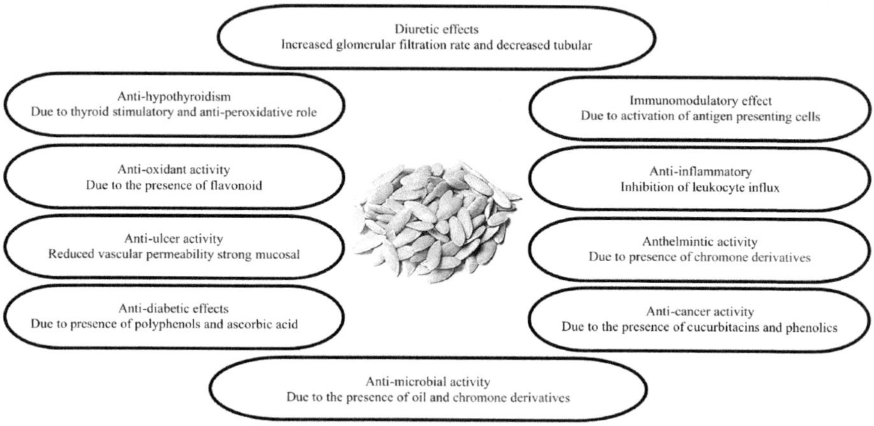

FIGURE 7.2 (See color insert.) Potential therapeutic benefits and mechanisms of bioactive compounds present in the melon seeds.

7.5.8 MISCELLANEOUS USES

Although melon seeds contain a significant amount of bioactive components with a healthy chemical composition, it also possesses a good taste and flavor. So in this way, it can serve as an excellent source of functional food or the source of functional ingredients.[11,20] Linoleic acid plays a significant metabolic role in the formation of prostaglandins.[45]

The α-glucosidase and α-amylase inhibitory effects were strongly correlated with the levels of triacylglyceride esterified with linoleic acid (trilinolein). These results suggest that oriental melon seed extracts contain TAG and UFAs, which are potent inhibitors of α-glucosidase and α-amylase.

7.6 SUMMARY

On the basis of available information, it is concluded that the *C. melo* seems to be rich in many bioactive and useful compounds. In spite of its nutritional value, *C. melo* has great potential for a therapeutic role due to the presence of different bioactive/functional compounds. The photochemistry of fruit and its various parts further elaborates that the variety and concentration of bioactive compounds are much higher in the melon seed. So, it is concluded that the sweet melon seed could be used for the treatment/management of various ailments/disorders with a number of beneficial potentials for human health. Moreover, we should adopt proper strategies for utilization and value addition of melon seed oil rather than dumping in the soil as waste. At the same time, there is also a need to explore more health potentials of melon especially of its seed which will be helpful for proper utilization of *C. melo* and its wastes/by-products.

KEYWORDS

- **melon seeds**
- **phytochemistry**
- **therapeutic value**
- **agroindustrial waste**
- **nutrients**
- **sweet melon *(Cucumis melo)***
- **functional food products**

REFERENCES

1. Mariod, A.; Matthäus, B. Fatty Acids, Tocopherols, Sterols, Phenolic Profiles and Oxidative Stability of *Cucumis melo* var. *agrestis* Oil. *J. Food Lipids* **2008,** *15*, 56–67.
2. ABF. Anuário Brasileiro de Fruticultura. Santa Cruz do Sul. *Gazeta Santa Cruz*, 2014, p 136.
3. Achu, M. B.; Fokou, E.; Kansil, G.; Martin, Fotso. Chemical Evaluation of Protein Quality and Phenolic Compound Levels of Some Cucurbitaceae Oilseeds from Cameroon. *Afr. J. Biotechnol.* **2013**, *12* (7), 735–743.
4. Ahamed, H. A. U.; Uddin, M. A.; Mannan, M. A.; Barua, S.; Hoque, M. A. Studies on the Isolation, Physico-chemical Characterization and Microbial Activities of Melon (*Cucumis melo*) Seed Oil. *Int. J. Innov. Sci. Res.* **2014,** *11*, 105–111.
5. Arora, R.; Kaur, M.; Gill, N. S. Antioxidant Activity and Pharmacological Evaluation of *Cucumis melo* var. *agrestis* Methanolic Seed Extract. *Phytochemistry* **2011,** *5* (3), 146–155.

6. Azhari, S.; Xu, Y. S.; Jiang, Q. X.; Xia, W. S. Physicochemical Properties and Chemical Composition of Seinat (*Cucumis melo* var. *tibish*) Seed Oil and Its Antioxidant Activity. *Gras. Y Aceit.* **2014**, *65*, e008.
7. Badifu, G. I. O. Effect of Processing on Proximate Composition, Antinutritional and Toxic Contents of Kernels from Cucurbitaceae Species Grown in Nigeria. *J. Food Compos. Anal.* **2001,** *14*, 153–161.
8. Bar-Nun, N.; Mayer, A. M. Cucurbitacin-Repressor of Induction of Laccase Formation. *Phytochemistry* **1989**, *28*, 1369–1371.
9. Besong, S. A.; Ezekwe, M. O.; Fosung, C. N.; Senwo, Z. N. Evaluation of Nutrient Composition of African Melon Oilseed (*Cucumeropsis mannii* Naudin) for Human Nutrition. *Int. J. Nutr. Metab.* **2011**, *3* (8), 103–108.
10. Bouazzaoui, N.; Drici, W.; Bouazzaoui, W.; Lemerini, W.; Arrar, Z.; Bendiabdellah, D.; Mulengi, J. K. Fatty Acids and Mineral Composition of Melon (*Cucumis melo* L. *inodorus*) Seeds from West Algeria. *Mediterr. J. Chem.* **2016**, *5* (1), 340–346.
11. Chan, K. T.; Li, K.; Liu, S. L.; Chu, K. H.; Toh, M.; Xie, W. D.; Cucurbitacin, B. Inhibits STAT3 and the Raf/MEK/ERK Pathway in Leukemia Cell Line K562. *Cancer Lett.* **2010**, *289* (1), 46–52.
12. Chen, L .; Kang, Y. H.; Suh, J. K. Roasting Processed Oriental Melon (*Cucumis melo* L. var. *makuwa* Makino) Seed Influenced the Triglyceride Profile and the Inhibitory Potential Against Key Enzymes Relevant for Hyperglycemia. *Food Res. Int.* **2014**, *56*, 236–242.
13. De Marino, S.; Festa, C.; Zollo, F.; Iorizzi, M. Phenolic Glycosides from *Cucumis melo* var. *inodorus* Seeds. *Phytochem. Lett.* **2009**, *2* (3), 130–133.
14. De Melo, M. L. S.; Narain, N.; Bora, P. S. Characterisation of Some Nutritional Constituents of Melon (*Cucumis melo* Hybrid AF-522) Seeds. *Food Chem.* **2000**, *68*, 411–414.
15. Debnath, B.; Uddin, J.; Patari, P.; Das, M.; Maiti, D.; Manna, K. Estimation of Alkaloids and Phenolics of Five Edible Cucurbitaceous Plants and Their Antibacterial Activity. *Int. J. Pharm. Pharm. Sci.* **2015**, *7* (12), 223–227.
16. Décordé, K.; Ventura, E.; Lacan, D.; Ramos, J.; Cristol, J. P.; Rouanet, J. M. An SOD Rich Melon Extract Extramel Prevents Aortic Lipids and Liver Steatosis in Diet-Induced Model of Atherosclerosis. *Nutr. Metab. Cardiovasc. Dis.* **2010**, *20*, 301–307.
17. Edem, D. O. Palm Oil: Biochemical, Physiological, Nutritional, Hematological and Toxicological Aspects: A Review. *Plant Foods Hum. Nutr.* **2002**, *57*, 319–341.
18. Fahamiya, N.; Aslam, M.; Siddiqui, A.; Shiffa, M. Review on *Cucumis melo*: Ethnobotany and Unani Medicine. *World J. Pharm. Pharm. Sci.* **2016**, *5* (12), 621–636.
19. Gill, N. S.; Bajwa, J.; Sharma, P.; Dhiman, K.; Sood, S.; Sharma, P. D.; Singh, B.; Bali, M. Evaluation of Therapeutic Potential of Traditionally Consumed *Cucumis melo* Seeds. *Asian J. Plant Sci.* **2011**, *10* (1), 86–91.
20. Gill, N. S.; Garg, M.; Bansal, R.; Sood, S.; Muthuraman, A.; Bali, M.; Sharma, P. D. Evaluation of Antioxidant and Antiulcer Potential of *Cucumis sativum* L. Seed Extract in Rats. *Asian J. Clin. Nutr.* **2009**, *1*, 131–138.
21. Gohain, N.; Prakash, A. An Ethnobotanical Survey of Anti-Malarial Plants in Some Highly Malaria. *Int. J. Pharm. Pharm. Sci.* **2015**, *7* (9), 147–152.
22. Ibrahim, S. R. New 2-(2-Phenylethyl)chromone Derivatives from the Seeds of *Cucumis melo* L. var. *reticulatus*. *Nat. Prod. Commun.* **2010**, *5* (3), 403–406.

23. Ismail, H. I.; Chan, K. W.; Mariad, A. A.; Ismail, M. Phenolic Content and Antioxidant Activity of Cantaloupe (*Cucumis melo*) Methanolic Extracts. *Food Chem.* **2010**, *119* (2), 643–647.
24. Lester, G. E. Antioxidant, Sugar, Mineral, and Phytonutrient Concentrations across Edible Fruit Tissues of Orange-Fleshed Honeydew Melon (*Cucumis melo* L.). *J. Agric. Food Chem.* **2008**, *56*, 3694–3698.
25. Lester, G. E.; Hodges, D. M. Antioxidants Associated with Fruit Senescence and Human Health: Novel Orange Fleshed Non-Netted Honey Dew Melon Genotype Comparisons Following Different Seasonal Productions and Cold Storage Durations. *Postharvest Biol. Technol.* **2008**, *48*, 347–354.
26. Mallek-Ayadi, S.; Bahloul, N.; Kechaou, N. Chemical Composition and Bioactive Compounds of *Cucumis melo* L. Seeds: Potential Source for New Trends of Plant Oils. *Process Saf. Environ. Protect.* **2018**, *113*, 68–77.
27. Mansouri, A.; Mirzabe, A. H.; Raufi, A. Physical Properties and Mathematical Modeling of Melon (*Cucumis melo* L.) Seeds and Kernels. *J. Saudi Soc. Agric. Sci.* **2017**, *16* (3), 218–226.
28. Mehra, M.; Pasricha, V.; Gupta, R. K. Estimation of Nutritional, Phytochemical and Antioxidant Activity of Seeds of Musk Melon (*Cucumis melo*) and Water Melon (*Citrullus lanatus*) and Nutritional Analysis of Their Respective Oils. *J. Pharm. Phytochem.* **2015**, *3* (6), 98–102.
29. Mian-Hao, H.; Yansong, A. Characteristics of Some Nutritional Composition of Melon (*Cucumis melo* Hybrid 'ChunLi') Seeds. *Int. J. Food Sci. Technol.* **2007**, *42* (12), 1397–1401.
30. Naito, Y.; Akagiri, S.; Uchiyama, K.; Kokura, S.; Yoshida, N.; Hasegawa, G. Reduction of Diabetes-Induced Renal Oxidative Stress by a Cantaloupe Melon Extract/Gliadin Biopolymers, Oxykine, in Mice. *BioFactors* **2005**, *23* (2), 85–95.
31. Nawirska-Olszanska, A.; Kita, A.; Biesiada, A.; Sokol-Letowska, A.; Kucharska, Z. Characteristics of Antioxidant Activity and Composition of Pumpkin Seed Oils in 12 Cultivars. *Food Chem.* **2013**, *139*, 155–161.
32. Parle, M.; Kulwant, S. Musk Melon in the Role of a Memory Melon. *Ann. Pharm. Pharm. Sci.* **2012**, *3* (1), 36–41.
33. Parmar, H. S.; Kar, A. Protective Role of *Mangifera indica*, *Cucumis melo* and *Citrullus vulgaris* Peel Extracts in Chemically Induced Hypothyroidism. *Chem. Biol. Interact.* **2009**, *177* (3), 254–258.
34. Patel, M.; Patel, P.; Patel, M. Aqueous Extract of *Ficus bengalensis* Linn. Bark for Inflammatory Bowel Disease. *J. Young Pharm.* **2010**, *2* (2), 130–136.
35. Petkova, Z.; Antova, G. Proximate Composition of Seeds and Seed Oils from Melon (*Cucumis melo* L.) Cultivated in Bulgaria. *Cog. Food Agric.* **2015**, *1*, 1018779.
36. Prajapati, N. D.; Purohit, S. S.; Sharma, A. K.; Kumar, T. *A Handbook of Medicinal Plants;* Agrobios: India, 2001; p 176.
37. Rashid, U.; Rehman, H. A.; Hussain, I.; Ibrahim, M.; Haider, M. S. Muskmelon (*Cucumis melo*) Seed Oil: A Potential Non-Food Oil Source for Biodiesel Production. *Energy* **2011,** *36*, 5632–5639.
38. Ravishankar, K.; Priya, V. P. S. V. Evaluation of Diuretic Effect of Ethanolic Seed Extracts of *Macrotyloma uniflorum* and *Cucumis melo* in Rats. *Int. J. Pharma Bio Sci.* **2012,** *3* (3), 251–255.

39. Siddeeg, A.; Xu, Y.; Qixing-Jiang.; Wenshui-Xia. Physicochemical and Functional Properties of Flour and Protein Isolates Extracted from Seinat (*Cucumis melo* var. *tibish*) Seeds. *Food Sci. Biotechnol.* **2014**, *23* (2), 345–353.
40. Silva, A. C. S.; Jorge, N. Bioactive Compounds of the Lipid Fractions of Agro-Industrial Waste. *Food Res. Int.* **2014**, *66*, 493–500.
41. New National Institute of Science Communication. *The Wealth of India*; New National Institute of Science Communication, Council of Science & Industrial Research: Delhi, 2001; vol II, pp 252–253.
42. Trease, G. E.; Evans, W. C. *Pharmacognosy*; W. B. Saunders: London, New York, 2002; p 473.
43. Vikasari, S. N.; Sukandar, E. Y.; Sutjiatmo, A. B.; Riyanti, S. Diuretic Effect of the Ethanol Extracts of *Phyllanthus acidus* (Skeels) Leaves in Wistar Rats. *Int. J. Pharm. Pharm. Sci.* **2005,** *7* (1), 120–123.
44. Wang, Y.; Jones, P. Dietary Conjugate Linoleic Acid and Body Composition. *Am. J. Clin. Nutr.* **2004**, *79*, 1153–1158
45. Warra, A. A. Production of Soap from an Indigenous *Moringa oliefera* Lam Seed Oil. *J. Raw Mater. Res.* **2012**, *7*, 26–35.
46. Wright, C. I.; Van-Buren, L.; Kroner, C. I.; Koning, M. M. Herbal Medicines as Diuretics: A Review of the Scientific Evidence. *J. Ethnopharmacol.* **2007**, *114* (1), 1–31.
47. Yanty, N. A. M.; Lai, M. A.; Osman, K.; Long, K.; Ghazali, H. M. Physicochemical Properties of *Cucumis melo* var. *inodorus* (Honeydew Melon) Seed and Seed Oil. *J. Food Lipids* **2008**, *15*, 42–55.

CHAPTER 8

LINALOOL: A NATURAL CHEMOTHERAPY AGAINST VARIOUS MALADIES

MUHAMMAD IMRAN[1], IAHTISHAM-UL-HAQ[2], AURANG ZEB[1], and HAFIZ ANSAR RASUL SULERIA[3,*]

[1]*Faculty of Allied Health Sciences, University Institute of Diet & Nutritional Sciences, The University of Lahore, Pakistan*

[2]*Department of Diet and Nutritional Sciences, Faculty of Health and Allied Sciences, Imperial College of Business Studies, Lahore, Pakistan*

[3]*School of Life and Environmental Sciences, Faculty of Science, Engineering and Built Environment, Centre for Chemistry and Biotechnology, Deakin University, 75 Pigdons Road, Geelong, VIC 3216, Australia*

**Corresponding author. E-mail: hafiz.suleria@uqconnect.edu.au*

ABSTRACT

Researchers have been mainly emphasized on nutritional and functional bioactive moieties, which have enormous health-endorsing perspectives. Linalool is a bioactive component and utilized as a functional and nutraceutical tool in different foodstuffs. Linalool (3,7-dimethyl-1,6-octadien-3-ol)-flavoring compound is volatile in nature and majorly present in flowers and other plant tissues including fruits and leaves. This compound and its esters are frequently employed in perfume industry, for example, petitgrain (bitter orange), and lavender-like odor is given by (*S*)- and (*R*)-linalool, respectively. Owing to their phytochemical profile, it prevents from lipid oxidation and enhances the shelf-life of food. Chemically, linalool is monoterpene alcohol that exists in numerous medicinal plants and offers various biological activities like anti-inflammatory, antioxidant, antimicrobial, antiviral, and antitumor effects. Besides, linalool possesses several sedating properties on central nervous systems (CNSs) and literature elucidated the potential perspectives of this compound. Furthermore, health-promoting potentials of linalool against various physiological threats like Alzheimer's disease, coronary atherosclerosis, aging processes, and carcinogenesis are also the limelight of the article.

8.1 INTRODUCTION

Phytochemicals are isolated from the plants to curtail various health disorders from ancient times. These phytochemicals are naturally present in the human diet being consumed since prehistoric era. The consumers of current era are well aware of the importance of diet in routine life, so they adopt selective materials as a diet-based therapy to cure various maladies. Researchers and scientists are diverting toward herbs and spices owing to their health-promoting attributes used as natural food preservatives in many food-based products. In addition, herbal medicines are grabbing equal importance in developed and developing countries as a possible health-caring strategy to combat numerous dysfunctions as these therapies are safer in nature and possess a wide range of biological activities.[99]

Among different herbs, coriander (*Coriandrum sativum* L.) has unique significance due to presence of essential bioactive compounds. It is fully enriched with diversified food constituents, for instance, protein, fat, minerals, fiber, carbohydrates, and water.[129] Moreover, coriander seeds are promising source of advantageous phytochemicals such as borneol, geraniol, carvone, elemol, limonene, camphor, and linalool. The essential and fatty oil contents were present as 0.03–2.6% and fatty oil content ranged from 9.9% to 27.7%, respectively.[70,101] The coriander seeds are composed of linalool, which varied from 50% to 70%, as well as used in creams, perfumes, detergents, surfactants, emulsifiers, and lotions.[4,22,54]

8.2 LINALOOL: CHEMISTRY, ABSORPTION, METABOLISM, AND EXCRETION

The linalool is a monoterpene compound and known as a vital flavoring compound having volatile nature. Its chemical structure is illustrated in Figure 8.1. It is widely distributed in various plant parts, especially in flowers, fruits, and leaves.[20,84] Primarily, linalool is isolated from the flowers, leaves, herbs, and wood and concentrated in the oils of petitgrain, rosewood, jasmine, rose, linaloe seed, lavender, coriander, and bergamot. Linalool has two imperative forms generally known as *S*(π)-linalool (coriandrol) and *R*(−)-linalool (licareol). Both these forms varied in different plants depending upon nature and agro-climatic conditions.[24] Linalool is also used in preparation of perfumed hygiene goods and cleaning agents, that is, detergents, lotions, shampoos, soaps, insecticides, and mosquito repellent.[24,75]

Diederichsen and Hammer[40] determined that out of 1% essential oil (EO), the principal component is *S*-(+)-linalool that accounts for 60–70%. However, other minor compounds were detected as monoterpenes hydrocarbons involving glimpnene, γ-terpinene, α-pinene, camphor, citronellol, *p*-cymene, geraniol, geraniol acetate, and borneol. Other heterocyclic components like pyrazine, furan, pyridine, thiazole and tetrahudrofuran derivatives, dihydrocoriandrin, coriandrons A–E, coriandrin, isocoumarins, neochidilide, digustilide phenolic acids, pthlides, and sterols were also present in the EO. It also passes the cytoplasmic membrane and cell wall, dislocates the structure of fatty acids, polysaccharides, and phospholipids, and permeabilizes them due to lipophilic nature.[7] This permeabilization of bacterial membranes is linked with reduction of membrane potential and loss of ions, diminution of adenosine triphosphate pool, and collapse of proton pump.39,124 Further, the EOs damage proteins and lipids and coagulate the cytoplasm. The damaged cell membrane and wall allows the macromolecules to leak through them and promotes cell lysis.[21,74,91]

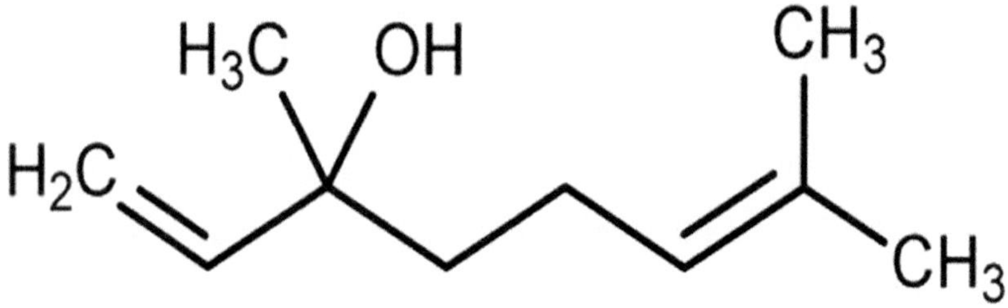

FIGURE 8.1 Chemical structure of linalool.

Chemically, linalool is a monoterpene alcohol with potent aromatic properties. It acquires two isomeric forms, that is, *S*(−)-linalool and *R*(−)-linalool. Both these isomers considerably contribute aromatic properties with different scents in various common plants.[117] Linalool and its esters are most frequently employed compounds in perfume industries. Amongst them, (*R*)- and (*S*)-linalool have been used as lavender and petitfrain (bitter orange) like odors, correspondingly.[9,118] In the presence of air, linalool can be oxidized at normal pressure and temperature. Under standard conditions, linalool concentration is reduced from 98% to 80% after oxidation for 10 weeks. Furthermore, 50% reduction in compound concentration has been observed after 30 weeks, whereas the linalool content has been found to fall to 4% after 80 weeks.[117]

In context of contact allergy, the greatest risk linked to linalool is its probable ability to form hydroperoxides. Commonly, linalool exhibits contact allergen properties only in its oxidized form as evident from clinical trials designed on patients prevailing consecutive dermatitis. Amongst allergic patients to oxidized linalool, 64% of individuals showed positive reactions to hydroperoxides.[83] Another study revealed nature of hydroperoxides produced during the oxidation process. They deduced that 7-hydroperoxy-3,7-dimethyl-octa-1,5-diene-3-ol was the dominating hydroperoxide formed during the oxidation process of linalool. However, the presence of 6-hydroperoxy-3,7-dimethyl-octa-1,7-diene-3-ol has also been observed together with 8-hydroperoxy-3,7-dimethyl-octa-1,6-diene-3-ol to a lesser extent. In addition, apparently analogous oxidation compound 5-hydroperoxy-3,7-dimethyl-octa-1,6-diene-3-ol was not detected in the investigation. Mechanistically, the chemical reaction of oxidation engages double bond proximity at C6–C7. Furthermore, larger class of chemical reactions is involved in the oxidation processes of unsaturated hydrocarbons.[117]

The repeated administration of linalool can result in onset of allylic oxidation. However, primary routes of metabolism of linalool involve glucuronic acid conjugation and excretion. Most of the linalool and its metabolic products are excreted via urine, whereas lower amounts are excreted from the body via expired air and feces. Alongside glucuronide conjugation, sulfation and hydroxylation of linalool have also been identified. Consequently, these findings suggest that linalool metabolism is quite rapid with quick absorption and excretion after subsequent metabolic changes from the body which depict its high biological compatibility in biological absorption, metabolism, and excretion process.[12,111]

8.3 LINALOOL: AUTO-OXIDATION

The content of linalool decreases over time due to oxidation on exposure to air. Hydroperoxides have been identified as the principal components of oxidation products from a complex mixture of compounds. The linalool lacks sensitizing property, whereas its oxidation mixture recovered after 10 weeks of oxidation on exposure to air showed sensitizing effects in guinea pigs in a Freund's Complete Adjuvant Test experiment.[67] About 20% of linalool is consumed upon exposure to air for 10 weeks in comparison to D-limonene (60%); hence, linalool oxidation is found to be comparatively slower. In case of linalool, the primary oxidation compounds appear more stable with lowered formation of polymers.[63]

Initially, the auto-oxidation of terpenes gives rise to hydroperoxides, which are comparatively unstable compounds and readily degrade to secondary oxidation products consequently forming polymeric material. This allylic hydroperoxide has been observed to be relatively stable in nature as elucidated by tertiary chemical structure of hydroperoxide.[11] Originally, the said hydroperoxide had been synthesized to assess the sensitization mechanism of allylic hydroperoxides but not as an aspirant of sensitizing chemical from oxidized linalool. One of the key properties of skin sensitizers is that the compound must be reactable to skin proteins. Hence, mechanistic understanding of formation of antigen is of cardinal significance for predictive modeling of allergenic effect of various compounds.[63]

Keeping in mind the relationship between structure and activity, linalool is not thought of a contact allergen because protein-reactive groups are lacking its molecular structure. However, hydroperoxides are recognized as contact allergens where colophonium[65] and D-limonene[63] have been recognized as oxidation products with strong allergen properties. It is worth mentioning that the exact mechanism of hydroperoxides binding with skin proteins is yet to be known but numerous reports have indicated the involvement of radical reaction depicted by antigen formation.[47,85]

In another investigation, linalool and its oxidized form have been tested as sensitizing agents in a model employing guinea pigs. The finding of the research indicated that linalool lacks sensitizer properties, whereas its oxidation products are responsible for inducing allergic responses in animal model. Further, linalool of commercial quality without any further purification showed insignificant response as sensitizer. A weak sensitizing property of linalool has been reported while employing a local lymph node assay. In available literature, purity of linalool used is rarely considered that might explain divergent results regarding sensitizing capacity of linalool in various reports. However, the concentration of oxidation products is of sufficient nature to evoke sensitized effects in animals.[107]

To report a chemical to be a sensitizer, it is necessary that a stimulation index (SI) of three or more may be attained at any concentration. Hence, while assessing SI of undiluted linalool, it resulted in 8.3 SI. Notably, the SI was dropped considerably (4.9) when linalool is purified and tested for SI. Nevertheless, the difference is within biologic variations; this finding supports the overall nonsensitizing behavior of linalool as the dropped SI after purity of the compound depicts a possible removal of oxidation products from the original mixture. Linalool demonstrated no sensitization effect in human maximization tests when administered at 8% and 20% concentration in 25 volunteers. Only a few clinical trials have shown linalool-related contact allergy.[110]

It has been seen that the degree of oxidation of products is naturally affected by handling and storage conditions that may consequently raise the risk of adverse side effects. In allergic contact dermatitis patients, frequent contact allergy of oxidized linalool can be observed undiagnosed because there is a wide use of linalool in aromatic industry, and screening of linalool hydroperoxides and oxidized linalool compound have not been performed. In a clinical trial by Karlberg and Dooms-Goossens,[66] 1–6% of individuals exhibited positive reactions with oxidized limonene and its hydroperoxides upon undergoing patch testing. In standard series, this frequency is in parallel to the most common contact allergens being prevailed.

8.4 HEALTH PERSPECTIVES

Numerous health-promoting perspectives have been studied by various scientific groups as highlighted in Table 8.1. The detailed discussion regarding these health aspects are summarized as follows.

8.4.1 ANTITUMOR ACTIVITY

Amongst hematological malignancies, leukemia is one of the serious conditions that is often found lethal.[133] Chemotherapy has been seen to be quite effective in reduction of most leukemia patients; however, degeneration and poor long-term endurance are often common. Currently, chemotherapy drugs being used nowadays like anthracyclines and nucleoside analogues are sensitive to flourish leukemia cells.[8,52,56] Linalool possesses potential biologic worth against Burkitt lymphoma cells P3HR1 and histiocytic lymphoma cells U937. The inhibitory concentrations (IC50) for the aforementioned cells were reported as 4.21 and 3.51 μg/mL, respectively, with SI as 494.1 and 592.6, correspondingly.[31] In human breast adenocarcinoma cells, linalool has also been identified to converse doxorubicin resistance[103] and showed antiproliferative properties against solid tumor cells including renal adenocarcinoma cells, amelanotic melanoma cells, Michigan cancer foundation (MCF)-7, and HepG2 cells.[92,126]

In a recent investigation, it has been observed that linalool (0–2.5 mM) inhibited cell proliferation in a dose-dependent fashion. In this context, inhibition of cell proliferation is attributed to downregulation of cyclin A and Cdk4, apoptosis characterized by poly ADP ribose polymerase (PARP) cleavage, DNA fragmentation, loss of mitochondrial membrane potential and caspase-3 activation, and upregulation of p21 and 27 inducing cell cycle arrest at G0/G1 in hepatocellular carcinoma HepG2 cells. Linalool reduced Akt activity and membrane-bound Ras at 1 mM concentration, whereas reactive oxygen species (ROS) generation and mammalian target of rapamycin inhibition along with Akt phosphorylation and mitogen-activated protein kinases (MAPKs) activation were triggered by higher linalool concentrations (2 mM). Anticancer activity of linalool potentiated by Akt phosphorylation and specific extracellular signal-regulated kinases (ERK) inhibitors pointed involvement of ERK and Akt activation as prosurvival mechanisms.[105]

Viability of U87-MG cells momentously is reduced by linalool in time- and concentration-dependent manner. The cells exposed to linalool showed increased terminal deoxynucleotidyl transferase dUTP nick end labeling -stained cells indicative of apoptotic cells death. The oxygen consumption rate of mitochondria and expression of Bcl-xl and Bcl-2 were reduced, whereas expression of Bak and Bax and caspase-3 and 9 activities were enhanced on exposure to linalool treatment that directed apoptosis in cells. It was further seen that linalool significantly increased acetylated superoxide dismutase-2 (SOD-2) expression while inhibited SIRT3 expression and mRNA alongside decrement in sirtuin-3 (SIRT3) and SOD2 interactions. SIR3 overexpression promisingly retarded linalool-induced apoptotic cell death and increased in mitochondrial ROS production besides reducing cell viability.[30,105]

Apoptosis was induced in cancer cells from human fibroblast cell line by treatment with linalool in vitro. In animal model, mice receiving high dose of linalool showed reduction in tumor-specific lipid peroxidation and xenograft tumor weight.[62] Similarly, lipid peroxidation, edema formation, hyperplasia, and antioxidant reduction in skin were prevented by linalool in acute ultraviolet B (UVB)-induced inflammation and photocarcinogenesis. Furthermore, ornithine decarboxylase and cyclooxygenase-2 overexpressions in mouse skin were prevented by linalool treatment that also prevented the expression of proliferative biomarkers subsequently lowering tumor incidence.[50] In another study by Cerchiara et al.,[26] linalool exhibited dose-dependent selective inhibitory response on melanoma cell growth that indicates potential utilization of linalool as therapeutic agent against melanoma. Furthermore, linalool from *Plantaginaceae* has shown cytotoxic effects presenting anticancer activity by cell cycle arrest in human cancer cell lines.[28]

A robust and rapid induction of apoptosis has been linked to administration of linalool in human leukemia cells in conjunction with activation of cyclin-dependent kinase inhibitors and p53. The compound has depicted potent cytotoxic effects in tumor cells in human hematopoietic malignancies comprising lymphoblastic leukemia (Molt-4, H-9), myeloid leukemia (Kasumi-1, HL-60), and lymphoma (Raji). It suggests that linalool acquires wide spectrum of antitumor activities. It is worth mentioning that the concentration (130 μM) of linalool required for killing tumor cells does not affect the normal hematopoietic cells and their growth, which is an additive effect to use this compound as a potential remedy against carcinomas. Although linalool provides sufficient window to persuade apoptosis in tumor cells, the requirement of high in vivo doses limits its pharmacological activity. Importantly, linalool can be optimized and modified as well for further development of anticarcinogenic compounds.[89]

Quiescent (nonproliferating) leukemia cells have been extensively reported to resist conventional cytotoxic agents, whereas linalool imparts positive impacts in this scenario.[8,52,56] The cell cycle is the primitive regulator by p53 that acquires powerful regulatory mechanisms and imparts critical role in prevention of improper cell proliferation and maintenance of genomic integrity caused by genotoxicity. Mechanistically, the activation of cyclin-dependent kinase inhibitors involving p27Kip, p21Waf1, and p57Kip2 and p53 tumor suppressor is triggered by treatment with linalool that further promotes induction of growth arrest and DNA damage (GADD) 45μJNK signaling pathway.[104,123,128] Apoptosis in quiescent leukemia stem cells can be induced by the upregulation of p53 triggered by linalool treatment[53] that opens a new horizon of patient's treatment relying on functional p53 strategically eliminating leukemia cells.[29] Conclusively, activation of p53 and cyclin-dependent kinase inhibitors induced by linalool is the core mechanism involved in providing protective role against carcinogenic events via regulation of apoptosis and cell cycle of proliferating and nonproliferating leukemia cells.[79]

8.4.2 HYPOCHOLESTEROLEMIC EFFECTS

Hypercholesterolemia is associated with building up cholesterols in arteries giving rise to formation of atheromas.[78] The earlier findings of Anum and Adera[5] determined that 10 mg/dL decrease in cholesterol levels in plasma lowers the death rate due to cardiovascular ailments up to 9%. The diet enriched with phytochemicals and fibers but free from the saturated fats can reduce the concentration of cholesterol.[10] The small aromatic terpenes from herbs and spices reduce the plasma cholesterol level by regulating 3-hydroxy-3-methylglutaryl-coenzyme A reductase expression.[32,33]

TABLE 8.1 Health Effects of Linalool.

Diseases	Functions	References
Hypercholesterolemia	Regulate HMGCR expression Inhibits LDL oxidation Decrease expression of HMGCR	[33, 94]
Oxidative stress	Enhanced level of antioxidant enzymes, that is, glutathione peroxidase (GSH-Px) and superoxide dismutase Scavenge free radicals	[6, 86]
Cancer	Inhibits HepG2 and MCF-7 cells Activate p53 and cyclin-dependent kinase inhibitors (P57Kip2, p27Kip1, and p21Waf1)	[89, 92, 126]
Microbial infections	Inhibits *Penicillium commune*, *Mucordimor phosphorus*, *Rhizopus azigosporum*, and *Fusarium solani* Inhibits *Helicobacter pylori* *Escherichia coli*, *Proteus vulgaris*, *Pseudomonas aeruginosa*, and *Klebsiella* species except *Corynebacterium diphtheriae* Inhibits *Salmonella* species	[60, 87, 115]
Anxiety	Inhibits excitatory glutamatergic Stimulate GABAergic pathways	[18, 45, 57, 113]
Inflammation	Possess peripheral analgesic action	[94]
Nociception	Inhibits carrageenan-induced edema Reduces acetic acid-induced pain responses	[94, 95]
Sedative hypnotic activity	Increases pentobarbital induced sleeping time Protect against picrotoxin, PTZ, and trans-corneal electroshock-induced convulsions Modulates in vitro expression of glutamate activation and delays in vivo *N*-methyl-D-aspartate-induced convulsions Blockade of intracerebroventricular quino-linic acid-induced convulsions	[18, 44, 46, 113]

GSH-Px, glutathione peroxidase; HMGCR, 3-Hydroxy-3-methylglutaryl-coenzyme A reductase; LDL, low-density lipoprotein; PTZ, pentylenetetrazol.Linalool suppresses the low-density lipoprotein (LDL) oxidation that increases the cholesterol uptake through macrophage scavenger receptors.[94] The treatment of linalool to high fructose diet lowered the total plasma and LDL cholesterol level in hepatic lipid levels whereas enhanced the high-density lipoprotein. The linalool in higher concentrations has been proven effective in reduction of plasma cholesterol levels in mice which was achieved via enhancing the expression of HMG co-A reductase.[17]

8.4.3 ANTIDIABETIC

The diabetes mellitus is linked with abnormal carbohydrate metabolism, an absolute impairment in insulin secretion, and peripheral resistance against action of insulin.[43] There are multiple activities such as poor lifestyle, physical inactivity, and overeating resulting in metabolic syndrome ultimately giving rise to prevalence of type-2 diabetes and cardiovascular ailments in both in vitro and in vivo studies.[120]

A study has shown promising effects of linalool in prophylaxis of streptozotocin-induced diabetes in rats. At a dose rate of 25 mg/kg, linalool prevented collagen accumulation, altered glucose metabolism, and improved NF-κB expression and transforming growth factor-β1 (TGF-β1) in kidney of STZ-induced diabetic rats. It was also observed that collagen content, glucose metabolizing enzymes, and glucose transporter-1 expression were restored upon treatment with linalool. It also prevented from nephron loss in diabetic rats and abolished oxidative stress-mediated inflammation by lowering NF-κB and TGF-β1 expressions alongside preserving normal histological tissue characters in kidneys of diabetic rats.[37]

8.4.4 OXIDATIVE STRESS

It is an imbalance between oxidation and antioxidants that produce the nitric oxide (NO) synthase activity, hydrogen peroxide in fasted rats via reducing the concentrations of glutathione, mitochondrial glutathione.[1] Testes hold vital significance due to their involvement in exchange of heredity and fertility source in male mammals. This organ is specifically sensitive to feeding and lifestyle practices. In extensive oxidative stress, the secretion of testosterone is compromised due to degeneration of Leydig cells.[108]

The data from other studies also revealed that administration of linalool at 120 mg/kg protected the cell membrane from the oxidative stress and enhanced the biochemical damage of reproductive organ such as testis via increasing the concentration of superoxide dismutase and glutathione peroxidase (GSH-Px).[6,86]

Recently, an investigation by Máté et al.[82] reported significant antioxidant capacity of linalool. They tested different dose rates of linalool and evaluated their acute toxicity effects on *Candida albicans* and deduced that it plays vital role in reducing the pathogen germ tube formation. It was seen that 20% and 30% reduction in colony formation ability were induced in an hour upon treatment with 0.7 and 1.4 mM of linalool, respectively, to 107 cells/mL. The reduction in colony-formation ability was noticed to be dose dependent alongside reduction in total reactive species and superoxide anion radicals. It is noteworthy that concentration of linalool showed an adaptive modification in antioxidant system. For an instance, a decrease in activities of catalase and superoxide dismutase was indicated on exposure of 07 mM linalool, whereas an upregulation of glutathione peroxidase, glutathione reductase, and catalase were exhibited when the cells were treated with 1.4 mM linalool concentration.

Oxidative imbalance can also be caused by UVB radiations that activate NF-κB and MAPK and signaling in human dermal fibroblasts adult (HDFa) cells. These radiations provoke numerous elements involved in cellular signaling. Linalool has been found to prevent the negative effects of acute UVB radiations that cause reduction in antioxidant enzyme activities in HDFa cells. Treatment with linalool (30 μM) significantly prohibited oxidative DNA damage as illustrated by lowered formation of 8-deoxy guanosine mediated by UVB that can be attributed to reduction in formation of reactive oxygen species in response to radiation exposure. Alongside, linalool treatment restored overall antioxidant balance in cellular system as reflected by overexpression of NF-κB and MAPK signaling induced by UVB. Mechanistically, linalool-suppressed radiations-induced phosphorylation of JNK, ERK1, and p38 proteins of MAPK. Further, IκBa was activated that inhibited NF-κB/p65 activation induced via radiation exposure. In addition, modulation of mitochondrial membrane potential -2 and 9, IL-6 and 10 and TNF-α was also noticed when the HDFa cells were treated with linalool which confirms that linalool may serve as photoprotective agent.[51]

Antioxidant properties of linalool have also been shown by Xu et al.[132] They narrated that administration of linalool (100 mg/kg) promisingly lowered malondialdehyde and acetylcholinesterase activity, whereas improved glutathione peroxidase (GPx) and superoxide dismutase functioning. Besides, protections against suppression of Nrf2 and heme oxygenase-1 expressions were evident upon treatment with this bioactive agent. Additionally, an improvement in expression of calmodulin-dependent protein kinase II (CaMKII), p-CaMKII, tropomyosin receptor

Kinase B, brain-derived neurotrophic factor, and synapse plasticity-related proteins in hippocampus was noticed after administration of linalool. All these findings firmly endorsed the protective effects of linalool against oxidative stress-induced malfunctions.

8.4.5 ANTIMICROBIAL ACTIVITY

Scientists have focused the research regarding the utilization of phytochemicals which have antimicrobial characters instead of conventional food preservatives.[127] More than 1340 plants exhibit the antimicrobial activity.[130] Spices are promising source of EOs and exhibited antimicrobial activity.[61,100] Boyle[14] was the first scientist who reported the antimicrobial property of the spices, including clove, mustard, and cinnamon, in 1880s. Their antimicrobial potential is mainly dependent on the type, concentration, and composition of spice, substrate composition, processing, and food storage conditions.[41,81] Earlier, Hulin et al.[58] determined the antimicrobial character of purified essentials oils from spices variety of microorganism such as *Shigella*.[34,58] Linalool shows the antimicrobial potential against various microorganism outbreaks due to the interaction between food matrix and phenolic compounds.[106,121]

The data from the research of Thyagaraja and Hosono[122] and O'Mahony et al.[87] implicated that linalool showed antifungal activity against *Mucordim orphosphorus*, *Rhizopus azigosporum*, *Fusarium solani*, *Penicillium commune*, and *Helicobacter pylori*. Likewise, Al-Jedah et al.[3] determined that coriander extract is effectual against Gram-positive and Gram-negative bacteria as well as foodborne pathogens like *Salmonella* species.[60] The studies of different scientists determined that linalool treatment exhibited the antimicrobial activity against *Escherichia coli*, *Staphylococcus aureus*, and *C. albicans*.[23] Similarly, Pattnaik et al.[93] examined the antibacterial perspectives of linalool against Gram-positive bacteria and fungi. It also showed the antibacterial properties against eight diverse human pathogens' Gram-positive and Gram-negative bacteria such as *Streptococcus haemolyticus*, *S. aureus*, *Bacillus subtilis*, *E. coli*, *Pseudomonas aeruginosa*, *Proteus vulgaris*, and *Klebsiella* species.[115] A study conducted by Lis-Balchin et al.[77] determined the effectiveness of linalool (500 ppm) against *Salmonella enteriditis*, *Zygosaccharomyces bailii*, *Saccharomyces ludwigii*, and *Listeria innocua*. The concentrations of coriander oil [less than 0.5% (vol/vol)] observed the minimum inhibitory concentration for *E. coli* (0.23%), *S. aureus* (0.4%), *Listeria monocytogenes* (0.47%), and *Saccharomyces cerevisiae* (0.13%).[38]

8.4.6 ANTIALLERGIC

Allergy especially dermatitis is associated with fragrances fragrance ingredients which are used in scents and other cosmetics products.[36,49] One of the research investigations examined 73 deodorants amongst which 97% contained linalool. In occupational and domestic products, D-limonene-2 and linalool are the most frequently used fragrances.[102] It is well known that linalool does not cause allergic reactions; however, possible allergic effects of linalool containing scented products may be due to the production of reactive substances as a result of oxidation during handling and storage. The nonallergic behavior of linalool is also confirmed from its chemical structure that resembles D-limonene that also lacks allergic activities but generates potent contact allergens upon exposure to air. Further, it has also been observed that diterpene abietic acid is also sensitive to air exposure which is the major constituent of colophonium.[64]

8.4.7 ANXIETY PREVENTION

In developing countries, anxiety disorders have emerged as public health concerns owing to their high prevalence. The situation is further worsened due to associated high-cost pharmacological treatment of anxiety disorders.[15,16,35] In addition, various problems are associated with treatment of anxiety using anxiolytic drugs. Hence, alternative treatments with safer approach are in the limelight of researcher that has driven a keen attention of scientific community to find possible solutions. Some of the therapies like aromatherapy have been found effective in curing chronic pain, anxiety, and depression.[2,16]

Aromatic plants generate strong odors because of EOs which are volatile secondary metabolites.[7,114] In aromatherapy and folk medicines, the use of EOs is quite extensive.[35,125] Amongst various EOs, certain compounds reduced anxiety-related behavior in both animals and humans.[13,16] Especially, linalool has been found to show sedative effects, induce hypothermia, lower ambulation, and elevate pentobarbital-induced sleep time in mice. Furthermore, linalool has also shown anxiolytic properties when examined in aggressive behavior during social interaction tests based on light/dark box test.[76]

One of important compounds present in some herbal EOs is linalool oxide. Chemically, it is monocyclic alcohol arising from natural oxidation of linalool or via synthetic procedures like biotransformation. In biotransformation, linalool is converted upon the action of a fungus known as *Aspergillus niger*.[55,108] Similar to the mechanistic involvement of anticonvulsant and anxiolytic compounds in effecting CNS, the molecular mechanisms of action of linalool are the same. It acts by stimulation of GABAergic pathways and inhibition of excitatory glutamatergic pathways.[18,45,57,113] The negative inflection of adenylate cyclase as affected by linalool has not yet been studied. However, protein kinase cyclic adenosine monophosphate (cAMP)-dependent phosphorylation-based positive modulation of m-methyl-D-aspartate glutamate receptors has been reported in relation to linalool administration.[116] There is only one report on rat cortex membranes which represented absence of linalool effects on cAMP via activation of Gs protein.[45]

8.4.8 ANTI-INFLAMMATORY

Conventionally, numerous traditional medicine systems use linalool, and linalyl acetate-producing species cure and alleviate chronic as well as acute ailments. These preventive properties of these plants are attributed to linalool and its ester linalyl acetate, which provide sufficient pharmacological activities.[94] It has been noticed that permeability biological tissues like mucus membranes and skin to allow drugs can be enhanced using terpenes, terpenoids, and linalool.[27,71,72] Only few reports are available suggesting the anti-inflammatory role of linalool; however, linalool producing plant species are endorsed to acquire virtuous peripheral analgesic action and anti-inflammatory activities.[96]

In an investigation, linalool has been demonstrated to possess significant activity to increase the antioxidant enzymes expression that is regulated by Nrf-2. Furthermore, numerous proinflammatory cytokines like interleukin (IL)-6 and TNF-α were downregulated in lung tissues.[131] Additionally, linalool acquires attenuation power to reduce in vitro and in vivo tumor formation in lipopolysaccharide (LPS)-stimulated RAW 264.7

cells. Linalool suppressed the production of IL-6 and TNF-α simultaneously induced by LPS. Moreover, it also blocked IκBα protein phosphorylation, c-Jun terminal kinase, p38, and ERK in abovementioned cells.[59]

8.4.9 ANTINOCICEPTIVE EFFECTS

Several studies have identified that linalool possess antinociceptive activity and anti-inflammatory role as indicated by experimental animal model trials. It has been found that linalool administration is correlated with inhibition of carrageenan-induced edema in rats[94] and lowered responsive pain induced by various impetuses, such as hot-line plate, formalin injection,[95] and acetic acid-induced pain responses.[98] Furthermore, protective effects against L-glutamate and prostaglandin E2 and hyperalgesia have also been demonstrated.[95]

Antinociceptive properties of linalool are delivered by positive interaction with opioid, muscarinic, and dopaminergic transmission. Additionally, it acquires antioxidant properties[25] and a local anesthetic effect.[48] Linalool positively manages the harmful effects of NO-induced cellular stress. The production and release of NO is triggered by intraplanar administration of formalin and carrageenan in animal models that enhances the nitrosative stress markers at site of injury.[90] Resultantly, NO elevates the production and release of proinflammatory mediators like prostanoids,[109] reactive oxygen species, and cytokines[80] with subsequent promotion of inflammatory responses. Furthermore, peripheral release of NO causes inflammation, hyperalgesia, and development of edema at the site of injury. These negative impacts are promisingly reduced by the administration of linalool due to its anti-inflammatory and antioxidant properties.[42]

8.4.10 SUPPRESSION OF VOLTAGE-GATED CURRENTS

Studies have suggested that linalool inhibits glutamatergic neurons in cerebral cortex of neurons and affects human brain β-wave.[45,119] The chemosensory information is conveyed from olfactory receptor cells (ORCs) to olfactory bulb and express different ionic channels like voltage-gated Na^+, Ca^{2+}, and K^+ channels on their somatic membranes.[112] The deodorants including amyl acetate, acetophenone, and limonene have lower lipid solubility than linalool. The ionic channels are affected by the interaction of somatic membranes, especially lipid molecules.[68,69] Additionally, these volatile substances block voltage-gated and ligand-gated channels including cyclic nucleotide-gated channels and glutamate-gated channels.[69] In neurons of retina, reports have shown that linalool inhibits different voltage-gated currents similar to the effect on ligand-gated currents by odorants in retinal neurons and ORCs.[69,88]

8.4.11 SEDATIVE HYPNOTIC ACTIVITY

Linalool exhibits sedative and hypnotic activity as indicated by increase in sleeping time induced by pentobarbital in comparison to saline-treated experimental subjects. It was found that linalool elevated the pentobarbital-induced sleeping time only at the concentration of 600 mg/kg.[46] Further, it has been noticed that dose-dependent sedative properties are delivered on CNS upon administration with linalool.[19] It also shows protective effects against picrotoxin, pentylenetetrazol (PTZ), transcorneal electroshock-induced convulsions, and hypothermic and hypnotic symptoms.[44,45]

Further reports suggested that it modulates in vivo and in vitro activation of glutamate expression. The in vitro expression activation of glutamate is delivered by competitive antagonism binding of L-[3*H*]–glutamate, whereas delayed blockade of intracerebroventricular quinolinic acid-induced convulsions and subcutaneous *N*-methyl-D-aspartate-induced convulsions.[18,113] In addition, various scientific reports exhibited its anesthetic activity due to its spasmolytic effect and action on nicotinic receptor-ion channel.[23,93,97] Moreover, behavioral expression of PTZ-kindling is significantly impaired by linalool; however, related elevation in L-[3*H*]-glutamate binding in cortical membranes is not modified.[44,73]

8.5 SUMMARY

Interference in the territory of nutraceutical and functional foods mainly focused on detection and isolation of bioactive moieties from plant origin that holds therapeutic prospective. Recently, linalool has been highlighted in assorted systematic researches, probing its therapeutic potential to authorize its remedial significance. Scientific investigations in the era of sustenance led to the detection of phenomena accredited for occurrence and pathogenesis of numerous health disparities. Linalool has high antioxidant activity owing to presence of phenolic compounds. Importance of natural products revitalized currently to alleviate such maladies. Linalool holds a promising place as it possesses valuable health-promoting benefits due to antioxidant, anticarcinogenic, anti-inflammatory, antibacterial, and anti-diabetic properties. Nevertheless, there is adequate evidence in favor of medicinal significance of linalool but needs further scientists' interest to explore its health-endorsing aspects for strengthening the claims as natural therapeutic stimulant.

KEYWORDS

- **bergamot**
- **coriander**
- **diet-based therapy**
- **functional foods**
- **linalool**
- **nutraceuticals**
- **volatile compounds**

REFERENCES

1. Abdelmegeed, M. A.; Moon, K.-H.; Hardwick, J. P.; Gonzalez, F. J.; Song, B.-J. Role of Peroxisome Proliferator-Activated Receptor-α in Fasting-Mediated Oxidative Stress. *Free Radic. Biol. Med.* **2009,** *47* (6), 767–778.
2. Abuhamdah, S.; Chazot, P. L. Lemon Balm and Lavender Herbal Essential Oils: Old and New Ways to Treat Emotional Disorders? *Curr. Anaesth. Crit. Care* **2008,** *19* (4), 221–226.
3. Al-Jedah, J.; Ali, M.; Robinson, R. The Inhibitory Action of Spices Against Pathogens That Might Be Capable of Growth in a Fish Sauce (Mehiawah) from the Middle East. *Int. J. Food Microbiol.* **2000,** *57* (1–2), 129–133.
4. Al-Mashhadani, E. H.; Al-Jaff, F. K.; Hamodi, S. J.; Al-Mashhadani, H. E. Effect of Different Levels of Coriander Oil on Broiler Performance and Some Physiological Traits under Summer Condition. *Pak. J. Nutr.* **2011,** *10* (1), 10–14.
5. Anum, E. A.; Adera, T. Hypercholesterolemia and Coronary Heart Disease in the Elderly: A Meta-analysis. *Ann. Epidemiol.* **2004,** *14* (9), 705–721.
6. Awad, S.; Constantin-Teodosiu, D.; Macdonald, I. A.; Lobo, D. N. Short-Term Starvation and Mitochondrial Dysfunction—A Possible Mechanism Leading to Postoperative Insulin Resistance. *Clin. Nutr.* **2009,** *28* (5), 497–509.

7. Bakkali, F.; Averbeck, S.; Averbeck, D.; Idaomar, M. Biological Effects of Essential Oils—A Review. *Food Chem. Toxicol.* **2008,** *46* (2), 446–475.
8. Barnes, D. J.; Melo, J. V. Primitive, Quiescent and Difficult to Kill: The Role of Non-Proliferating Stem Cells in Chronic Myeloid Leukemia. *Cell Cycle* **2006,** *5* (24), 2862–2866.
9. Bauer, K.; Garbe, D.; Surburg, H. *Common Fragrance and Flavor Materials: Preparation, Properties and Uses*; John Wiley and Sons: Hoboken, NJ, 2008.
10. Behall, K. M.; Scholfield, D. J.; Hallfrisch, J. Diets Containing Barley Significantly Reduce Lipids in Mildly Hypercholesterolemic Men and Women. *Am. J. Clin. Nutr.* **2004,** *80* (5), 1185–1193.
11. Bezard, M.; Karlberg, A.-T.; Montelius, J.; Lepoittevin, J.-P. Skin Sensitization to Linalyl Hydroperoxide: Support for Radical Intermediates. *Chem. Res. Toxicol.* **1997,** *10* (9), 987–993.
12. Boutin, J. A.; Thomassin, J.; Siest, G.; Cartier, A. Heterogeneity of Hepatic Microsomal UDP-Glucuronosyltransferase Activities: Conjugations of Phenolic and Monoterpenoid Aglycones in Control and Induced Rats and Guinea Pigs. *Biochem. Pharmacol.* **1985,** *34* (13), 2235–2249.
13. Bowers, L. To What Extent Does Aromatherapy Use in Palliative Cancer Care Improve Quality of Life and Reduce Levels of Psychological Distress? A Literature Review. *Int. J. Aromather.* **2006,** *16* (1), 27–35.
14. Boyle, W. Spices and Essential Oils as Preservatives. *Am. Perfum. Essent. Oil Rev.* **1955,** *66*, 25–28.
15. Bradley, B.; Starkey, N.; Brown, S.; Lea, R. Anxiolytic Effects of *Lavandula angustifolia* Odour on the Mongolian Gerbil Elevated Plus Maze. *J. Ethnopharmacol.* **2007,** *111* (3), 517–525.
16. Bradley, B. F.; Starkey, N.; Brown, S.; Lea, R. The Effects of Prolonged Rose Odor Inhalation in Two Animal Models of Anxiety. *Physiol. Behav.* **2007,** *92* (5), 931–938.
17. Brown, M. S.; Goldstein, J. L. Multivalent Feedback Regulation of HMG CoA Reductase, a Control Mechanism Coordinating Isoprenoid Synthesis and Cell Growth. *J. Lipid Res.* **1980,** *21* (5), 505–517.
18. Brum, L. S.; Emanuelli, T.; Souza, D.; Elisabetsky, E. Effects of Linalool on Glutamate Release and Uptake in Mouse Cortical Synaptosomes. *Neurochem. Res.* **2001,** *26* (3), 191–194.
19. Buchbauer, G.; Jirovetz, L.; Jäger, W. Aromatherapy: Evidence for Sedative Effects of the Essential Oil of Lavender After Inhalation. *Zeitsch. Naturforsch. C* **1991,** *46* (11–12), 1067–1072.
20. Burdock, G. *Fenaroli's Handbook of Flavor Ingredients*; 5th ed.; CRC Press: London, 2005.
21. Burt, S. Essential Oils: Their Antibacterial Properties and Potential Applications in Foods—A Review. *Int. J. Food Microbiol.* **2004,** *94* (3), 223–253.
22. Cabuk, M.; Alcicek, A.; Bozkurt, M.; Imre, N. Antimicrobial Properties of the Essential Oils Isolated from Aromatic Plants and Using Possibility as Alternative Feed Additives. II. In *National Animal Nutrition Congress*, 2003; pp 18–20.
23. Carson, C.; Riley, T. Antimicrobial Activity of the Major Components of the Essential Oil of *Melaleuca alternifolia*. *J. Appl. Microbiol.* **1995,** *78* (3), 264–269.

24. Casabianca, H.; Graff, J.; Faugier, V.; Fleig, F.; Grenier, C. Enantiomeric Distribution Studies of Linalool and Linalyl Acetate. A Powerful Tool for Authenticity Control of Essential Oils. *J. High Resolut. Chromatogr.* **1998,** *21* (2), 107–112.
25. Çelik, S.; Ozkaya, A. Effects of Intraperitoneally Administered Lipoic Acid, Vitamin E, and Linalool on the Level of Total Lipid and Fatty Acids in Guinea Pig Brain with Oxidative Stress Induced by H_2O_2. *BMB Rep.* **2002,** *35* (6), 547–552.
26. Cerchiara, T.; Straface, S.; Brunelli, E.; Tripepi, S.; Gallucci, M. C.; Chidichimo, G. Antiproliferative Effect of Linalool on RPMI 7932 Human Melanoma Cell Line: Ultrastructural Studies. *Nat. Prod. Commun.* **2015,** *10* (4), 547–549.
27. Ceschel, G.; Maffei, P.; Moretti, M.; Demontis, S.; Peana, A. *In vitro* Permeation through Porcine Buccal Mucosa of *Salvia desoleana* Atzei & Picci Essential Oil from Topical Formulations. *Int. J. Pharm.* **2000,** *195* (1–2), 171–177.
28. Chang, M.-Y.; Shieh, D.-E.; Chen, C.-C.; Yeh, C.-S.; Dong, H.-P. Linalool Induces Cell Cycle Arrest and Apoptosis in Leukemia Cells and Cervical Cancer Cells through CDKIs. *Int. J. Mol. Sci.* **2015,** *16* (12), 28169–28179.
29. Chen, Q.; Xie, W.; Kuhn, D. J.; Voorhees, P. M.; Lopez-Girona, A.; Mendy, D.; Corral, L. G.; Krenitsky, V. P.; Xu, W.; Moutouh-De Parseval, L. Targeting the p27 E3 Ligase SCFSkp2 Results in p27- and Skp2-Mediated Cell-Cycle Arrest and Activation of Autophagy. *Blood* **2008,** *111* (9), 4690–4699.
30. Cheng, Y.; Dai, C.; Zhang, J. SIRT3–SOD2–ROS Pathway Is Involved in Linalool-Induced Glioma Cell Apoptotic Death. *Acta Biochim. Pol.* **2017,** *64* (2), 343–350.
31. Chiang, L.-C.; Chiang, W.; Chang, M.-Y.; Ng, L.-T.; Lin, C.-C. Antileukemic Activity of Selected Natural Products in Taiwan. *Am. J. Chin. Med.* **2003,** *31* (01), 37–46.
32. Chung, M. J.; Kang, A.-Y.; Park, S.-O.; Park, K.-W.; Jun, H.-J.; Lee, S.-J. The Effect of Essential Oils of Dietary Wormwood (*Artemisia princeps*), with and without Added Vitamin E, on Oxidative Stress and Some Genes Involved in Cholesterol Metabolism. *Food Chem. Toxicol.* **2007,** *45* (8), 1400–1409.
33. Chung, M. J.; Park, K. W.; Kim, K. H.; Kim, C.-T.; Baek, J. P.; Bang, K.-H.; Choi, Y.-M.; Lee, S.-J. Asian Plantain (*Plantago asiatica*) Essential Oils Suppress 3-Hydroxy-3-Methyl-Glutaryl-*co*-Enzyme A Reductase Expression In Vitro and In Vivo and Show Hypocholesterolaemic Properties in Mice. *Br. J. Nutr.* **2008,** *99* (1), 67–75.
34. Davidson, P.; Naidu, A. *Natural Food Antimicrobial Systems: Phytophenols*; CRC Press: Boca Raton, FL, 2000.
35. De Almeida, R. N.; Motta, S. C.; De Brito Faturi, C.; Catallani, B.; Leite, J. R. Anxiolytic-Like Effects of Rose Oil Inhalation on the Elevated Plus-Maze Test in Rats. *Pharmacol. Biochem. Behav.* **2004,** *77* (2), 361–364.
36. De Groot, A. C.; Nater, J. P.; Weyland, J. W. *Unwanted Effects of Cosmetics and Drugs Used in Dermatology*; Elsevier: Amsterdam, 1994.
37. Deepa, B.; Venkatraman Anuradha, C. Effects of Linalool on Inflammation, Matrix Accumulation and Podocyte Loss in Kidney of Streptozotocin-Induced Diabetic Rats. *Toxicol. Mech. Methods* **2013,** *23* (4), 223–234.
38. Delaquis, P. J.; Stanich, K.; Girard, B.; Mazza, G. Antimicrobial Activity of Individual and Mixed Fractions of Dill, Cilantro, Coriander and Eucalyptus Essential Oils. *Int. J. Food Microbiol.* **2002,** *74* (1–2), 101–109.

39. Di Pasqua, R.; Hoskins, N.; Betts, G.; Mauriello, G. Changes in Membrane Fatty Acids Composition of Microbial Cells Induced by Addiction of Thymol, Carvacrol, Limonene, Cinnamaldehyde, and Eugenol in the Growing Media. *J. Agric. Food Chem.* **2006,** *54* (7), 2745–2749.
40. Diederichsen, A.; Hammer, K. Vielfalt von koriander im Weltsortiment der Genbank Gatersleben. *Drogenreport* **1994,** *7* (11), 13–17.
41. Dorman, H.; Deans, S. G. Antimicrobial Agents from Plants: Antibacterial Activity of Plant Volatile Oils. *J. Appl. Microbiol.* **2000,** *88* (2), 308–316.
42. Dudhgaonkar, S. P.; Kumar, D.; Naik, A.; Devi, A. R.; Bawankule, D. U.; Tandan, S. K. Interaction of Inducible Nitric Oxide Synthase and Cyclooxygenase-2 Inhibitors in Formalin-Induced Nociception in Mice. *Eur. J. Pharmacol.* **2004,** *492* (2–3), 117–122.
43. Eckel, R. H.; York, D. A.; Rössner, S.; Hubbard, V.; Caterson, I.; Jeor, S. T. S.; Hayman, L. L.; Mullis, R. M.; Blair, S. N. Prevention Conference VII: Obesity, a Worldwide Epidemic Related to Heart Disease and Stroke: Executive Summary. *Circulation* **2004,** *110* (18), 2968–2975.
44. Elisabetsky, E.; Brum, L. S.; Souza, D. Anticonvulsant Properties of Linalool in Glutamate-Related Seizure Models. *Phytomedicine* **1999,** *6* (2), 107–113.
45. Elisabetsky, E.; Coelho De Souza, G.; Dos Santos, M.; Siquieira, I.; Amador, T.; Nunes, D. Sedative Properties of Linalool. *Fitoterapia* **1995,** *66* (5), 407–414.
46. Emamghoreishi, M.; Khasaki, M.; Aazam, M. F. *Coriandrum sativum*: Evaluation of Its Anxiolytic Effect in the Elevated Plus-Maze. *J. Ethnopharmacol.* **2005,** *96* (3), 365–370.
47. Gafvert, E.; Shao, L. P.; Karlberg, A.-T.; Nilsson, U.; Nilsson, J. L. G. Contact Allergy to Resin Acid Hydroperoxides. Hapten Binding via Free Radicals and Epoxides. *Chem. Res. Toxicol.* **1994,** *7* (2), 260–266.
48. Ghelardini, C.; Galeotti, N.; Matucci, R.; Bellucci, C.; Gualtieri, F.; Capaccioli, S.; Quattrone, A.; Bartolini, A. Antisense 'Knockdowns' of M1 Receptors Induces Transient Anterograde Amnesia in Mice. *Neuropharmacology* **1999,** *38* (3), 339–348.
49. Groot, A. C.; Frosch, P. J. Adverse Reactions to Fragrances. *Contact Derm.* **1997,** *36* (2), 57–86.
50. Gunaseelan, S.; Balupillai, A.; Govindasamy, K.; Muthusamy, G.; Ramasamy, K.; Shanmugam, M.; Prasad, N. R. The Preventive Effect of Linalool on Acute and Chronic UVB-Mediated Skin Carcinogenesis in Swiss Albino Mice. *Photochem. Photobiol. Sci.* **2016,** *15* (7), 851–860.
51. Gunaseelan, S.; Balupillai, A.; Govindasamy, K.; Ramasamy, K.; Muthusamy, G.; Shanmugam, M.; Thangaiyan, R.; Robert, B. M.; Nagarajan, R. P.; Kandan Ponniresan, V. Linalool Prevents Oxidative Stress Activated Protein Kinases in Single UVB-Exposed Human Skin Cells. *PLoS ONE,* **2017,** *12* (5), e0176699.
52. Guzman, M. L.; Rossi, R. M.; Karnischky, L.; Li, X.; Peterson, D. R.; Howard, D. S.; Jordan, C. T. The Sesquiterpene Lactone Parthenolide Induces Apoptosis of Human Acute Myelogenous Leukemia Stem and Progenitor Cells. *Blood* **2005,** *105* (11), 4163–4169.
53. Guzman, M. L.; Swiderski, C. F.; Howard, D. S.; Grimes, B. A.; Rossi, R. M.; Szilvassy, S. J.; Jordan, C. T. Preferential Induction of Apoptosis for Primary Human Leukemic Stem Cells. *Proc. Natl. Acad. Sci.* **2002,** *99* (25), 16220–16225.

54. Hamodi, S. J.; Al-Mashhadani, E. H.; Al-Jaff, F. K.; Al-Mashhadani, H. E. Effect of Coriander Seed (*Coriandrum sativum* L.) as Diet Ingredient on Broilers Performance under High Ambient Temperature. *Int. J. Poultry Sci.* **2010,** *9* (10), 968–971.
55. Hilmer, J. M.; Gatfield, I. L. Process for the Preparation of Linalool Oxide or of Linalool Oxide-Containing Mixtures. Google Patents, 2004.
56. Holtz, M.; Forman, S.; Bhatia, R. Nonproliferating CML CD34+ Progenitors Are Resistant to Apoptosis Induced by a Wide Range of Proapoptotic Stimuli. *Leukemia* **2005,** *19* (6), 1034.
57. Hossain, S. J.; Aoshima, H.; Koda, H.; Kiso, Y. Fragrances in Oolong Tea That Enhance the Response of GABAA Receptors. *Biosci., Biotechnol. Biochem.* **2004,** *68* (9), 1842–1848.
58. Hulin, V.; Mathot, A.-G.; Mafart, P.; Dufosse, L. Les Proprietés Anti-microbiennes des Huiles Essentielles et Composés D'arômes. *Sci. Alim.* **1998,** *18* (6), 563–582.
59. Huo, M.; Cui, X.; Xue, J.; Chi, G.; Gao, R.; Deng, X.; Guan, S.; Wei, J.; Soromou, L. W.; Feng, H. Anti-inflammatory Effects of Linalool in RAW 264.7 Macrophages and Lipopolysaccharide-Induced Lung Injury Model. *J. Surg. Res.* **2013,** *180* (1), e47–e54.
60. Isao, K.; Ken-Ichi, F.; Aya, K.; Ken-Ichi, N.; Tetsuya, A. Antibacterial Activity of Coriander Volatile Compounds against *Salmonella choleraesuits*. *J. Agric. Food Chem.* **2004,** *52*, 3329–3332.
61. Isman, M. B. Plant Essential Oils for Pest and Disease Management. *Crop Protect.* **2000,** *19* (8–10), 603–608.
62. Iwasaki, K.; Zheng, Y.-W.; Murata, S.; Ito, H.; Nakayama, K.; Kurokawa, T.; Sano, N.; Nowatari, T.; Villareal, M. O.; Nagano, Y. N. Anticancer Effect of Linalool via Cancer-Specific Hydroxyl Radical Generation in Human Colon Cancer. *World J. Gastroenterol.* **2016,** *22* (44), 9765.
63. Karlberg, A.-T.; Magnusson, K.; Nilsson, U. Influence of an Anti-Oxidant on the Formation of Allergenic Compounds During Auto-oxidation of d-Limonene. *Ann. Occup. Hyg.* **1994,** *38* (2), 199–207.
64. Karlberg, A. T.; Bergstedt, E.; Boman, A.; Bohlinder, K.; Lidén, C.; Lars, J.; Nilsson, G.; Wahlberg, J. E. Is Abietic Acid the Allergenic Component of Colophony? *Contact Derm.* **1985,** *13* (4), 209–215.
65. Karlberg, A. T.; Bohlinder, K.; Boman, A.; Hacksell, U.; Hermansson, J.; Jacobsson, S.; Nilsson, J. L. G. Identification of 15-Hydroperoxyabietic Acid as a Contact Allergen in Portuguese Colophony. *J. Pharmacy Pharmacol.* **1988,** *40* (1), 42–47.
66. Karlberg, A. T.; Dooms-Goossens, A. Contact Allergy to Oxidized d-Limonene among Dermatitis Patients. *Contact Derm.* **1997,** *36* (4), 201–206.
67. Karlberg, A. T.; Magnusson, K.; Nilsson, U. Air Oxidation of d-Limonene (the Citrus Solvent) Creates Potent Allergens. *Contact Derm.* **1992,** *26* (5), 332–340.
68. Kawai, F. Odorants Suppress T-and L-Type Ca^{2+} Currents in Olfactory Receptor Cells by Shifting Their Inactivation Curves to a Negative Voltage. *Neurosci. Res.* **1999,** *35* (4), 253–263.
69. Kawai, F.; Miyachi, E.-I. Direct Suppression by Odorants of Cyclic Nucleotide-Gated Currents in the NEWT Photoreceptors. *Brain Res.* **2000,** *876* (1–2), 180–184.

70. Kaya, N.; Yılmaz, G.; Telci, İ. Farklı zamanlarda ekilen kişniş (*Coriandrum sativum* L.) populasyonlarının agronomik ve teknolojik özellikleri. *Turk. J. Agric. Forest.* **2000,** *24*, 355–364.
71. Kommuru, T. R.; Khan, M. A.; Reddy, I. K. Racemate and Enantiomers of Ketoprofen: Phase Diagram, Thermodynamic Studies, Skin Permeability, and Use of Chiral Permeation Enhancers. *J. Pharm. Sci.* **1998,** *87* (7), 833–840.
72. Kunta, J. R.; Goskonda, V. R.; Brotherton, H. O.; Khan, M. A.; Reddy, I. K. Effect of Menthol and Related Terpenes on the Percutaneous Absorption of Propranolol across Excised Hairless Mouse Skin. *J. Pharm. Sci.* **1997,** *86* (12), 1369–1373.
73. Kuroda, K.; Inoue, N.; Ito, Y.; Kubota, K.; Sugimoto, A.; Kakuda, T.; Fushiki, T. Sedative Effects of the Jasmine Tea Odor and (*R*)-(–)-Linalool, One of Its Major Odor Components, on Autonomic Nerve Activity and Mood States. *Eur. J. Appl. Physiol.* **2005,** *95* (2–3), 107–114.
74. Lambert, R.; Skandamis, P. N.; Coote, P. J.; Nychas, G. J. A Study of the Minimum Inhibitory Concentration and Mode of Action of Oregano Essential Oil, Thymol and Carvacrol. *J. Appl. Microbiol.* **2001,** *91* (3), 453–462.
75. Lewinsohn, E.; Schalechet, F.; Wilkinson, J.; Matsui, K.; Tadmor, Y.; Nam, K.-H.; Amar, O.; Lastochkin, E.; Larkov, O.; Ravid, U. Enhanced Levels of the Aroma and Flavor Compound S-Linalool by Metabolic Engineering of the Terpenoid Pathway in Tomato Fruits. *Plant Physiol.* **2001,** *127* (3), 1256–1265.
76. Linck, V.; Da Silva, A.; Figueiró, M.; Caramão, E.; Moreno, P.; Elisabetsky, E. Effects of Inhaled Linalool in Anxiety, Social Interaction and Aggressive Behavior in Mice. *Phytomedicine* **2010,** *17* (8–9), 679–683.
77. Lis-Balchin, M.; Buchbauer, G.; Hirtenlehner, T.; Resch, M. Antimicrobial Activity of *Pelargonium* Essential Oils Added to a Quiche-Filling as a Model Food System. *Lett. Appl. Microbiol.* **1998,** *27* (4), 207–210.
78. Lusis, A. J. Atherosclerosis. *Nature* **2000,** *407*, 233–241.
79. Mantena, S. K.; Sharma, S. D.; Katiyar, S. K. Berberine Inhibits Growth, Induces G1 Arrest and Apoptosis in Human Epidermoid Carcinoma A431 Cells by Regulating CDKI–CDK–Cyclin Cascade, Disruption of Mitochondrial Membrane Potential and Cleavage of Caspase 3 and PARP. *Carcinogenesis* **2006,** *27* (10), 2018–2027.
80. Marcinkiewicz, J.; Grabowska, A.; Chain, B. Nitric Oxide Up-regulates the Release of Inflammatory Mediators by Mouse Macrophages. *Eur. J. Immunol.* **1995,** *25* (4), 947–951.
81. Marino, M.; Bersani, C.; Comi, G. Impedance Measurements to Study the Antimicrobial Activity of Essential Oils from Lamiaceae and Compositae. *Int. J. Food Microbiol.* **2001,** *67* (3), 187–195.
82. Máté, G.; Kovács, D.; Gazdag, Z.; Pesti, M.; Szántó, Á. Linalool-Induced Oxidative Stress Processes in the Human Pathogen *Candida albicans*. *Acta Biol. Hung.* **2017,** *68* (2), 220–231.
83. Matura, M.; Goossens, A.; Bordalo, O.; Garcia-Bravo, B.; Magnusson, K.; Wrangsjö, K.; Karlberg, A. T. Patch Testing with Oxidized R-(+)-Limonene and Its Hydroperoxide Fraction. *Contact Derm.* **2003,** *49* (1), 15–21.
84. Mitić-Ćulafić, D.; Žegura, B.; Nikolić, B.; Vuković-Gačić, B.; Knežević-Vukčević, J.; Filipič, M. Protective Effect of Linalool, Myrcene and Eucalyptol against *t*-Butyl

Hydroperoxide Induced Genotoxicity in Bacteria and Cultured Human Cells. *Food Chem. Toxicol.* **2009,** *47* (1), 260–266.
85. Mutterer, V.; Arnau, E. G.; Karlberg, A.-T.; Lepoittevin, J.-P. Synthesis and Allergenic Potential of a 15-Hydroperoxyabietic Acid-Like Model: Trapping of Radical Intermediates. *Chem. Res. Toxicol.* **2000,** *13* (10), 1028–1036.
86. Nygren, J. The Metabolic Effects of Fasting and Surgery. *Best Pract. Res. Clin. Anaesthesiol.* **2006,** *20* (3), 429–438.
87. O'mahony, R.; Al-Khtheeri, H.; Weerasekera, D.; Fernando, N.; Vaira, D.; Holton, J.; Basset, C. Bactericidal and Anti-adhesive Properties of Culinary and Medicinal Plants against *Helicobacter pylori*. *World J. Gastroenterol.* **2005,** *11* (47), 7499.
88. Ohkuma, M.; Kawai, F.; Miyachi, E. Direct Suppression by Odorants of Ionotropic Glutamate Receptors in NEWT Retinal Neurons. *J. Neur. Transm.* **2002,** *109* (11), 1365–1371.
89. Olagnier, D.; Costes, P.; Berry, A.; Linas, M.-D.; Urrutigoity, M.; Dechy-Cabaret, O.; Benoit-Vical, F. Modifications of the Chemical Structure of Terpenes in Antiplasmodial and Antifungal Drug Research. *Bioorg. Med. Chem. Lett.* **2007,** *17* (22), 6075–6078.
90. Omote, K.; Hazama, K.; Kawamata, T.; Kawamata, M.; Nakayaka, Y.; Toriyabe, M.; Namiki, A. Peripheral Nitric Oxide in Carrageenan-Induced Inflammation. *Brain Res.* **2001,** *912* (2), 171–175.
91. Oussalah, M.; Caillet, S.; Lacroix, M. Mechanism of Action of Spanish Oregano, Chinese Cinnamon, and Savory Essential Oils against Cell Membranes and Walls of *Escherichia coli* O157:H7 and *Listeria monocytogenes*. *J. Food Prot.* **2006,** *69* (5), 1046–1055.
92. Paik, S.-Y.; Koh, K.-H.; Beak, S.-M.; Paek, S.-H.; Kim, J.-A. The Essential Oils from *Zanthoxylum schinifolium* Pericarp Induce Apoptosis of HepG2 Human Hepatoma Cells through Increased Production of Reactive Oxygen Species. *Biol. Pharm. Bull.* **2005,** *28* (5), 802–807.
93. Pattnaik, S.; Subramanyam, V.; Bapaji, M.; Kole, C. Antibacterial and Antifungal Activity of Aromatic Constituents of Essential Oils. *Microbios* **1997,** *89* (358), 39–46.
94. Peana, A. T.; D'aquila, P. S.; Panin, F.; Serra, G.; Pippia, P.; Moretti, M. D. L. Anti-Inflammatory Activity of Linalool and Linalyl Acetate Constituents of Essential Oils. *Phytomedicine* **2002,** *9* (8), 721–726.
95. Peana, A. T.; De Montis, M. G.; Sechi, S.; Sircana, G.; Paolo, S.; Pippia, P. Effects of (−)-Linalool in the Acute Hyperalgesia Induced by Carrageenan, l-Glutamate and Prostaglandin E2. *Eur. J. Pharmacol.* **2004,** *497* (3), 279–284.
96. Peana, A. T.; Moretti, M. D. Pharmacological Activities and Applications of *Salvia sclarea* and *Salvia desoleana* Essential Oils. *Studies in Natural Products Chemistry*; Elsevier: Amsterdam, 2002.
97. Peana, A. T.; Moretti, M. D.; Juliano, C. Chemical Composition and Antimicrobial Action of the Essential Oils of *Salvia desoleana* and *S. sclarea*. *Planta Med.* **1999,** *65* (8), 752–754.
98. Peana, A. T.; Paolo, S. D.; Chessa, M. L.; Moretti, M. D.; Serra, G.; Pippia, P. (−)-Linalool Produces Antinociception in Two Experimental Models of Pain. *Eur. J. Pharmacol.* **2003,** *460* (1), 37–41.

99. Potwale, S. E.; Mantri, R. A.; Luniya, K. P.; Mehta, U. K.; Waseem, M.; Sadiq, M.; Dhalawat, H. J.; Deshmukh, R. S. *Camellia sinensis*: An Ethnopharmacological Review. *Pharmacol. Online* **2008,** *3*, 1–25.
100. Prasad, M.; Seenayya, G. Effect of Spices on the Growth of Red Halophilic Cocci Isolated from Salt Cured Fish and Solar Salt. *Food Res. Int.* **2000,** *33* (9), 793–798.
101. Ramadan, M. F.; Moersel, J. T. Screening of the Antiradical Action of Vegetable Oils. *J. Food Compos. Anal.* **2006,** *19* (8), 838–842.
102. Rastogi, S. C.; Heydorn, S.; Johansen, J.; Basketter, D. Fragrance Chemicals in Domestic and Occupational Products. *Contact Derm.* **2001,** *45* (4), 221–225.
103. Ravizza, R.; Gariboldi, M. B.; Molteni, R.; Monti, E. Linalool, a Plant-Derived Monoterpene Alcohol, Reverses Doxorubicin Resistance in Human Breast Adenocarcinoma Cells. *Oncol. Rep.* **2008,** *20* (3), 625–630.
104. Renouf, B.; Hollville, E.; Pujals, A.; Tetaud, C.; Garibal, J.; Wiels, J. Activation of p53 by MDM2 Antagonists Has Differential Apoptotic Effects on Epstein–Barr Virus (EBV)-Positive and EBV-Negative Burkitt's Lymphoma Cells. *Leukemia* **2009,** *23* (9), 1557.
105. Rodenak-Kladniew, B.; Castro, A.; Stärkel, P.; De Saeger, C.; De Bravo, M. G.; Crespo, R. Linalool Induces Cell Cycle Arrest and Apoptosis in HepG2 Cells through Oxidative Stress Generation and Modulation of Ras/MAPK and Akt/mTOR Pathways. *Life Sci.* **2018,** *199*, 48–59.
106. Ross, Z.; O'gara, E. A.; Hill, D. J.; Sleightholme, H.; Maslin, D. J. Antimicrobial Properties of Garlic Oil Against Human Enteric Bacteria: Evaluation of Methodologies and Comparisons with Garlic Oil Sulfides and Garlic Powder. *Appl. Environ. Microbiol.* **2001,** *67* (1), 475–480.
107. Ryan, C. A.; Gerberick, G.; Cruse, L.; Basketter, D. A.; Lea, L.; Blaikie, L.; Dearman, R. J.; Warbrick, E.; Kimber, I. Activity of Human Contact Allergens in the Murine Local Lymph Node Assay. *Contact Derm.* **2000,** *43* (2), 95–102.
108. Santos, F.; Silva, R.; Campos, A.; De Araujo, R.; Júnior, R. L.; Rao, V. 1,8-Cineole (Eucalyptol), a Monoterpene Oxide Attenuates the Colonic Damage in Rats on Acute TNBS-Colitis. *Food Chem. Toxicol.* **2004,** *42* (4), 579–584.
109. Sautebin, L.; Ialenti, A.; Ianaro, A.; Rosa, M. Modulation by Nitric Oxide of Prostaglandin Biosynthesis in the Rat. *Br. J. Pharmacol.* **1995,** *114* (2), 323–328.
110. Schaller, M.; Korting, H. Allergic Airborne Contact Dermatitis from Essential Oils Used in Aromatherapy. *Clin. Exp. Dermatol.* **1995,** *20* (2), 143–145.
111. Scheline, R. R. *Handbook of Mammalian Metabolism of Plant Compounds*; CRC Press: Boca Raton, FL, 1991.
112. Schild, D.; Restrepo, D. Transduction Mechanisms in Vertebrate Olfactory Receptor Cells. *Physiol. Rev.* **1998,** *78* (2), 429–466.
113. Silva Brum, L.; Elisabetsky, E.; Souza, D. Effects of Linalool on [3*H*] MK801 and [3*H*] Muscimol Binding in Mouse Cortical Membranes. *Phytother. Res.* **2001,** *15* (5), 422–425.
114. Silva, M. I. G.; Silva, M. a. G.; De Aquino Neto, M. R.; Moura, B. A.; De Sousa, H. L.; De Lavor, E. P. H.; De Vasconcelos, P. F.; Macêdo, D. S.; De Sousa, D. P.; Vasconcelos, S. M. M. Effects of Isopulegol on Pentylenetetrazol-Induced Convulsions in Mice: Possible Involvement of GABAergic System and Antioxidant Activity. *Fitoterapia* **2009,** *80* (8), 506–513.

115. Singh, G.; Kapoor, I.; Pandey, S.; Singh, U.; Singh, R. Studies on Essential Oils: Part 10; Antibacterial Activity of Volatile Oils of Some Spices. *Phytother. Res.* **2002,** *16* (7), 680–682.
116. Skeberdis, V. A.; Chevaleyre, V.; Lau, C. G.; Goldberg, J. H.; Pettit, D. L.; Suadicani, S. O.; Lin, Y.; Bennett, M. V.; Yuste, R.; Castillo, P. E. Protein Kinase A Regulates Calcium Permeability of NMDA Receptors. *Nat. Neurosci.* **2006,** *9* (4), 501.
117. Sköld, M.; Börje, A.; Harambasic, E.; Karlberg, A.-T. Contact Allergens Formed on Air Exposure of Linalool. Identification and Quantification of Primary and Secondary Oxidation Products and the Effect on Skin Sensitization. *Chem. Res. Toxicol.* **2004,** *17* (12), 1697–1705.
118. Sugawara, Y.; Hara, C.; Aoki, T.; Sugimoto, N.; Masujima, T. Odor Distinctiveness between Enantiomers of Linalool: Difference in Perception and Responses Elicited by Sensory Test and Forehead Surface Potential Wave Measurement. *Chem. Sens.* **2000,** *25* (1), 77–84.
119. Sugawara, Y.; Hara, C.; Tamura, K.; Fujii, T.; Nakamura, K.-I.; Masujima, T.; Aoki, T. Sedative Effect on Humans of Inhalation of Essential Oil of Linalool: Sensory Evaluation and Physiological Measurements Using Optically Active Linalools. *Anal. Chim. Acta* **1998,** *365* (1–3), 293–299.
120. Tahraoui, A.; El-Hilaly, J.; Israili, Z.; Lyoussi, B. Ethnopharmacological Survey of Plants Used in the Traditional Treatment of Hypertension and Diabetes in South-Eastern Morocco (Errachidia Province). *J. Ethnopharmacol.* **2007,** *110* (1), 105–117.
121. Tassou, C.; Koutsoumanis, K.; Nychas, G.-J. Inhibition of *Salmonella enteritidis* and *Staphylococcus aureus* in Nutrient Broth by Mint Essential Oil. *Food Res. Int.* **2000,** *33* (3–4), 273–280.
122. Thyagaraja, N.; Hosono, A. Effect of Spice Extract on Fungal Inhibition. *LWT—Food Sci. Technol.* **1996,** *29* (3), 286–288.
123. Tovar, C.; Rosinski, J.; Filipovic, Z.; Higgins, B.; Kolinsky, K.; Hilton, H.; Zhao, X.; Vu, B. T.; Qing, W.; Packman, K. Small-Molecule MDM2 Antagonists Reveal Aberrant p53 Signaling in Cancer: Implications for Therapy. *Proc. Nat. Acad. Sci. USA.* **2006,** *103* (6), 1888–1893.
124. Turina, A. D. V.; Nolan, M.; Zygadlo, J.; Perillo, M. Natural Terpenes: Self-Assembly and Membrane Partitioning. *Biophys. Chem.* **2006,** *122* (2), 101–113.
125. Umezu, T.; Ito, H.; Nagano, K.; Yamakoshi, M.; Oouchi, H.; Sakaniwa, M.; Morita, M. Anticonflict Effects of Rose Oil and Identification of Its Active Constituents. *Life Sci.* **2002,** *72* (1), 91–102.
126. Usta, J.; Kreydiyyeh, S.; Knio, K.; Barnabe, P.; Bou-Moughlabay, Y.; Dagher, S. Linalool Decreases HepG2 Viability by Inhibiting Mitochondrial Complexes I and II, Increasing Reactive Oxygen Species and Decreasing ATP and GSH Levels. *Chem.-Biol. Interact.* **2009,** *180* (1), 39–46.
127. Valero, M.; Salmeron, M. Antibacterial Activity of 11 Essential Oils against *Bacillus cereus* in Tyndallized Carrot Broth. *Int. J. Food Microbiol.* **2003,** *85* (1–2), 73–81.
128. Vassilev, L. T.; Vu, B. T.; Graves, B.; Carvajal, D.; Podlaski, F.; Filipovic, Z.; Kong, N.; Kammlott, U.; Lukacs, C.; Klein, C. *In Vivo* Activation of the p53 Pathway by Small-Molecule Antagonists of MDM2. *Science* **2004,** *303* (5659), 844–848.

129. Vasudevan, K.; Vembar, S.; Veeraraghavan, K.; Haranath, P. Influence of Intragastric Perfusion of Aqueous Spice Extracts on Acid Secretion in Anesthetized Albino Rats. *Indian J. Gastroenterol.* **2000,** *19* (2), 53–56.
130. Velluti, A.; Sanchis, V.; Ramos, A.; Egido, J.; Marın, S. Inhibitory Effect of Cinnamon, Clove, Lemongrass, Oregano and Palmarose Essential Oils on Growth and Fumonisin B1 Production by *Fusarium proliferatum* in Maize Grain. *Int. J. Food Microbiol.* **2003,** *89* (2–3), 145–154.
131. Wu, Q.; Yu, L.; Qiu, J.; Shen, B.; Wang, D.; Soromou, L. W.; Feng, H. Linalool Attenuates Lung Inflammation Induced by *Pasteurella multocida* via Activating Nrf-2 Signaling Pathway. *Int. Immunopharmacol.* **2014,** *21* (2), 456–463.
132. Xu, P.; Wang, K.; Lu, C.; Dong, L.; Gao, L.; Yan, M.; Aibai, S.; Yang, Y.; Liu, X. The Protective Effect of Lavender Essential Oil and Its Main Component Linalool Against the Cognitive Deficits Induced by d-Galactose and Aluminum Trichloride in Mice. *Evid. Based Complem. Altern. Med.* **2017,** *2017*, 11.
133. Xu, R.; Yu, Y.; Zheng, S.; Zhao, X.; Dong, Q.; He, Z.; Liang, Y.; Lu, Q.; Fang, Y.; Gan, X. Overexpression of Shp2 Tyrosine Phosphatase Is Implicated in Leukemogenesis in Adult Human Leukemia. *Blood* **2005,** *106* (9), 3142–3149.

CHAPTER 9

GARLIC: FROM NUTRITIONAL TO NUTRACEUTICAL VIEWPOINT

KANZA AZIZ AWAN[1], MASOOD SADIQ BUTT[1], HUSSAN MUNIR[2], and HAFIZ ANSAR RASUL SULERIA[3,*]

[1]*National Institute of Food Science and Technology, Faculty of Food, Nutrition & Home Sciences, University of Agriculture, Faisalabad, Pakistan*

[2]*Al-Shaheer Foods, Karachi, Pakistan*

[3]*School of Life and Environmental Sciences, Faculty of Science, Engineering and Built Environment, Centre for Chemistry and Biotechnology, Deakin University, 75 Pigdons Road, Geelong, VIC 3216, Australia*

[*]*Corresponding author. E-mail: hafiz.suleria@uqconnect.edu.au*

ABSTRACT

Lifestyle-related malfunctions, the "diseases of affluence/civilization," are emerging largely because of changing living pattern. Due to inflation in medical care cost, nutritionists and health-care professionals have been forced to explicate the notion of functional foods and nutraceuticals. Nutraceuticals are substances considered as food or components of food-imparting medicinal or health benefits encompassing prevention or management of diseases. Recent research has shown beneficial therapeutic properties of various traditional plants including garlic, ginger, green tea, black cumin, and onion. The diet-based therapies can be helpful in mitigating various lifestyle-related disparities. In this context, garlic is a powerhouse of nutraceuticals known to possess certain beneficial compounds, such as allicin, that are helpful in ameliorating numerous physiological threats. This chapter includes the potential applications and benefits of garlic and its bioactive moieties. The extraction technologies and conditions have also been discussed for better isolation of biologically active agents. Moreover, the prophylactic role of garlic and its major nutraceutical agent allicin against hyperglycemia and dyslipidemia have also been discussed at the end.

9.1 INTRODUCTION

9.1.1 *DIET AND HEALTH INTERACTION*

Modern era is characterized by health dysregulation due to behavioral and environmental changes. Dietary habits play an imperative role in the occurrence of various physiologic malfunctions like cardiovascular ailments, hyperlipidemia, diabetes, atherosclerosis, thrombosis, and hypertension.[21] The vast diversity in human diet cannot be apprehended due to the socioeconomic, ethnic, regional, and environmental factors. With the rapidly increasing occurrence rates of acute as well as chronic disparities, nutritional therapies are emerging to a great extent. Food, as a biological system, in itself constitutes an array of bioactive molecules that interact with human biology in several ways. The prevention of chronic diseases requires a careful knowhow of nutritional environment as well as genes interaction.

Food and nutrients are the original medicines and the shoulders on which contemporary pharmaceutical drugs stand. Due to the hypnotizing impact of profit-oriented market economy, diet and medicine have taken divergent pathways. The natural curing properties of food have been diminished in the wake of present technological advancements. Nevertheless, the role of diet and nutrition is progressively becoming an important public health concern these days.[3] Diet remains a feasible and logical approach to impart favorable conditions for normal body functioning.[96] The use of natural medicinal products and herbs has gained immense popularity and is backed by solid scientific evidences.[51,93,116] Nutritional therapy has turned out to be a core element of disease-managing regime. Herbs and spices, having strong role in cultural heritage and food appreciation, also hold pivotal health linkages. Demonstrating the food benefits is still a challenge, especially when compared with pharmaceuticals. These agents are of small molecular weight, purified as well as concentrated to provide immediate relief, whereas food is eaten in combination form often in large quantities. Therefore, purifying and concentrating food components, making a pharma food would be more effective and comparable to the chemical entities consumed in the form of medicines.[112,123] Unhealthy diet is the primary contributor of lifestyle-related ailments.[27] Lifestyle behaviors, deskbound living, increased tobacco consumption, moving away from natural foods, elevated stress levels, and obesity are some major pandemics of morbidity and premature death.[38,45] Various medicinal plants are nowadays being used for health promotion and disease prevention. The bioactive entities in these natural compounds are worth considering due to their prophylactic properties. In this millennium, natural products are gaining more popularity for their effectiveness in averting physiological threats owing to their safety and fewer side effects as compared to pharmaceutical counterparts. Diet-induced infirmities such as hyperglycemia, hypercholesterolemia, obesity, and related disorders can be effectively controlled using beneficial diet-based therapies involving the use of natural foods possessing ample amounts of phytonutrients to ensure a healthy living.[6] For instance, herbs and spices possess a number of health-boosting phytomedicinal compounds and onions contain quercetin, a disease-remedial agent. Similarly, ginger constitutes gingerol, that is, helpful in fighting various ailments. Likewise, garlic has revealed magical phytotherapeutic properties. It lowers the levels of cholesterol and blood pressure, prevents heart disease, delays progression of atherosclerosis, prevents cancer, reduces infections, and improves the blood circulation.[56]

9.2 GARLIC: A NATURAL PANACEA

Garlic, scientifically known as *Allium sativum* from the family Alliaceae, is presumed to have its origin from Central Asia that extended to other parts of the world.[48] The word "garlic" is derived from the Anglo-Saxon "gar-leac" or spear plant, while the genus *Allium* originates from "celtic all" that means scorching, burning, hot, or pungent and *sativum* meaning cultivated.[2] It is a perennial herb; most commonly used part is the bulb, which is divided into various fleshy segments, the cloves. These cloves are exploited for disease preventive and culinary purposes. Garlic imparts significant flavor and taste to the final product.[36]

Garlic is being used since primeval times as a traditional medicine. Sanskrit histories depict its therapeutic applications for about 5000 years ago, whereas in China, it is being used for at least 3000 years ago. The early Greeks, Egyptians, Romans, and Babylonians used garlic for healing purposes.[63] Pasteur discovered the antibacterial activity of garlic in 1858. During the world wars, garlic was used as an antiseptic agent for the prevention of gangrene.[85,124] Garlic, in the prehistoric era, was sliced, grounded, and applied to wounds to suppress the infections.[5] Thus, garlic acquired the folklore position of many civilizations over decades as a well-recognized therapeutic and prophylactic disease-remedial agent.

Garlic possesses health-promoting properties, which have been recognized globally. In ancient times, it was used as a flavoring condiment due to its preventive and curative properties. In the recent epochs, there is an increase in the production and consumption of garlic due to its health-enhancing nature.[107] Epidemiological investigations have revealed close relation among the dietary habits, garlic consumption, and manifestation of illnesses.[4] Worldwide garlic production has increased about 35% due to the changes in eating habits of the consumer that resulted in 18% increase in cropped area with 13% increase in the yield. Intact garlic as well as different commercial preparations are available. Due to health claims of garlic, it is also used in the form of dietary supplements.[29]

9.2.1 NUTRITIONAL FACTS AND BIOACTIVE MOIETIES

Health promoting properties of garlic have been recognized worldwide. It contains moisture 65%, carbohydrates 30%, dietary fiber 1.5%, and organosulfur compounds 2.3%.[89,101] The protein content varies between 2.6% and 3.0% depending upon the cultivar. On an average, about 2.13% of free amino acids are present in garlic. It also contains 103.1 mg/100 g of total tocopherols. Moreover, garlic is a promising source of trace elements. In 100 g of fresh garlic, manganese 446.9 μg, zinc 556.1 μg, selenium 5.5 μg, iodine 2.5 μg, and copper 143.3 μg are present. The total polyphenols and ascorbic acid levels are 1.9 and 73.6 mg in 100 g, respectively. About 70 fatty acids are present in garlic with 46–53% linoleic acid, 20–23% palmitic acid, 4–13% oleic acid, and 3–7% α-linoleic acid that account for 80% of total lipids.[126]

The prophylactic properties of garlic are due to the presence of its nutraceutical components. These biomolecules mainly composed of sulfur containing complexes like diallyl sulfide (DAS), diallylthiosulfate (allicin), *S*-allyl-L-cysteine sulfoxides (alliin), and glutamyl-*S*-allyl-L-cysteine.[28] About 70% of total sulfur in garlic is due to γ-glutamyl peptides and alkylcysteine sulfoxides. The alkyl cysteine sulfoxides, alliin, isoalliiin, and methiin are degraded by the enzyme alliinase. Due to the enzymatic action, tissues are disrupted to release pungent volatiles that are responsible for the characteristic flavor and odor of the *Alliums*. Among these flavor precursors, the predominant one is the alliin.[57] The chief sulfur-containing compound identified in the human breath on consuming garlic is allyl mercaptan, which with its metabolites is among the plausible causes for the systemic effectivity of diallyldisulfide, alliin, and allicin. Garlic has substantial antioxidative capacities that are irrefutable due to the supportive scientific evidences covering/laboratory as well as clinical studies.[70] Consequently, allicin, alliin, γ-glutamyl, diallysulfide, cysteine, and diallyldisulfide have shown in vitro antioxidant potential.[111] Nevertheless, feed model studies regarding the free radical and lipid peroxidation inhibitory activity of garlic and its various preparations necessitates researcher's attention to deduce the consequences of cell culture system in disease inhibition.

9.2.1.1 ALLICIN

Allicin [$C_3H_5SS(O)C_3H_5$] is known as diallyl thiosulfinate and considered to be a major nutraceutical compound present in garlic extract, and for this reason, garlic is called pharma food mentioned in Figure 9.1. In garlic bulbs, 1.7% alliin and 2.8% alliinase are present on dry weight basis. Allicin is sequestered from the enzyme alliinase. When the cell membranes are ruptured by chopping, crushing, or chewing, alliinase rapidly converts alliin to allicin, an odoriferous component with a pungent smell that characterizes garlic. It contains about 60–80% of the entire garlic thiosulfinates.[50,133] It is among the most debatable compounds as numerous studies reveal it as an antioxidant having significant free radical quenching ability. It also ceases the hepatic homogenate lipid peroxidation by scavenging hydroxyl radicals in a dosage-reliant manner.[69] Allicin metabolizes into numerous active compounds and reacts with glutathione and L-cysteine to form *S*-allyl-mercaptoglutathione and *S*-allyl-mercaptocysteine, correspondingly. *S*-allyl-mercaptocysteine controls the cytochrome P450 2E1 in hepatocytes and provides protection against acetaminophen-induced hepatic injuries. The *S*-allyl-mercaptoglutathione possesses considerable antioxidant activity by inhibiting the reactions amongst the hydroxyl radicals.[98] Various scientific records have shown that the physiological effect of allicin in the biological system is mainly accredited to its interactions with the thiol containing protein instead of direct antioxidant action. In a recent scientific investigation, it was found that allicin is the sole component accountable for antioncogenic activity.[72] Allicin- and allicin-derived thiosulfates are also believed to be the major compounds responsible for the antithrombotic activity.[30] Congenital high reactivity and high membrane permeability of allicin work favorably in tumorigenesis. A number of researchers have generated allicin in situ to induce cancer cell apoptosis in the bioevaluation trials.[8,82]

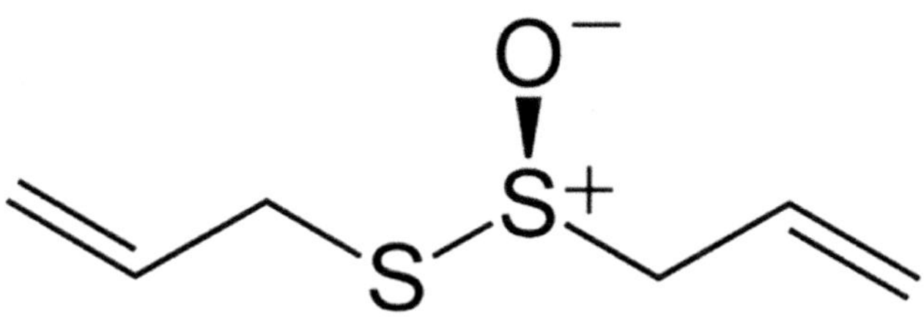

FIGURE 9.1 Chemical structure of allicin.

9.3 EXTRACTION AND QUANTIFICATION OF ALLICIN

The acidic, pungent, oxidizing composites in fresh garlic can be isolated, modified, or eliminated by numerous extraction methods. For the quantification of bioactive moieties, novel analytical procedures are required for extraction and identification of various sulfur-containing compounds.[5,9] Due to health-enhancing potential of garlic and its active constituent, allicin, garlic extract can be used as nutraceutical agent in functional or designer foods or dietary supplements. The extraction from garlic can be obtained using such techniques as solvent extraction, steam distillation, solid phase extraction, and supercritical fluid extraction (SFE). Ajoene is identified as the principle bioactive compound if the extract is prepared using ether as a solvent, whereas allicin is obtained when extracted using ethanol, methanol, ethyl acetate, and hexane.[126] Garlic is also subjected to aging process, which is responsible for decreasing the allicin content in the extract as allicin is converted to other organosulfur compounds.[64] For the quantification of the compounds obtained in extraction, several techniques as high-performance liquid chromatography (HPLC), gas chromatography (GC)–mass spectroscopy, and supercritical fluid (SF) chromatography are in vogue.[23]

There is great variation in allicin yield and botanical traits of different garlic varieties. The HPLC technique is used to quantify allicin content and yields from different varieties obtained. Moreover, the climatic differences also influence the garlic potential regarding pharmaceutical and agronomical traits. Consequently, for consistent production, different garlic varieties need to be optimized.[17]

Usually, chromatographic approaches are employed to explore garlic bioactives that may pose effect on acceptability of the resultant food or pharmaceutical product. Numerous sulfur-containing molecules have been investigated including allylmethylsulfide, 3,3′-thiobis-1-propene, allyl mercaptan, and diallyl disulfide (DADS). The DADSs are noticeable components for all garlic-based preparations. The pharmacological activities of chief garlic moieties like pectin, lectins (a sort of garlic protein), adenosine, fructans, prostaglandins, glycolipids, phospholipids, vitamins, fatty acids, and essential amino acids have been explored broadly. The pharmacological effects of garlic as an antitumor, antifungal, antithrombotic agent, and dyslipidemic potential of steroid saponins and sapogenins as β-chlorogenin have been reported as well.[68]

9.3.1 STEAM DISTILLATION

Steam distillation is commonly employed to isolate and concentrate the volatile organosulfur compounds.[130] It primarily encompasses diallyltrisulfide, diallyldisulfide, allyl propyl disulfide, and disulfide with a small quantity of diallylpolysulfide.[94] It is deliberated as an effectual aqueous extract possessing an array of pharmacological effects including hypoglycemic and antimutagenic properties.[2] However, its food application is very limited owing to its strong odor, low physicochemical stability, and water insolubility.[37]

9.3.2 HIGH-HYDROSTATIC PRESSURE

The high hydrostatic pressure (HHP) extraction system is also used to obtain extracts. An important advantage of this technique is to separate active entities from herbs or plants at ambient temperatures.[73] When the cell membranes are ruptured, a number of biologically active moieties are extracted easily from the cells. HHP technology disturbs the volatile substances in garlic, thereby reducing the alliinase enzyme activity, hence affecting the allicin formation.[60,77]

9.3.3 CONVENTIONAL SOLVENT EXTRACTION

Numerous solvents are used to extract different phytochemical entities. Conventional solvent extraction (CSE) is a traditional and obsolete technique employed for obtaining extracts. Solvents such as ethanol, methanol, hexane, ethyl acetate, or ethyl ether are most commonly used for this purpose. The extracts obtained using these solvents often contain solvent traces along with the required phenolics. Samples are macerated with solvents and allowed to stand at ambient temperature. Finally, solvent is evaporated using evaporators. Some solvents are more effective than the others. The residual solvent must be removed from the extract that is a critical process. There is regulatory level recommended for every solvent; it is necessary to minimize the residual solvent traces below the recommended regulatory levels to ensure safety and effectiveness of the material. Moreover, care needs to be taken to remove only the solvent from the extract without degradation or loss of the required components. The choice of extraction solvent is also dependent on the safety regulations. In an investigation, methanol was considered as an ideal solvent for extraction of bioactive moieties from garlic, and maximum antioxidant activity was also reported for methanolic extracts followed by ethyl acetate and hexane.[16]

Regardless of the benefits, CSE offers certain disadvantages as well. Flammable and hazardous solvents are used which can be life-threatening and pose an environmental hazard as various toxins are emitted during extraction. Moreover, for effectual separation, high purity solvents are required which are quite costly. It is a laborious process and nonselective extraction technology; as we cannot extract one component alone, an array of components is extracted, which is further quantified using HPLC. The ideal extraction system requires the process to be environment-friendly releasing no pollutant to increase the pollution, thereby damaging the ozone layer. CSE is not considered an ideal system; it is a contributor in ozone damage due to the release of hazardous and toxic solvent fumes. Recently, novel techniques, such as SFE, for extracting specific bioactive moieties from complex matrices, have been designed.[108]

9.3.4 SUPERCRITICAL FLUID EXTRACTION—GREEN TECHNOLOGY

SFE being the technological cutting-edge system employs the use of gases, usually carbon dioxide, which is then compressed to a dense liquid as shown in Figure 9.2. Afterward, the liquid is pumped through a stainless steel cylinder having the raw material to be isolated. The resultant extract-laden liquid is pumped to a separation assembly, where the extract is separated from the gaseous phase. Solvent properties of carbon dioxide can be adjusted and manipulated by varying the operating temperature and pressure. The versatility of the SFE system makes it advantageous to extract specific constituents from a given raw material. High initial cost of the system is often a hindrance in the adoption of this technology.[42] SFs possess gas-like properties of diffusion, surface tension, and viscosity as well as liquid-like density and solvating power. These liquid and gas-like properties make it suitable to make changes in temperature and pressure to form a unique medium that is beneficial to execute the extraction process for getting the valuable nutraceuticals. The conventional system used for obtaining bioactive compounds is solvent extraction, while SFE is a novel technique that has certain benefits over this system.[40] Earlier, del Valle et al.[39] reported 9.3% extraction yield using SFE technology. SFs are proved to be efficient solvents with enhanced diffusivity, mass transfer properties, and penetration capacity than the organic solvents such as acetone, ethanol, methanol, hexane, etc. Altering the extraction conditions affects the overall yield as well as selectivity of the extraction process. The use of organic solvents causes high-energy cost and safety hazards leaving the solvent residues behind.[117]

SFE is regarded as an environment-friendly and energy-efficient method. The critical point of CO_2, the most commonly used SF, is 7.4 MPa which allows the extraction at low temperature posing no harm to heat labile constituents and avoiding thermal degradation of the product.[79] Subsequently, to attain high levels of labile allicin, mild extraction processes are being used by the nutraceutical industries. In a study conducted by del Valle, Glatzel, and Martínez,[40] extraction selectivity for allicin along with the oleoresin yield in SFE system was optimized. It was observed that oleoresin yield was slightly improved by increasing the extraction pressure (15–45 MPa), whereas it was dramatically enhanced by the process temperature (35–65°C). However, a decrease in allicin concentration is observed as the temperature increases. Therefore, an optimum temperature/pressure combination is selected for higher yields. Based on results, 4-h extraction procedure at 55°C temperature and 30 MPa pressure using 55 kg/kg CO_2/substrate is suggested for large-scale nutraceutical extraction. Moreover, agglomeration may pose a problem in extraction that can be avoided by lowering the sample moisture content.

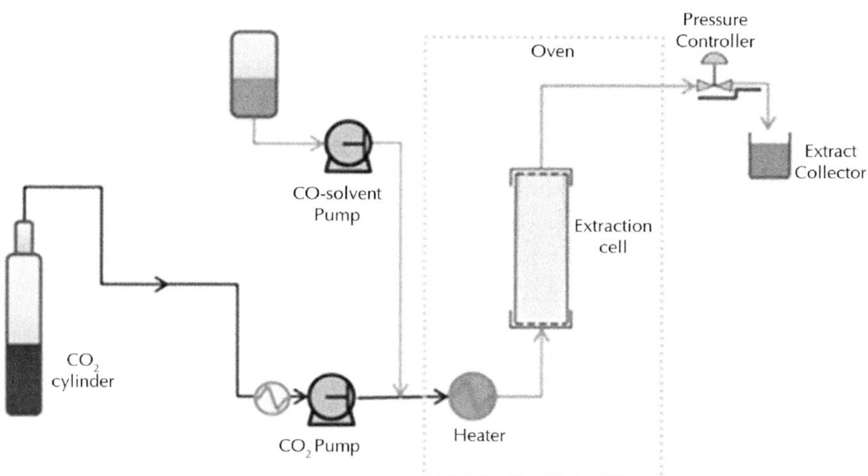

FIGURE 9.2 (See color insert.) A typical supercritical fluid extraction system.

9.3.4.1 *BENEFITS OF SFE OVER CSE*

There are several reports on benefits of SFE over CSE. The high selectivity, efficiency, and speed make SFE suitable for extracting even the minor constituents from a food matrix. It offers low operating temperature and minimal postextraction manipulation posing no toxic effect to consumer and environment. Extraction pressure has positive influence with fluid density, that is, higher the extraction pressure, more will be the fluid density, which ultimately increases the solvating power. The interaction between the matrix and analytes decreases by increasing the extraction temperatures resulting in improved extraction efficiency.[105] The main plus point of SFE is the tremendous quality of the resulting product.[103]

In a comparative study between CSE and SFE of garlic, the yield and quality of extracts obtained were analyzed. Yield improves 0.65–1% by increasing the pressure of the system at constant temperature. Yield also rises with extraction pressure (150–400 bar) at a constant temperature of 50°C. On the basis of quality of extracts and yield obtained, the most suitable extraction conditions are 300–400 bar pressure at 35–50°C temperatures; 5.5% yield is obtained using conventional extraction system employing ethanol as the solvent but it is nonselective.[39]

Thiosulfinates, primarily allicin and its counterparts, are heat labile and unstable. Their stability can be improved by dissolving or diluting in water or other binary solvents. Therefore, allicin quantification is problematic because of its unstable nature. In GC, allicin characterization is difficult due to the oven temperature. Even at moderate temperature, it readily undergoes decomposition to other compounds. The most suitable method for allicin quantification is reversed-phase HPLC.[71] The peak area of the HPLC graph is proportional to the quantity of a certain substance that has passed the HPLC detector, while the retention time is an indicator of the size of the particular molecule.

9.4 ILLNESS-FIGHTING ANTIOXIDANTS

Phytonutrients in garlic may act synergistically to exert antioxidant potential by scavenging reactive oxygen species (ROS),[25] augmenting the cellular antioxidant enzyme systems (glutathione peroxidase, catalase, and superoxide dismutase) and increasing the glutathione concentration in cells. Garlic, due to its rich phytochemistry, lessens the peril of cardio- and cerebro-vascular conditions by inhibiting low-density lipoproteins (LDL) oxidation and thiobarbituraic acid-reactive species formation.[5] Previously, it was also deduced that plasma antioxidant levels of rats significantly increase up to 14.7% on administration with raw garlic and 13.8% when boiled garlic at 100°C for 20 min is given. An increase in Trolox values is also observed.[51]

Many degenerative disorders of aging result from the long-term exposure to chemical entities that are destructive in nature leading to free radical production. These molecules are highly reactive and energized due to the presence of free unpaired electrons. These may be a consequence of human metabolism or may come from outside the body. Garlic extract also contributes in antiaging, its preparations promote neural cells survival, memory, longevity, and cognitive abilities. Nevertheless, meta-analysis is desirable to validate the evidences.[121] For this reason, neurotrophic potential, primarily attributed to the garlic extract, is a natural substitute for averting neurodegenerative ailments like Alzheimer's disease.[33] The promising antioxidants present in various garlic extracts regulate the ROS concentrations in ischemia, favor the pr-survival pathways, and mitigate mitochondrial dysfunction.[31]

Earlier, the antioxidant capacities of mature and immature garlic bulbs were investigated in terms of radical scavenging activity and the effect on lipid peroxidation. The IC_{50} 1, 1-diphenyl-2-picrylhydrazyl (DPPH) activity ranges from 1.03 to 6.01 mg/mL. Strong lipid peroxidation inhibition is observed in the samples. The total phenolic contents range from 0.05 to 0.98 mg gallic acid/g of extract.[26] The color change of garlic extract in DPPH scavenging method from deep violet to yellow also indicates the strong antioxidant activity.[99] Scientific evidences support the fact that by the intake of dietary antioxidants in various plant-based formulations, the quality of life can be improved by delaying the onset and reducing the threat of degenerative diseases associated with it.[52]

9.5 NUTRIFIED GARLIC-BASED FORMULATIONS

Garlic and its preparations such as aqueous extract, garlic juice, garlic powder, and oil are being employed in the formulation of different food products like bread, sauces, mayonnaise, and dips. It is an important culinary ingredient in the subcontinent.[15,48] Freshly sliced garlic contains limited amounts of allicin; hence, the allicin content can be adjusted in the commercial products for better bioavailabilty of the bioactive agent. Garlic preparations such as aqueous extract, essential oil, oil macerate, and powder extract can be used in the development of numerous nutraceutical and phytopharamaceutical products.[22]

Some important garlic preparations are also used as dietary supplements containing effective, highly stable, and odorless compounds. Aged garlic extract (AGE) is more effective and safe as compared to other-related garlic preparations. AGE has been vigorously studied on account of its affirmative health effects.[20] Garlic paste is a widely used ingredient in our traditional dishes. In this regard, garlic storage temperatures and preservatives affect the sensory, chemical, physical, and microbiological characteristics of the final paste. Greenish pigment formation can be avoided by storing the fresh bulbs at 25–40°C. Furthermore, skin removal is facilitated by heating the bulbs. The garlic paste is processed and thermally treated at 85°C for 5 min. The color was affected by the additives, temperature, and the storage time. Moreover, the apparent viscosity and consistency index of the paste decreased by increasing the temperatures.[35]

Various nutrified garlic-based formulations are being prepared to utilize the beneficial properties of the clove. Raw garlic homogenate is the most common garlic preparation that is being consumed all around the globe. This preparation is characterized by the presence of allicin in it. Garlic oil extracted through steam distillation technique is also widely used. It yields about 2.5–3.0 g/kg of fresh garlic. In garlic oil, DAS, DADS, and diallyl trisulfide differ in number of sulfur atoms.[126] Garlic cloves are also dehydrated and pulverized to generate garlic powder. Allicin is often degraded due to thermal treatment during the dehydration process. Alliin is the active constituent present in garlic powder due to the alliinase deactivation with small amounts of oil soluble sulfurous compounds. For the consumers who do not like garlic because of its odor and pungency, AGE is prepared by soaking the cloves in vinegar or alcohol for a year or more. This process removes various pungent sulfur components along with stabilizing unstable components.[83] AGE contains the S-allylmercaptocysteine and S-allyl cysteine (SAC) and trace quantities of oil soluble compounds. However, fresh garlic and garlic powder and aged extract are nowadays among the popular supplements in market.[126]

Aqueous garlic extract prevents the production of lipid peroxidation product, that is, malondialdehyde (MDA).[95] Furthermore, raw garlic homogenate increases the endogenous antioxidants and also reduces the lipid peroxidation at 125, 250, and 500 mg/kg body weight.[21] Additionally, chronic administration of garlic homogenate orally protects heart from adriamycin-induced oxidative stress and ischemic reperfusion injury in vitro.[20,84]

Heat processing inactivates the alliinase enzyme that catalyzes the alliin conversion into allicin; thus, thermally treated garlic chiefly contains alliin (*S*-allyl-cysteine sulfoxide). When the cloves are dehydrated and pulverized to form garlic powder, it demonstrates similar alliinase activity as that of the fresh garlic. Nevertheless, the temperature should not exceed from 60°C; otherwise, it would inactivate alliinase. Garlic powder exhibits considerable antioxidant potential; it is capable of interacting with 1,1-diphenyl-2-picrylhydroxyl stable free radicals.[67] AGE is produced when garlic cloves are soaked in the ethanol solution for several months. The AGE may quench the ROS and improve the cellular antioxidant capacity by reducing the endogenous glutathione and other antioxidant enzymes such as superoxide dismutase, catalase, and glutathione peroxidase.[128] Besides, laboratory and clinical studies are convenient to curb out the prophylactic role of AGE regarding various physiological threats as cardiovascular ailments, aging, oxidant-mediated brain cell damage, and allied health issues leading to Alzheimer's disease.

Currently, garlic and onion treatment are also used for preventing oxidative damages in irradiated raw beef. With the addition of garlic, the sulfurous compounds increase and the amount of volatiles also change significantly. On the other hand, the irradiation aroma of treated beef is similar to nontreated one that means that the interaction of garlic volatiles with beef has masked the aroma. It was then concluded that $<0.01\%$ garlic is necessary to avert or mask the irradiation odor in irradiated raw ground beef.[131]

Several garlic-based preparations are added in commercial food products for numerous reasons. As a nutraceutical agent, they play an imperative role in ailment prevention and mitigation, while as an additive, these preparations improve the shelf stability of the final product by decreasing the oxidative damage, thereby increasing the antioxidant capacity of the product. Furthermore, these preparations also act as antimicrobial agents or synergistically with such agents. Previously, these effects were deliberated using different garlic formulations such as fresh and powdered garlic and garlic oil on raw chicken sausage during the storage period. It has been concluded that fresh garlic has the highest antioxidant activity followed by powdered garlic and garlic oil. Similarly, fresh garlic imparts a strong flavor in sausages. Accordingly, fresh and powdered garlic can be used in the preservation of meat products.[109] Addition of supercritical garlic extract in chicken nuggets improved their storage stability as well as overall acceptability alongside increasing the antioxidant potential of the product.[14,15]

Aqueous garlic addition in yogurt was studied by Hassan et al.[54] Different concentrations of the extract ranging from 0.1% to 0.5% were used. However, 0.1% garlic extract yogurt was considered as the best treatment organoleptically. Enhancement in the processing technologies and combinatorial use can make garlic-based formulations as an effective tool in the prevention and management of various metabolic syndromes.[6]

9.6 PROPHYLACTIC PERSPECTIVES OF GARLIC EXTRACT

Prophylaxis is a Greek word, which means to prevent or guard beforehand. A medical or public health procedure intends to prevent rather than treat or cure a particular syndrome. Generally, these prophylactic measures are divided into two main categories: primary prophylaxis (prevention of disease development) and secondary prophylaxis (protection after disease development from deterioration in condition). Scientific investigations have revealed diverse biological properties of garlic bioactive components including the regulation blood glucose, ameliorating plasma lipid profile, anticarcinogenic, antihypertensive, antithrombotic, insecticidal, antimicrobial, and antioxidant activities.[1,81]

9.6.1 IMMUNOMODIFICATION THROUGH GARLIC

Advances in industrialization have led to various autoimmune syndromes in this technological era. To cope with such disparities, there is a need to boost up the immune system. This can be done using immune booster foods. Foods rich in antioxidants also pose immunomodulatory effects. Antioxidants alter the cellular "redox" status. They influence the production of APC cytokine, which plays a substantial role in the development of type 1-T helper cells and regulatory T cells.[96] The declining intake of polyphenols and decreasing antioxidant levels is leading to devastating health problems. The immunomodulatory properties of many dietary nutrients make them potential adjunctive strategies in coping with health crisis. Garlic or allicin intake is helpful in boosting up the immune system. It potentially induces the macrophage, phagocytosis, and lymphocyte proliferation, whilst activates the infiltration of lymphocytes and macrophages in tumors. It also stimulates the release of interlukin-2, interferon gamma, and tumor necrosis factor alpha and improves the lymphokine activated killer cell and natural killer cell activity. These effects efficaciously stimulate immune responses. Thus, immunomodification through garlic is helpful in the treatment and prevention of immune dysfunctions.[32,34]

9.6.2 ANTIPROLIFERATIVE PROPERTIES

Cancer is a disease of complex etiology that results from the uncontrolled cell growth. The transformation of normal cell to cancerous cell involves three phases: initiation, promotion, and finally progression of cancerous cells.[114] When normal cell functions are interrupted by some factor, uncontrolled growth and abnormal cellular development or neoplasms begin to occur. The neoplasmic mass often forms a clumping of cells known as tumor, which can be benign or malignant depending upon the type and severity of invasion. Phytochemicals in numerous plant-based foods help cancer-fighting enzymes to purge carcinogens in cells.[12] Garlic is considered as one of the earliest plants used for the treatments of tumors over 3500 years ago by the Egyptians.[114] The chemopreventive activity of garlic is due to the organosulfur compounds.[58] In an ecological study in China, it was reported that 74.7% women and 82.3% men are at low risk due to three times more garlic consumption per week.[113]

Uncontrolled activities of microorganisms can prove catastrophic to man. Increased intake of antibiotics has made bacteria more resistant to them; hence, there is a huge demand for new antimicrobial agents for controlling infections.[13] The important antimicrobial constituent of garlic is its oxygenated sulfur compound, thio-2-propene-1-sulfinic acid *S*-allyl ester, that is, allicin formed after crushing the raw garlic. Allicin reacts rapidly with free thiol groups, via thiol-disulfide exchange. It is known to be potent antimicrobial agent because of its interaction with thiol-containing enzymes such as alcohol dehydrogenases and cysteine proteases. These enzymes are essential for bacterial metabolism and nutrition.[92] Aqueous garlic extract is administered for its ability to inhibit growth of some oral bacterial species. About 57.1% garlic extract containing 220 μg/mL allicin has been found effective for inhibiting and killing the putative periodontal pathogens.[18]

The hypoglycemic and hypocholestrolemic potential of garlic requires worthful considerations. Owing to the sulfur-rich components, it is regarded as an effective antihyperglycemic and dyslipidemic agent.

9.6.3 HYPOGLYCEMIC ACTIVITY

Diabetes mellitus is a condition involving homeostasis of carbohydrate where lipid metabolism is improperly regulated by the pancreatic hormone, insulin, and released via pancreatic β cells. According to the World Health Organization (WHO), the predicted diabetic cases by the year 2030 will be around 366 million. Diabetes is divided into two main categories depending upon the insulin metabolism.[12] The type 1 diabetes mellitus is insulin dependent (IDDM), while the type II is noninsulin-dependent diabetes mellitus. However, in accordance with WHO, diabetes mellitus is an etiologically and clinically diverse group of infirmities in which hypoglycemia is a commonly occurring condition.[129] Hyperglycemia is an endocrine disorder affecting the human physiological system.[134] Poorly controlled blood glucose levels lead to the development of diabetic complications.[62] A higher than optimal glucose level is now acknowledged as more common than anticipated and involves long-term effects on metabolism specially cardiovascular health.[59] Diabetes could also be a factor or root cause for several other complications including cardiovascular complaints, oxidative stress mediated diseases, and immune dysfunction.[100] The increased cholesterol level and LDL oxidation trigger events that initiate atherosclerosis.[7,127] As a result of various health malfunctions, there is an onset of autoimmune disorders and immune dysfunction.[136] About 33% of all diabetics take diet-based regimes that are effectual; among these, garlic is among the frequently used dietary regimen.[87] Its treatment is helpful in lowering and controlling the blood glucose level. The hypoglycemic activity of garlic is primarily due to the presence of allicin, which enhances the serum insulin concentrations by combining with substances such as cysteine, thus preventing the insulin inactivation.[125] Some studies suggest that garlic alleviates hyperglycemia by increasing the insulin secretion from pancreatic β cells, enhancement of insulin sensitivity, or release of bound insulin.[24,78]

Diabetes also increases the oxidative stress due to free radical production and disturbing the natural antioxidant mechanisms.[43] Garlic extract reduces serum glucose level, cholesterol, and triglyceride (TG) levels in hyperglycemic rabbits.[120] Hyperglycemia during diabetes leads to cognition impairment. Different garlic extracts act as neuroprotective agents in diabetic rats. Both fresh and aged extracts improve learning in male diabetic rats, whereas cooked garlic reverses impaired memory in both female as well as male rats. Garlic is thus beneficial in lowering blood glucose, inhibiting production, and scavenging free radicals in brain acting as a neuroprotector.[112]

Type I diabetes or IDDM is an autoimmune disease in which pancreatic β cells are targeted and destroyed by an abnormal host immune system. Both cellular and humoral branches of the immune system are involved. Autoantibodies also work against islet cell antigens.[41] Allicin administration intraperitonealy (8–16 mg/kg body weight) for about 30 days in type 1 diabetes-induced rats effectively lowers the anti-islet cell antibody indices, which are the major antibodies produced during this autoimmune syndrome. Moreover, elevated pan B-cell (CD19), pan T-cell (CD90), and pan innate cell (CD11b) markers that increase during the disease significantly decreased due to the allicin administration. The same study indicated that insulin level reduces because of the damage to Islet's of Langerhans cells that rise in serum due to tissue repair process after allicin treatment.[91] The bioactive components of garlic like allicin and SAC sulfoxide are potent antioxidants that exert affirmative impact on insulin secretion, remove abnormalities from the β cell, and detoxify the body, hence alleviating the risk of associated complications.[21,118]

Extraction solvents also affect the efficacy of the extract. Pure bioactive components have better therapeutic efficacy. In alloxan-induced diabetic rabbits, about 0.25 g/kg of ethyl ether, petroleum ether, and ethanolic extracts of garlic reduce the blood sugar about 26.2%, 17.9%, and 18.9%, respectively.[102]

Garlic extract oral administration for a period of 14 days was studied on hyperglycemic rats in comparison to glibenclamide, an antidiabetic drug. It was indicated that garlic extract considerably lowers serum glucose, TGs, total cholesterol (TC), uric acid, urea, creatinine, alanine aminotransferase, and aspartate aminotransferase indices, while serum insulin concentration increases in diabetic rats. It was found that 0.1–0.5 g/kg of garlic extract is more helpful in managing diabetes than 600 μg/kg of commercial antidiabetic drug.[44] Moreover, garlic extract can also be used as an adjunct to pharmaceticals to enhance their effectiveness.

9.6.4 HYPOLIPIDEMIC POTENTIAL

Hyperlipidemia management includes optimal nutrition and diet therapy combined with appropriate exercise. The nutritional supplements are cornerstone for therapy in low-to-moderate risk patients. However, in high-risk patients, pharmacological interventions are required. The combinatorial therapy involving the use of pharmacological agents and lipid-lowering nutraceuticals can reduce LDL, increase their particle size, decrease the TG concentrations, and increase high-density lipoproteins (HDL), thereby improving the blood lipid profile.[66] The intake of excessive calories aggravates the serum TG levels, while diet containing cholesterol and fatty acids leads to hypercholesterolemia.[90] Spices, including garlic, have various chemopreventive and antioxidative agents that are helpful in mitigating various lifestyle-related disorders such as dyslipidemia. Low-density lipoprotein is the chief atherogenic lipoprotein, and its levels associate well with atherosclerosis and coronary heart diseases (CHD) risk. It has been recommended that each percentage decrease in LDL-cholesterol (LDL-C) can reduce CHD risk by 2%.[86]

Garlic extract treatment alleviates blood cholesterol level and consequently mitigates dyslipidemia. It also affects lipid peroxidation indicators by increasing the oxidation resistance, blood antioxidant potential, and reducing the MDA level.[43] The equilibrium between peroxidants and antioxidants is disrupted by oxidative stress that leads to free radical generation and lipid peroxidation. MDA level is considered to be the biomarker of lipid peroxidation.[76] Garlic extract treatment increases the catalase and superoxide dismutase activities. Alliin in garlic prevents the reduction of these enzyme activities in diabetic rats. Raw and boiled garlic also have impact on plasma antioxidant indices and lipid concentrations in hyperlipidemic rats. Raw garlic as well as garlic boiled for 20 min at 100°C contains highly comparable amounts of polyphenolic fractions and possess sufficient antioxidant activities. Likewise, they are helpful in alleviating plasma lipid levels in rats fed on cholesterol-containing diets. Thus, slightly boiled garlic can also be consumed and preferable if it is to be added in dishes, it should be added in the last 20 min.[51] Aqueous extract of raw garlic scavenges superoxide anions and hydroxyl radicals and inhibits lipid peroxidation, formation of hydroperoxides, and Cu^{+2}-induced lipoprotein oxidation.[88] It has been reported that garlic considerably increases the antioxidant enzyme activity and lowers the plasma and erythrocyte MDAs in elderly subjects.[11]

Heidarian et al.[55] elucidated the lipid lowering potential of garlic and concluded that garlic supplementation (4%) in hypercholestrolemic rats is effective in reducing the TG levels. In the same study, it was observed that supplementation with garlic resulted in reduction in phosphatidic acid phosphatase (PAP) activity which decreases the serum lipid profile. More or less similar results were obtained in some other studies.[115,122] The hypocholestrolemic and hypolipidemic properties of garlic on biomodels can be accredited to the presence of allicin and its derivatives. Garlic imparts its lipid-lowering impact by limiting liver cholesterol biosynthesis.[115]

Garlic is regarded as an antiatherogenic food, beside others, hemodynamic and hemostatic activities.[51] Its cholesterol-lowering effect is primarily due to decrease in LDL-C that is attributed to the inhibition of hepatic reductase enzyme (hydroxymethylglutaryl-CoA reductase) by alliin and allicin.[110,126] The relationship between lipid indices and haematological parameters was studied by administration of aqueous garlic extracts to rats at 0.5–1.5 mg /kg body weight for 28 days. The results showed that the extract lowered the TGs and TC concentrations and indicated a direct correlation with the hematological indices.[47]

Aqueous garlic extract reduces cholesterol biosynthesis up to 75% without cellular toxicity mediated by sterol 4-alpha-methyl oxidase. It is also effective in reducing coronary calcium progression.[132] High blood pressure or hypertension is a major risk factor for cardiovascular complications. Hypertension is affecting almost 1 billion individuals worldwide and is becoming a major cause of morbidity and mortality, especially in developed countries. Blood pressure ameliorating properties of garlic are accredited to its allicin content and production of hydrogen sulfide.[106] This effect was observed in a trial where the normotensive rats fed on 2% wild garlic leaf powder for a period of 8 weeks resulted in lowering the blood pressure.[80] Garlic is effective in blocking hypoxic pulmonary hypertension as allicin acts as a pulmonary vasodilator.[75] In another study, hypertensive rats were treated with 80 mg/kg/day of pure allicin. It showed reduction in TG levels from 96 to 71 mg/dL. Moreover, a significant decrease in serum blood pressure from 190 to 168 mmHg was also observed.[46]

High blood cholesterol also leads to obesity. It is a state of disease in which excess body fat accumulates to the extent that health may be impaired. The underlying denominators are positive energy balance and gain in weight. It can also be characterized by a number of proinflammatory and antiinflammatory factors such as cytokines, adipokines, and chemokines.[49] Obesity arises from excess energy consumption over energy expenditure.[53] It is becoming a major concern worldwide due to its prevalence and association with metabolic syndromes. Dietary garlic has been found helpful in reducing body mass and accumulation of several white adipose tissues.[73] Dietary garlic supplementation causes reduction in body weight with an alleviation in uncoupling protein mRNA and a reduction in expression of adipogenic genes in obesity induced mice.[60] Accumulation of cholesterol is common with increased age and reduced physical activity. A raised blood cholesterol level (above 150 mg/100 mL plasma) is the main risk factor of cardiovascular disorders. When combined with other factors like smoking, likelihood of dying from heart diseases increases.[10] The foam cells, "lipid-laden" atherosclerosis, are macrophages are responsible for the production of oxygen radicals in their microbicidal role. They cause oxidation of lipoprotein and lipids. The oxidized sterols produced by the monocyte macrophages lead to the development of atherosclerotic lesion and necrosis that are most dangerous inducers of CHD.[43]

Epidemiological deliberations suggest that garlic consumption may have defensive effects against cardiovascular complications. Being herbal remedy, garlic treatment mitigates multitude risk factors that hold an important position in the genesis and progression of cardiac abnormalities. These effects comprise reduction in LDL, serum TG, TC, and fibrinogen concentration, and an increase in high-density lipoproteins. It also ameliorates the organ perfusion and lowers the arterial blood pressure that results in the augmentation in fibrinolysis, platelet aggregation inhibition, and plasma viscosity diminution, thus playing a key role in overall cardioprotection.[19]

Garlic and its preparations are used to avert various cardiovascular complications. Whole and AGEs are effective in ameliorating serum cholesterol, LDL, and TG levels in hyperlipidemic subjects. Similarly, garlic powder also improves serum cholesterol level in hypercholestrolemic subjects.[61] Garlic and its constituents have the ability to decrease serum cholesterol and triacylglycerol concentration by inhibiting the enzymes involved in cholesterol and fatty acid biosynthesis.[74] Platelet aggregation and gradual thrombus formation are reduced by utilizing garlic and its components, whilst fibrinolysis increases because of dissolution of clots and thrombi.[119]

In a study, Joo et al.[60] explicated the effect of hydrodistillated garlic extract and aqueous garlic extract on obesity and dyslipidemia. They observed that hydrodistilled garlic supplementation results decrease in body weight due to the reduction in mass of adipose tissue as compared to the aqueous garlic extract supplementation. Serum TC, TGs, and LDL cholesterol also decrease in the hydro-distilled supplemented group; less reduction was observed in the aqueous extract supplemented group.

Previously, in a placebo-controlled single blind investigation, lipid indices of dyslipidemic subjects (150) were observed. They were distributed into three groups and were given enteric-coated garlic powder tablets (1 mg allicin) two times a day, anethum tablets (650 mg) twice, and placebo tablet. After 6 weeks administration, lipid profiles were assessed. In garlic tablet supplemented group, a 12.1% decrease was observed in TC and 17.3% reduction in LDL, while HDL increased up to 15.7%. Nonetheless, in anethum-supplemented group, TC reduced by 0.4% and the LDL-C by 6.3%. The study revealed that garlic supplementation is helpful in the management of dyslipidemia.[65]

Various studies suggest that garlic alleviates TC levels by 10% approximately by altering the LDL/HDL ratios. Moreover, some researches also support the effectiveness of garlic as a mild antihypertensive by lowering the blood pressure (5–7%). It also enhances the fibrionolytic activity by reducing the clots on damaged endothelium.[75] In a recent meta-analysis, data collected from 29 trials indicated that garlic supplementation is helpful in ameliorating blood cholesterol and triacylglyceride levels from 0.33 to 0.06 mmol/L and 0.19 to 0.06 mmol/L, respectively.[104]

One of the researcher groups, Zeng et al.[135] conducted a meta-analysis and concluded that the therapeutic use of garlic is more effectual if employed for a long period with higher baseline cholesterol levels. They found that garlic powder and AGE are effectual in the reduction of serum cholesterol levels. Garlic can be used as an adjuvant with lipid-lowering medicines, but for using it as main therapeutic for lowering lipid levels, more meta-analysis should be conducted using standardized preparations.[97]

9.7 SUMMARY

Conclusively, garlic can be regarded as a natural panacea owing to its diverse health-benefiting properties. It encompasses a variety of phenolics and antioxidants that prevent the occurrence and progression of life-threating ailments arising from changing lifestyle, poor diet intake, and related factors. Allicin, being the chief organosulfur component, can be effectively extracted using traditional and novel extraction techniques. Allicin can be supplemented in designer foods to curtail the threat of diet-related malfunctions in vulnerable segments. Moreover, garlic consumption on daily basis can ensure a healthy life.

ACKNOWLEDGMENTS

The authors are grateful to Functional and Nutraceutical Food Research Section, National Institute of Food Science and Technology, University of Agriculture, Faisalabad, Pakistan. This research was partially supported by Higher Education Commission, Pakistan under Pak-US Science and Technology Cooperation Program Phase IV (Project Grant No. 10/01/10-09/30/12), project entitled "Establishment of Functional and Nutraceutical Food Research Section at the National Institute of Food Science and Technology, University of Agriculture, Faisalabad, Pakistan."

KEYWORDS

- **garlic**
- ***Allium sativum***
- **allicin**
- **functional food**
- **diet-based therapy**
- **hyperlipidemia**
- **diabetes**

REFERENCES

1. Abdalla, F.; Bellé, L.; De Bona, K.; Bitencourt, P.; Pigatto, A.; Moretto, M. *Allium sativum* L. Extract Prevents Methyl Mercury-Induced Cytotoxicity in Peripheral Blood Leukocytes (ls). *Food Chem. Toxicol.* **2010,** *48* (1), 417–421.

2. Agarwal, K. C. Therapeutic Actions of Garlic Constituents. *Med. Res. Rev.* **1996,** *16* (1), 111–124.
3. Ahmad, R. S.; Butt, M. S.; Sultan, M. T.; Mushtaq, Z.; Ahmad, S.; Dewanjee, S.; De Feo, V.; Zia-ul-Haq, M. Preventive Role of Green Tea Catechins from Obesity and Related Disorders Especially Hypercholesterolemia and Hyperglycemia. *J. Transl. Med.* **2015,** *13* (1), 79.
4. Amagase, H. Clarifying the Real Bioactive Constituents of Garlic. *J. Nutr.* **2006,** *136* (3), 716S–725S.
5. Amagase, H.; Petesch, B. L.; Matsuura, H.; Kasuga, S.; Itakura, Y. Intake of Garlic and Its Bioactive Components. *J. Nutr.* **2001,** *131* (3), 955S–962S.
6. Ambati, S. Garlic Derivatives: Role in Obesity and Related Disorders. *Biotechnology* **2013,** *14* (2 (1)), 1–5.
7. Andican, G.; Seven, A.; Uncu, M.; Cantaşdemir, M.; Numan, F.; Burçak, G. Oxidized LDL and Anti-OXLDL Antibody Levels in Peripheral Atherosclerotic Disease. *Scand. J. Clin. Lab. Invest.* **2008,** *68* (6), 473–478.
8. Arditti, F. D.; Rabinkov, A.; Miron, T.; Reisner, Y.; Berrebi, A.; Wilchek, M.; Mirelman, D. Apoptotic Killing of B-Chronic Lymphocytic Leukemia Tumor Cells by Allicin Generated In Situ Using a Rituximab–Alliinase Conjugate. *Mol. Cancer Ther.* **2005,** *4* (2), 325–332.
9. Arnault, I.; Auger, J. Seleno-Compounds in Garlic and Onion. *J. Chromatogr. Anal.* **2006,** *1112* (1), 23–30.
10. Aslam, M. N.; Malik, F. A.; Humera, J.; Anwar, E.; Awan, J. A.; Khan, Y. H.; Iqbal, M. Relation of Smoking and Elevated Homocysteine in Patients Presenting with Acute Coronary Syndrome. *Pak. J. Cardiol.* **2006,** *17* (2–3), 48–56.
11. Avcı, A.; Atlı, T.; Ergüder, İ. B.; Varlı, M.; Devrim, E.; Aras, S.; Durak, İ. Effects of Garlic Consumption on Plasma and Erythrocyte Antioxidant Parameters in Elderly Subjects. *Gerontology* **2008,** *54* (3), 173–176.
12. Awan, J. A. *Elements of Food and Nutrition*; Unitech Communications: Faisalabad Pakistan, 2011.
13. Awan, J. A.; Anjum, F. M. *Food Toxicology*; Unitech Communications, Faisalabad, Pakistan, 2010.
14. Awan, K.; Butt, M.; Haq, I.; Ashfaq, F.; Suleria, H. Storage Stability of Garlic Fortified Chicken Bites. *J. Food Chem. Nanotechnol.* **2017,** *3* (2), 80–85.
15. Awan, K. A.; Butt, M. S.; Awan, J. A.; Zahoor, T. Sensory Profile of Conventional and Supercritical Garlic Extract-Nutrified Chicken Bites. *Pak. J. Food Sci.* **2015,** *25* (2), 92–97.
16. Awan, K. A.; Butt, M. S.; Haq, I. U.; Suleria, H. A. R. Investigating the Antioxidant Potential of Garlic (*Allium sativum*) Extracts Through Different Extraction Modes. *Curr. Bioact. Compounds* **2017,** *14*, 1–6.
17. Baghalian, K.; Ziai, S. A.; Naghavi, M. R.; Badi, H. N.; Khalighi, A. Evaluation of Allicin Content and Botanical Traits in Iranian Garlic (*Allium sativum* L.) Ecotypes. *Sci. Hort.* **2005,** *103* (2), 155–166.
18. Bakri, I.; Douglas, C. Inhibitory Effect of Garlic Extract on Oral Bacteria. *Arch. Oral Biol.* **2005,** *50* (7), 645–651.

19. Baluchnejadmojarad, T.; Roghani, M.; Homayounfar, H.; Hosseini, M. Beneficial Effect of Aqueous Garlic Extract on the Vascular Reactivity of Streptozotocin-Diabetic Rats. *J. Ethnopharmacol.* **2003,** *85* (1), 139–144.
20. Banerjee, S.; Mukherjee, P. K.; Maulik, S. Garlic as an Antioxidant: The Good, the Bad and the Ugly. *Phytother. Res.* **2003,** *17* (2), 97–106.
21. Banerjee, S. K.; Maulik, S. K. Effect of Garlic on Cardiovascular Disorders: A Review. *Nutr. J.* **2002,** *1* (1), 1.
22. Bayan, L.; Koulivand, P. H.; Gorji, A. Garlic: A Review of Potential Therapeutic Effects. *Avi. J. Phytomed.* **2014,** *4* (1), 1–14.
23. Benkeblia, N.; Lanzotti, V. *Allium* Thiosulfinates: Chemistry, Biological Properties and Their Potential Utilization in Food Preservation. *Food* **2007,** *1* (2), 193–201.
24. Bokaeian, M.; Nakhaee, A.; Moodi, B.; Farhangi, A.; Akbarzadeh, A. Effects of Garlic Extract Treatment in Normal and Streptozotocin Diabetic Rats Infected with *Candida albicans*. *Indian J. Clin. Biochem.* **2010,** *25* (2), 182–187.
25. Borek, C. Antioxidant Health Effects of Aged Garlic Extract. *J. Nutr.* **2001,** *131* (3), 1010S–1015S.
26. Bozin, B.; Mimica-Dukic, N.; Samojlik, I.; Goran, A.; Igic, R. Phenolics as Antioxidants in Garlic (*Allium sativum* L., Alliaceae). *Food Chem.* **2008,** *111* (4), 925–929.
27. Burniston, J.; Eftekhari, F.; Hrabi, S.; Worsley, R.; Dean, E. Health Behaviour Change and Lifestyle-Related Condition Prevalence: Comparison of Two Epochs Based on Systematic Review of the Physical Therapy Literature. *Hong Kong Physiother. J.* **2012,** *30* (2), 44–56.
28. Butt, M. S.; Sultan, M. T.; Butt, M. S.; Iqbal, J. Garlic: Nature's Protection against Physiological Threats. *Crit. Rev. Food Sci. Nutr.* **2009,** *49* (6), 538–551.
29. Cardelle-Cobas, A.; Soria, A. C.; Corzo, N.; Villamiel, M. A Comprehensive Survey of Garlic Functionality. In *Garlic Consumption and Health*; Pacurar, M., Krejci, G., Eds.; Nova Science Publishers: New York, 2010; pp 1–60.
30. Cavagnaro, P. F.; Camargo, A.; Galmarini, C. R.; Simon, P. W. Effect of Cooking on Garlic (*Allium sativum* L.) Antiplatelet Activity and Thiosulfinates Content. *J. Agric. Food Chem.* **2007,** *55* (4), 1280–1288.
31. Cervantes, M. I.; de Oca Balderas, P. M.; de Jesús Gutiérrez-Baños, J.; Orozco-Ibarra, M.; Fernández-Rojas, B.; Medina-Campos, O. N.; Espinoza-Rojo, M.; Ruiz-Tachiquín, M.; Ortiz-Plata, A.; Salazar, M. I. Comparison of Antioxidant Activity of Hydroethanolic Fresh and Aged Garlic Extracts and Their Effects on Cerebral Ischemia. *Food Chem.* **2013,** *140* (1), 343–352.
32. Chandrashekar, P. M.; Prashanth, K. V. H.; Venkatesh, Y. P. Isolation, Structural Elucidation and Immunomodulatory Activity of Fructans from Aged Garlic Extract. *Phytochemistry* **2011,** *72* (2), 255–264.
33. Chauhan, N. Multiplicity of Garlic Health Effects and Alzheimer's Disease. *J. Nutr. Health Aging* **2004,** *9* (6), 421–432.
34. Clement, F.; Pramod, S. N.; Venkatesh, Y. P. Identity of the Immunomodulatory Proteins from Garlic (*Allium sativum*) with the Major Garlic Lectins or Agglutinins. *Int. Immunopharmacol.* **2010,** *10* (3), 316–324.
35. Constenla, D.; Lozano, J. Effect of Pretreatments and Processing Conditions on the Chemical, Physical, Microbiological and Sensory Characteristics of Garlic Paste. *J. Food Process Eng.* **2005,** *28* (3), 313–329.

36. Coppi, A.; Cabinian, M.; Mirelman, D.; Sinnis, P. Antimalarial Activity of Allicin, a Biologically Active Compound from Garlic Cloves. *Antimicrob. Agents Chemother.* **2006,** *50* (5), 1731–1737.
37. Corzo-Martínez, M.; Corzo, N.; Villamiel, M. Biological Properties of Onions and Garlic. *Trends Food Sci. Technol.* **2007,** *18* (12), 609–625.
38. Dean, E. Physical Therapy in the 21st Century (Part II): Evidence-Based Practice Within the Context of Evidence-Informed Practice. *Physiother. Theory Pract.* **2009,** *25* (5–6), 354–368.
39. del Valle, J.; Mena, C.; Budinich, M. Extraction of Garlic with Supercritical CO_2 and Conventional Organic Solvents. *Braz. J. Chem. Eng.* **2008,** *25* (3), 532–542.
40. del Valle, J. M.; Glatzel, V.; Martínez, J. L. Supercritical CO_2 Extraction of Allicin from Garlic Flakes: Screening and Kinetic Studies. *Food Res. Int.* **2012,** *45* (1), 216–224.
41. Diabetes, A. A. T. F. o. I. American College of Endocrinology and American Diabetes Association Consensus Statement on Inpatient Diabetes and Glycemic Control. *Endocr. Pract.* **2006,** *12* (4), 458–468.
42. Doughari, J. H. *Phytochemicals: Extraction Methods, Basic Structures and Mode of Action as Potential Chemotherapeutic Agents*; INTECH Open Access Publisher, 2012.
43. Durak, I.; Kavutcu, M.; Aytaç, B.; Avcı, A.; Devrim, E.; Özbek, H.; Öztürk, H. S. Effects of Garlic Extract Consumption on Blood Lipid and Oxidant/Antioxidant Parameters in Humans with High Blood Cholesterol. *J. Nutr. Biochem.* **2004,** *15* (6), 373–377.
44. Eidi, A.; Eidi, M.; Esmaeili, E. Antidiabetic Effect of Garlic (*Allium sativum* L.) in Normal and Streptozotocin-Induced Diabetic Rats. *Phytomedicine* **2006,** *13* (9), 624–629.
45. Elahi, H. H.; Awan, J. A.; Huma, N. Our Lifestyle, Food and Health. *Pak. J. Food Sci.* **1997,** *7* (1–2), 33–37.
46. Elkayam, A.; Peleg, E.; Grossman, E.; Shabtay, Z.; Sharabi, Y. Effects of Allicin on Cardiovascular Risk Factors in Spontaneously Hypertensive Rats. *Israel Med. Assoc. J.* **2013,** *15* (3), 170–173.
47. Emmanuel, U. C.; James, O. Comparative Effects of Aqueous Garlic (*Allium sativum*) and Onion (*Allium cepa*) Extracts on Some Haematological and Lipid Indices of Rats. *Annu. Rev. Res. Biol.* **2011,** *1*, 37–44.
48. Etoh, T.; Watanabe, H.; Iwai, S. Rapd Variation of Garlic Clones in the Center of Origin and the Westernmost Area of Distribution. *Memoirs Fac. Agric. Kogoshima Univ.* **2001,** *37*, 21–27.
49. Fantuzzi, G. Adipose Tissue, Adipokines, and Inflammation. *J. Allergy Clin. Immunol.* **2005,** *115* (5), 911–919.
50. Fugh-Berman, A. Herb–Drug Interactions. *Lancet* **2000,** *355* (9198), 134–138.
51. Gorinstein, S.; Leontowicz, H.; Leontowicz, M.; Drzewiecki, J.; Najman, K.; Katrich, E.; Barasch, D.; Yamamoto, K.; Trakhtenberg, S. Raw and Boiled Garlic Enhances Plasma Antioxidant Activity and Improves Plasma Lipid Metabolism in Cholesterol-Fed Rats. *Life Sci.* **2006,** *78* (6), 655–663.
52. Gupta, P.; Saraogi, R.; Khare, M. Free Redicals Scavenging Activity of Polyherbal Formulation (Diasweta). *Int. J. Phytopharmacol.* **2013,** *4* (1), 24–28.

53. Haslam, D.; James, W. Obesity. *Lancet* **2005,** *1*, 366.
54. Hassan, H. A.; Hafez, H. S.; Zeghebar, F. E. Garlic Oil as a Modulating Agent for Oxidative Stress and Neurotoxicity Induced by Sodium Nitrite in Male Albino Rats. *Food Chem. Toxicol.* **2010,** *48* (7), 1980–1985.
55. Heidarian, E.; Jafari-Dehkordi, E.; Seidkhani-Nahal, A. Effect of Garlic on Liver Phosphatidate Phosphohydrolase and Plasma Lipid Levels in Hyperlipidemic Rats. *Food Chem. Toxicol.* **2011,** *49* (5), 1110–1114.
56. Helou, L.; Harris, I. M. Garlic. *Herbal Products*; Springer: Berlin, 2007, pp 123–149.
57. Hughes, J.; Tregova, A.; Tomsett, A.; Jones, M.; Cosstick, R.; Collin, H. Synthesis of the Flavour Precursor, Alliin, in Garlic Tissue Cultures. *Phytochemistry* **2005,** *66* (2), 187–194.
58. Islam, M.; Kusumoto, Y.; Al-Mamun, M. A. Cytotoxicity and Cancer (Hela) Cell Killing Efficacy of Aqueous Garlic (*Allium sativum*) Extract. *J. Sci. Res.* **2011,** *3* (2), 375–382.
59. Isolauri, E.; Kalliomäki, M.; Rautava, S.; Salminen, S.; Laitinen, K. Obesity—Extending the Hygiene Hypothesis. *Microbial–Host Interaction: Tolerance versus Allergy*; Karger Publishers, 2009; Vol. 64, pp 75–89.
60. Joo, H.; Kim, C.-T.; Kim, I.-H.; Kim, Y. Anti-Obesity Effects of Hot Water Extract and High Hydrostatic Pressure Extract of Garlic in Rats Fed a High-Fat Diet. *Food Chem. Toxicol.* **2013,** *55*, 100–105.
61. Kannar, D.; Wattanapenpaiboon, N.; Savige, G. S.; Wahlqvist, M. L. Hypocholesterolemic Effect of an Enteric-Coated Garlic Supplement. *J. Am. Coll. Nutr.* **2001,** *20* (3), 225–231.
62. Khayatnouri, M.; Bahari, K.; Safarmashaei, S. Study of the Effect of Gliclazide and Garlic Extract on Blood Sugar Level in STZ-Induced Diabetic Male Mice. *Adv. Environ. Biol.* **2011,** *5*, 1751–1756.
63. Koch, H. P.; Lawson, L. D. *Garlic: The Science and Therapeutic Application of Allium sativum L. and Related Species*; Williams & Wilkins: Baltimore, MD, 1996; Vol xv, 329 p.
64. Kodera, Y.; Suzuki, A.; Imada, O.; Kasuga, S.; Sumioka, I.; Kanezawa, A.; Taru, N.; Fujikawa, M.; Nagae, S.; Masamoto, K. Physical, Chemical, and Biological Properties of S-Allylcysteine, an Amino Acid Derived from Garlic. *J. Agric. Food Chem.* **2002,** *50* (3), 622–632.
65. Kojuri, J.; Vosoughi, A. R.; Akrami, M. Effects of *Anethum graveolens* and Garlic on Lipid Profile in Hyperlipidemic Patients. *Lipids Health Dis* **2007,** *6* (1), 1.
66. Kota, S. K.; Jammula, S.; Kota, S. K.; Krishna, S. V. S.; Meher, L. K.; Rao, E. S.; Modi, K. D. Nutraceuticals in Dyslipidemia Management. *J. Med. Nutr. Nutraceut.* **2013,** *2* (1), 26.
67. Kourounakis, P.; Rekka, E. Effect on Active Oxygen Species of Alliin and *Allium sativum* (Garlic) Powder. *Res. Commun. Chem. Pathol. Pharmacol.* **1991,** *74* (2), 249–252.
68. Lanzotti, V.; Barile, E.; Antignani, V.; Bonanomi, G.; Scala, F. Antifungal Saponins from Bulbs of Garlic, *Allium sativum* L. var. *voghiera*. *Phytochemistry* **2012,** *78*, 126–134.
69. Lau, B. H. Suppression of LDL Oxidation by Garlic. *J. Nutr.* **2001,** *131* (3), 985S–988S.

70. Lawson, L. D. Garlic: A Review of Its Medicinal Effects and Indicated Active Compounds. *Blood* **1998,** *179*, 62.
71. Lawson, L. D.; Wood, S. G.; Hughes, B. G. HPLC Analysis of Allicin and Other Thiosulfinates in Garlic Clove Homogenates. *Planta Med.* **1991,** *57* (3), 263–270.
72. Lee, J.; Gupta, S.; Huang, J.-S.; Jayathilaka, L. P.; Lee, B.-S. HPLC–MTT Assay: Anticancer Activity of Aqueous Garlic Extract Is from Allicin. *Anal. Biochem.* **2013,** *436* (2), 187–189.
73. Lee, M.-S.; Kim, I.-H.; Kim, C.-T.; Kim, Y. Reduction of Body Weight by Dietary Garlic Is Associated with an Increase in Uncoupling Protein MRNA Expression and Activation of AMP-Activated Protein Kinase in Diet-Induced Obese Mice. *J. Nutr.* **2011,** *141* (11), 1947–1953.
74. Liu, L.; Yeh, Y.-Y. Water-Soluble Organosulfur Compounds of Garlic Inhibit Fatty Acid and Triglyceride Syntheses in Cultured Rat Hepatocytes. *Lipids* **2001,** *36* (4), 395–400.
75. Londhe, V. Role of Garlic (*Allium sativum*) in Various Diseases—An Overview. *J. Pharm. Res. Opin.* **2014,** *1* (4), 129–134.
76. Lykkesfeldt, J. Malondialdehyde as Biomarker of Oxidative Damage to Lipids Caused by Smoking. *Clin. Chim. Acta* **2007,** *380* (1), 50–58.
77. Ma, Y.; Song, D.; Wang, Z.; Jiang, J.; Jiang, T.; Cui, F.; Fan, X. Effect of Ultrahigh Pressure Treatment on Volatile Compounds in Garlic. *J. Food Process Eng.* **2011,** *34* (6), 1915–1930.
78. Mahesar, H.; Bhutto, M.; Khand, A. Garlic Used as an Alternative Medicine to Control Diabetic Mellitus in Alloxan-Induced Male Rabbits. *Pak. J. Physiol.* **2010,** *6* (1), 39–41.
79. McHugh, M.; Krukonis, V. *Supercritical Fluid Extraction: Principles and Practice*; Elsevier: Amsterdam, 2013.
80. McRae, M. P. A Review of Studies of Garlic (*Allium sativum*) on Serum Lipids and Blood Pressure Before and After 1994: Does the Amount of Allicin Released from Garlic Powder Tablets Play a Role? *J. Chiropract. Med.* **2006,** *4* (4), 182–190.
81. Meriga, B.; Mopuri, R.; MuraliKrishna, T. Insecticidal, Antimicrobial and Antioxidant Activities of Bulb Extracts of *Allium sativum. Asian Pac. J. Trop. Med.* **2012,** *5* (5), 391–395.
82. Miron, T.; Rabinkov, A.; Mirelman, D.; Wilchek, M.; Weiner, L. The Mode of Action of Allicin: Its Ready Permeability Through Phospholipid Membranes May Contribute to Its Biological Activity. *Biochim. Biophys. Acta—Biomembranes* **2000,** *1463* (1), 20–30.
83. Morihara, N.; Sumioka, I.; Ide, N.; Moriguchi, T.; Uda, N.; Kyo, E. Aged Garlic Extract Maintains Cardiovascular Homeostasis in Mice and Rats. *J. Nutr.* **2006,** *136* (3), 777S–781S.
84. Mukherjee, A. K.; Basu, S.; Sarkar, N.; Ghosh, A. C. Advances in Cancer Therapy with Plant Based Natural Products. *Curr. Med. Chem.* **2001,** *8* (12), 1467–1486.
85. Murray, M. *The Healing Power of Herbs: The Enlightened Person's Guide to the Wonders of Medicinal Plants. Revised and Expanded Second Edition*; Prima Publishing: Rocklin, CA, 1995; Vol xiv, p 410.

86. Nijjar, P. S.; Burke, F. M.; Bloesch, A.; Rader, D. J. Role of Dietary Supplements in Lowering Low-Density Lipoprotein Cholesterol: A Review. *J. Clin. Lipidol.* **2010,** *4* (4), 248–258.
87. Noguchi, H. Stem Cells for the Treatment of Diabetes. *Endocr. J.* **2007,** *54* (1), 7–16.
88. Nuutila, A. M.; Puupponen-Pimiä, R.; Aarni, M.; Oksman-Caldentey, K.-M. Comparison of Antioxidant Activities of Onion and Garlic Extracts by Inhibition of Lipid Peroxidation and Radical Scavenging Activity. *Food Chem.* **2003,** *81* (4), 485–493.
89. Odebunmi, E.; Oluwaniyi, O.; Bashiru, M. Comparative Proximate Analysis of Some Food Condiments. *J. Appl. Sci. Res.* **2010,** *6* (3), 272–274.
90. Ortega, R. M.; Palencia, A.; López-Sobaler, A. M. Improvement of Cholesterol Levels and Reduction of Cardiovascular Risk via the Consumption of Phytosterols. *Br. J. Nutr.* **2006,** *96* (S1), S89–S93.
91. Osman, M.; Adnan, A.; Salmah Bakar, N.; Alashkham, F. Allicin Has Significant Effect on Autoimmune Anti-Islet Cell Antibodies in Type 1 Diabetic Rats. *Pol. J. Pathol.* **2012,** *63* (4), 248–254.
92. Pérez-Giraldo, C.; Cruz-Villalón, G.; Sánchez-Silos, R.; Martínez-Rubio, R.; Blanco, M.; Gómez-García, A. In Vitro Activity of Allicin against *Staphylococcus epidermidis* and Influence of Subinhibitory Concentrations on Biofilm Formation. *J. Appl. Microbiol.* **2003,** *95* (4), 709–711.
93. Pierre, S.; Crosbie, L.; Duttaroy, A. K. Inhibitory Effect of Aqueous Extracts of Some Herbs on Human Platelet Aggregation In Vitro. *Platelets* **2005,** *16* (8), 469–473.
94. Pranoto, Y.; Salokhe, V. M.; Rakshit, S. K. Physical and Antibacterial Properties of Alginate-Based Edible Film Incorporated with Garlic Oil. *Food Res. Int.* **2005,** *38* (3), 267–272.
95. Prasad, K.; Laxdal, V. A.; Yu, M.; Raney, B. L. Evaluation of Hydroxyl Radical-Scavenging Property of Garlic. *Mol. Cell. Biochem.* **1996,** *154* (1), 55–63.
96. Prescott, S. L. Role of Dietary Immunomodulatory Factors in the Development of Immune Tolerance. In *Microbial Host-Interaction: Tolerance versus Allergy*; Karger Publishers: Basel, 2009; Vol 64.
97. Qidwai, W.; Ashfaq, T. Role of Garlic Usage in Cardiovascular Disease Prevention: An Evidence-Based Approach. *Evid.-Based Complement. Alternat. Med.* **2013**, *2013*, 1–9.
98. Rabinkov, A.; Miron, T.; Mirelman, D.; Wilchek, M.; Glozman, S.; Yavin, E.; Weiner, L. S-Allylmercaptoglutathione: The Reaction Product of Allicin with Glutathione Possesses SH-Modifying and Antioxidant Properties. *Biochem. Biophys. Acta—Mol. Cell Res.* **2000,** *1499* (1), 144–153.
99. Rahman, M.; Fazlic, V.; Saad, N. Antioxidant Properties of Raw Garlic (*Allium sativum*) Extract. *Int. Food Res. J.* **2012,** *19* (2), 589–591.
100. Rana, J.; Nieuwdorp, M.; Jukema, J.; Kastelein, J. Cardiovascular Metabolic Syndrome—An Interplay of, Obesity, Inflammation, Diabetes and Coronary Heart Disease. *Diab., Obes. Metab.* **2007,** *9* (3), 218–232.
101. Ranjani, R.; Raju, M. Anticancer Properties of *Allium sativum:* A Review. *Asian J. Biochem. Pharm. Res.* **2012,** *3* (2), 190–196.
102. Rashed, K. N. Z. Medicinal Plants as a Potential Source for Treatment of Diabetes Mellitus. *World J. Pharm. Sci.* **2013,** *1*, 1–2.

103. Raventós, M.; Duarte, S.; Alarcón, R. Application and Possibilities of Supercritical CO_2 Extraction in Food Processing Industry: An Overview. *Food Sci. Technol. Int.* **2002,** *8* (5), 269–284.
104. Reinhart, K. M.; Talati, R.; White, C. M.; Coleman, C. I. The Impact of Garlic on Lipid Parameters: A Systematic Review and Meta-Analysis. *Nutr. Res. Rev.* **2009,** *22* (1), 39–48.
105. Reverchon, E.; Caputo, G.; De Marco, I. Role of Phase Behavior and Atomization in the Supercritical Antisolvent Precipitation. *Ind. Eng. Chem. Res.* **2003,** *42* (25), 6406–6414.
106. Ried, K.; Toben, C.; Fakler, P. Effect of Garlic on Serum Lipids: An Updated Meta-Analysis. *Nutr. Rev.* **2013,** *71* (5), 282–299.
107. Rivlin, R. S. Historical Perspective on the Use of Garlic. *J. Nutr.* **2001,** *131* (3), 951S–954S.
108. Sahena, F.; Zaidul, I.; Jinap, S.; Karim, A.; Abbas, K.; Norulaini, N.; Omar, A. Application of Supercritical CO_2 in Lipid Extraction: A Review. *J. Food Eng.* **2009,** *95* (2), 240–253.
109. Sallam, K. I.; Ishioroshi, M.; Samejima, K. Antioxidant and Antimicrobial Effects of Garlic in Chicken Sausage. *LWT—Food Sci. Technol.* **2004,** *37* (8), 849–855.
110. Sangeetha, T.; Quine, S. D. Preventive effect of S-Allyl Cysteine Sulfoxide (Alliin) on Cardiac Marker Enzymes and Lipids in Isoproterenol-Induced Myocardial Injury. *J. Pharm. Pharmacol.* **2006,** *58* (5), 617–623.
111. Santhosha, S.; Jamuna, P.; Prabhavathi, S. Bioactive Components of Garlic and Their Physiological Role in Health Maintenance: A Review. *Food Biosci.* **2013,** *3*, 59–74.
112. Sarkaki, A.; Chehardacheric, S. V.; Farbood, Y.; Mansouri, S. M. T.; Naghizadeh, B.; Basirian, E. Effects of Fresh, Aged and Cooked Garlic Extracts on Short- and Long-Term Memory in Diabetic Rats. *Avicen. J. Phytomed.* **2013,** *3* (1), 45.
113. Sengupta, A.; Ghosh, S.; Bhattacharjee, S.; Das, S. Indian Food Ingredients and Cancer Prevention: An Experimental Evaluation of Anticarcinogenic Effects of Garlic in Rat Colon. *Asian Pac. J. Cancer Prevent.* **2004,** *5* (2), 126–132.
114. Shukla, Y.; Kalra, N. Cancer Chemoprevention with Garlic and Its Constituents. *Cancer Lett.* **2007,** *247* (2), 167–181.
115. Singh, D. K.; Porter, T. D. Inhibition of Sterol 4α-Methyl Oxidase Is the Principal Mechanism by Which Garlic Decreases Cholesterol Synthesis. *J. Nutr.* **2006,** *136* (3), 759S–764S.
116. Slekovec, M.; Goessler, W. Accumulation of Selenium in Natural Plants and Selenium Supplemented Vegetable and Selenium Speciation by HPLC–ICPMS. *Chem. Spec. Bioavail.* **2005,** *17* (2), 63–73.
117. Sovilj, M. N.; Nikolovski, B. G.; Spasojević, M. Đ. Critical Review of Supercritical Fluid Extraction of Selected Spice Plant Materials. *Macedonian J. Chem. Chem. Eng.* **2011,** *30* (2), 197–220.
118. Srinivasan, K. Plant Foods in the Management of Diabetes Mellitus: Spices as Beneficial Antidiabetic Food Adjuncts. *Int. J. Food Sci. Nutr.* **2005,** *56* (6), 399–414.
119. Steiner, M.; Li, W. Aged Garlic Extract, a Modulator of Cardiovascular Risk Factors: A Dose-Finding Study on the Effects of Age on Platelet Functions. *J. Nutr.* **2001,** *131* (3), 980S–984S.

120. Suleria, H. A. R.; Sadiq Butt, M.; Muhammad Anjum, F.; Saeed, F.; Batool, R.; Nisar Ahmad, A. Aqueous Garlic Extract and Its Phytochemical Profile: Special Reference to Antioxidant Status. *Int. J. Food Sci. Nutr.* **2012,** *63* (4), 431–439.
121. Sumi, S.-i.; Tsuneyoshi, T.; Matsuo, H.; Yoshimatsu, T. Isolation and Characterization of the Genes Up-regulated in Isolated Neurons by Aged Garlic Extract (Age). *J. Nutr.* **2001,** *131* (3), 1096S–1099S.
122. Sumioka, I.; Hayama, M.; Shimokawa, Y.; Shiraishi, S.; Tokunaga, A. Lipid-Lowering Effect of Monascus Garlic Fermented Extract (MGFE) in Hyperlipidemic Subjects. *Hiroshima J. Med. Sci.* **2006,** *55* (2), 59–64.
123. Tapsell, L. C.; Hemphill, I.; Cobiac, L.; Sullivan, D. R.; Fenech, M.; Patch, C. S.; Roodenrys, S.; Keogh, J. B.; Clifton, P. M.; Williams, P. G. Health Benefits of Herbs and Spices: The Past, the Present, the Future. *Med J Aust.* **2006,** *185* (4 Suppl.), S4–S24.
124. Tattelman, E. Health Effects of Garlic. *Am. Fam. Physician* **2005,** *72* (1), 103–106.
125. Thomson, M.; Al-Amin, Z. M.; Al-Qattan, K. K.; Shaban, L. H.; Ali, M. Antidiabetic and Hypolipidaemic Properties of Garlic (*Allium sativum*) in Streptozotocin-Induced Diabetic Rats. *Int. J. Diab. Metab.* **2007,** *15*, 108–115.
126. Tsai, C.-W.; Chen, H.-W.; Sheen, L.-Y.; Lii, C.-K. Garlic: Health Benefits and Actions. *BioMedicine* **2012,** *2* (1), 17–29.
127. Wahle, K. W.; Heys, S. D. Atherosclerosis: Cell Biology and Lipoproteins. *Curr. Opin. Lipidol.* **2008,** *19* (4), 435–437.
128. Wei, Z.; Lau, B. H. Garlic Inhibits Free Radical Generation and Augments Antioxidant Enzyme Activity in Vascular Endothelial Cells. *Nutr. Res.* **1998,** *18* (1), 61–70.
129. WHO. *Report of a WHO/IDF Consultation—Definition and Diagnosis of Diabetes mellitus and Intermediate Hyperglycemia*; WHO Document Production Services: Geneva, Switzerland, 2006, 5.
130. Wu, C.-C.; Sheen, L.-Y.; Chen, H.-W.; Kuo, W.-W.; Tsai, S.-J.; Lii, C.-K. Differential Effects of Garlic Oil and Its Three Major Organosulfur Components on the Hepatic Detoxification System in Rats. *J. Agric. Food Chem.* **2002,** *50* (2), 378–383.
131. Yang, H. S.; Lee, E. J.; Moon, S.; Paik, H. D.; Ahn, D. U. Effect of Garlic, Onion, and Their Combination on the Quality and Sensory Characteristics of Irradiated Raw Ground Beef. *Anim. Ind. Rep.* **2013,** *659* (1), 16.
132. Yeh, Y.-Y.; Liu, L. Cholesterol-Lowering Effect of Garlic Extracts and Organosulfur Compounds: Human and Animal Studies. *J. Nutr.* **2001,** *131* (3), 989S–993S.
133. Yu, L.; Xu, S.-Y. Preparation of Garlic Powder with High Allicin Content. *Agric. Sci. China* **2007,** *6* (7), 890–898.
134. Zakir, S.; Sarwar, M.; Allen, J.; Butt, M.; Arshad, U.; Javaid, A. Impact of Sweet Potato Cultivars on Blood Glucose Level in Diabetic and Healthy Participants. *Int. J. Agric. Biol.* **2008,** *10* (3), 316–320.
135. Zeng, T.; Guo, F. F.; Zhang, C. L.; Song, F. Y.; Zhao, X. L.; Xie, K. Q. A Meta-Analysis of Randomized, Double-Blind, Placebo-Controlled Trials for the Effects of Garlic on Serum Lipid Profiles. *J. Sci. Food Agric.* **2012,** *92* (9), 1892–1902.
136. Zhang, T.; Li, Y.; Wang, Y.-J.; Wang, X.; Young, M.; Douglas, S. D.; Ho, W.-Z. Natural Killer Cell Inhibits Human Immunodeficiency Virus Replication in Chronically Infected Immune Cells. *Antivir. Res.* **2007,** *73* (2), 132–139.

CHAPTER 10

HEALTH PERSPECTIVES OF PERSIMMON

MUHAMMAD IMRAN[1], ANEES AHMED KHALIL[1], MUHAMMAD SAJID ARSHAD[2], and HAFIZ ANSAR RASUL SULERIA[3,*]

[1]*Faculty of Allied Health Sciences, University Institute of Diet & Nutritional Sciences, The University of Lahore, Pakistan*

[2]*Institute of Home & Food Sciences, Government College University, Faisalabad, Pakistan*

[3]*School of Life and Environmental Sciences, Faculty of Science, Engineering and Built Environment, Centre for Chemistry and Biotechnology, Deakin University, 75 Pigdons Road, Geelong, VIC 3216, Australia*

**Corresponding author. E-mail: hafiz.suleria@uqconnect.edu.au*

ABSTRACT

In the current era, the consumption and utilization of fruits and vegetables is gaining significant importance as an effective tool to maintain human health. In this context, phytochemicals and bioactive molecules from fruits and vegetables are also becoming chemo-preventive agent against various maladies. Among these, persimmon (*Diospyros kaki* L.) fruit belongs to the family Ebenaceae and is used as a medicinal plant since many years to cure different human disorders, such as cancer, diabetes, cardio-vascular, obesity, and so on. Persimmon fruit has significant protective effects against various types of human syndromes. Their effectual role is mainly owing to the presence of significant amounts of antioxidants such as phenolic acids, flavonoids, anthocyanins, vitamins, and other phenolic compounds. These bioactive compounds have the potential to scavenge and neutralize the free radical chain reaction before causing any deleterious effects to the body. Research-based evidences strongly assert that application of persimmon ingredients provides protection against hyperlipidemia and hyperglycemia. Conclusively, persimmon and its components have potential as one of the effective modules in diet-based therapy.

10.1 INTRODUCTION

Evidence-based research pertaining to the therapeutic role of food and its bioactive ingredients has prompted new horizon of significant discoveries in the field of nutraceutical and designer foods to improve human health. Fruits and vegetables are considered as vital part of daily diet and are known to be a cache of biologically active phytochemicals possessing health-promoting perspectives. Ameliorating effect of fruits and vegetables against various metabolic syndromes is mainly associated with the presence of these phytomolecules.[73] Keeping in view the medicinal potential, various tropical and subtropical fruits and vegetables have acquired the attention of scientists. Among these, persimmon is nowadays the limelight of investigators owing to its potent antioxidative properties and unique taste.[31]

Persimmon (botanical name: *Diospyros kaki*; family: Ebenaceae) is a fleshy fibrous tropical, deciduous fruit; it is widely cultivated in warm regions of the world including China, Japan, Korea, Turkey, Brazil, and Italy. It is an edible fruit named as *Diospyros kaki* belonging to family *Ebenaceae*.[102] Among genus Diospyros, species *Diospyros kaki* is of most promising nature as more than 400 species of oriental persimmon are grown globally.[4,6,110] Fruits and leaves of persimmon plant are a cache of abundant phytochemicals, predominantly proanthocyanidins, catechins, tannins, carotenoids, phenolic, and organic acids.[3,7] On the other hand, Jung et al.[8] also documented polyphenols (160–250 mg/100 g), carotenoids (2 mg/100 g), and proteins (640–1300 mg/100 g) in dry persimmon residues. Nutritional profile of persimmon fruit is discussed in Table 10.1. Polyphenolic characterization reveals that persimmon fruit possesses a rich source of gallic acid, epicatechin, p-coumaric acid, ferulic acid, vanillic acid, and protocatechuic acid. Owing to these phenolic compounds (1150 mg/100 g) and fiber contents (63,480 mg/100 g), persimmon leaves are known for their medicinal perspectives.[54] Usually, the whole persimmon is edible if consumed in fresh form, but in some parts of the world, it is utilized in dry form also. On drying, peel of persimmon is removed physically; if not, it imparts bitter taste due to its astringency. Various persimmon-based products including juices, puree, syrups, jam, and jellies being commercialized in the market nowadays are prepared either from dried (whole/sliced) persimmon or fresh persimmon pulp. In spite of this, persimmon fruit, as a whole (unpeeled), is also being industrially subjected to fermentation for production of persimmon-based wine and vinegar.[3]

Habitats of Asian countries are quite aware of sensory characteristics and health benefits associated with consumption of persimmon fruit as its native to this continent. The demand for this fruit is increasing in European countries nowadays mainly due to the extensive research on its phytochemistry and allied medicinal properties. Phenolic profiling of persimmon has opened new opportunities for researchers to investigate the potential of this fruit in diet-based regime to augment numerous metabolic maladies. Among these, phytonutrients of persimmon possess antihypertensive, hypolipidemic, antidiabetic, neuroprotective, anticancer, antioxidative, cardio- and hepatoprotective, and the like, properties. The current review is an attempt to explicate the correlation between persimmon's phytochemistry and its curbing potential against various diseases.[25]

TABLE 10.1 Nutritional Value per 100 g (3.5 oz) of Persimmon Fruit (*Diospyros kaki*, Raw).

Parameters	Values
Energy	293 kJ
Fat	0.19 g
Saturated	0.02 g
Riboflavin (vitamin B2)	2.5 mg
Vitamin C	7.5 mg
Folate (vitamin B9)	8 μg
Protein	0.58 g
Sodium	1 mg
Iron	0.15 mg
Calcium	8 mg
Carbohydrates	18.59 g
Sugars	12.53 g
Dietary fiber	3.6 g

Source: Adapted with permission from USDA (1998).

10.2 BIOACTIVE COMPOUNDS IN PERSIMMON

Dietary uptake of phytochemicals in the human body is directly reliant on the administration of fruits and vegetables in our daily diet. Modulating effect of ingested food materials against various metabolic diseases is mainly due to the occurrence of natural antioxidants (polyphenols; carotenoids, tannins, catechins, proanthocyanidins, tocopherols, etc.) within the food commodity. However, to authenticate the use of these functional ingredients, scientific evidence is required to validate their potential use as nutraceutical foods in food supply chain system.[11] Biologically active phytoceutics present in persimmon fruit, including peel, pulp, and seed are discussed herein and illustrated in Figure 10.1.

Researchers have documented the fact that Persimmon fruit is an excellent source of plant secondary metabolites known as polyphenols. These naturally occurring compounds are identified and detected in Persimmon fruit both in free and bound form.[78] Phenolic compounds in persimmon fruit could be classified into two, that is, low-molecular-weight (hydrolyzable tannins, phenolic acids, and catechins) and high-molecular-weight (condensed tannins/proanthocyanidins) polyphenols depending upon their structure and molecular weight.[13]

While investigating the compositional analysis of persimmon along with phloroglucinol, epigallocatechin-3-O-gallate-phloroglucinol was found to be the most abundantly present condensed tannin in persimmon.[2] Tannins present in fruits and vegetables could be further classified into two classes depending upon their chemical properties, that is, nonhydrolyzable and hydrolyzable. Furthermore, they could be split into condensed and complex tannins depending upon their structural configurations.[10] Persimmon pulp constitutes of various other condensed tannins, such as flavonols, interflavan, flavan-3-O-galloylated, and flavan-3-ols, possessing higher molecular weights .[13] Phenolic compounds present in persimmon fruit are either simpler or highly polymerized constituents consisting of an aromatic ring comprising hydroxyl units. Persimmon phenolic compounds could be further characterized as extractable [EPP] and non-extractable phenolics [NEPP]. Metabolism and bioavailability of both types of phenolics differ in the human digestive system. EPP is up-taken directly via intestines, whereas NEPP does not interact with the metabolic system and are excreted from the body in the form of feces.[15,83] Sakanaka et al.[16] reported that fresh persimmon fruit consists of total phenolics (1.45 mg/100 g) and gallic acid contents (190–252 μg/100 g) in ample amount.

Phenolic compounds are a broader family consisting of various subclasses such as tannins, phenolic acids, and flavonoids. Various activities of phenolic compounds such as antioxidative potential widely depend upon glycosylation patterns and structural variations of these compounds. Persimmon-based tannins have reported minimizing the chance of stroke in hypertensive animal models.[96] This property was ascribed mainly due to elevated antioxidant activities of persimmon tannins, that is, 20 times more as compared to tocopherol (vitamin E). Profiling of persimmon tannins revealed its composition as epigallocatechin-3-O-gallate, epicatechin-3-O-gallate, epicatechin, and epigallocatechin.[18]

Protective external layers (peel and hull) on fruits encompass a broad spectrum of nutraceutical and therapeutic agents. Likewise, persimmon peel, an agro-waste is gaining importance of researchers due to its unique chemical profiling.[19] Main components of persimmon peel are dietary fiber, total carotenoids, total phenols, and antioxidants (Vit-C).[111] Among phenolic constituents, ample amount of gallic acid, caffeic acid, ferulic acid, and p-coumaric acids are present in persimmon. While among condensed tannins, proanthocyanidins are responsible for elevated antioxidative properties. These compounds are present abundantly in persimmon peel in comparison to its pulp.[21,22] Raised content of carotenoids (~340 mg/100 g β-carotene equivalents) are present in the peel of persimmon as compared to banana and apple peel (Kim et al., 2002). Among family carotenoids, xanthrophils (β-cryptoxanthin [50%]>zeaxanthin [5%] = lutein [5%]) are present in high concentration as compared to carotenes (β-carotene [10%]). Owing to the inimitable nutritional profiling of persimmon peel, ingestion of whole fruit must be preferred.[24,109]

Other than this, edible part of persimmon fruit possesses a rich percentage of micronutrients like vitamins C and A, calcium, and iron as 70, 65, 9, and 0.2 mg/100 g, respectively.[107] The rapid increase in total carotenoid contents is observed depending upon the variations in maturity index (green to orange-red) of persimmon fruit. Although lycopene and lutein content minimizes as fruit achieves its full maturity. These carotenes (α, β, and γ-carotene) and xanthophylls (zeaxanthin, lutein, and astaxanthin) are well known among scientists due to their ability to inhibit lipid peroxidation. Persimmon seeds contain high amount (70–78% of total fatty acid content) of linoleic, palmitic, and oleic acid as main fatty acids.[97]

Oleic acid has been reported as an anticancer agent, and this property is evidently supported by many earlier investigations.[27] Furthermore, ω-6 FA (omega-6 fatty acid) reduces the chances of cardiovascular complications. Along with this, persimmon leaves have also been reported to comprehend various bioactive compounds such as tatarine-C, saponin, hyperin, 40-dihydroxy-a-truxillic acid, kampferol, myricetin, quercetin, trifolin, isoquercetin, astragalin, annulatin, and kakispyrone.[28]

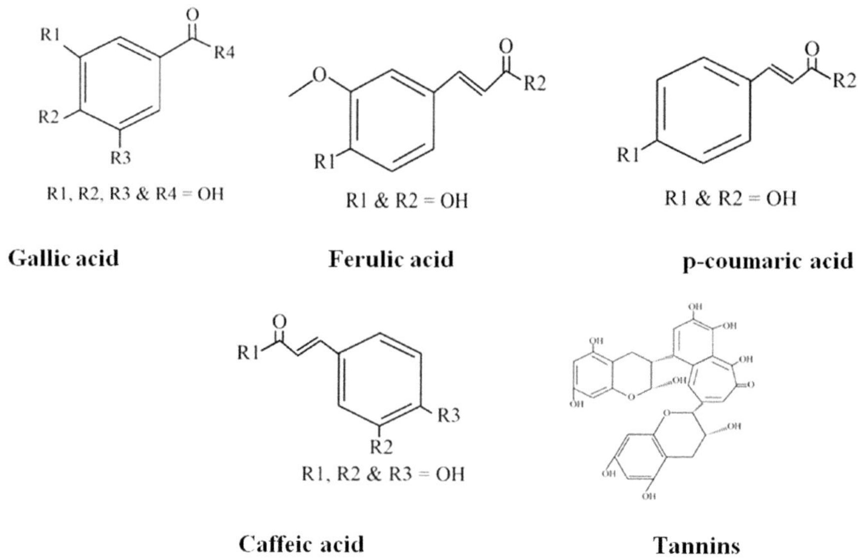

FIGURE 10.1 Chemical structures of various bioactive compounds in persimmon.

10.3 HEALTH PERSPECTIVES

10.3.1 ANTICANCER

Anticancer role regarding ethanolic extract of persimmon fruit has been explored by Park et al..[100] They examined the proliferative potential of persimmon fruit on various cancerous cell lines (HT-29, SW-480, LoVo, and HCT116) and stated its anticancer activity due to depression of cyclin-D1 expression in protein. The extract possesses the bioactive compounds that have an inhibitory action on the expression of β-catenin/TCF-dependent luciferase activity, catenin, and TCF4.[29] Apoptotic conditions were achieved due to the elevated permeability of mitochondrial membrane in prostate cancer cell lines due to the administration of rutin and quercetin. These bioactive compounds not only activated downstream apoptosis signaling but also deactivated Bcl-2 expression, released cytochrome-c, and upregulated BAX protein expression. As a result, both these persimmon leaves flavonoids suppressed proliferation and caused apoptosis of human prostate cell line (PC-3) mainly due to activation of oxidative stress conditions and mitochondria-related apoptosis.[23]

Flavonoids innate to persimmon fruit have also been known to act as an anticancer agent against the propagation of various cells like MCF-7, TK-10, and UACC-62.[31,91] Administration of acetone extract (100 μg/mL) of persimmon leaves in human leukemia (HL)-60 cells enhanced the differentiation of HL-60 cells.[47] Methanolic extraction of persimmon fruit polyphenols subjected to Ehrlich ascites carcinoma (EAC)-bearing albino mice (Swiss) demonstrated a significant increase in surviving rate of experimented mice due to a remarkable reduction in all examined tumor growth parameters.[19]

Furthermore, leaf extract of persimmon momentously enhanced the cytotoxic action of DOX (doxorubicin) drug in adenocarcinoma (A-549) cells. Reduction in phosphorylation of p-53 and ataxia telangiectasia mutated (ATM) cells was noticed in a concentration-dependent manner.[38] 24-hydroxyursolic acid showed inhibitory effect on cell proliferation, activation of AMPK (activated protein kinase), and mediation of potent anticancer properties due to down-regulation of cyclooxygenase-2 (COX-2) expression in human leukemia cells (HT-29). Moreover, cellular apoptotic conditions were induced due to activation of caspase-3 and poly-ADP ribose polymerase at Ser-15. In HT-29 cells, 24-hydroxyursolic acid also prompted DNA fragmentation and thus, expressively retarded colony propagation of experimented cancer cells on soft agar. It has also been documented that due to the application of 24-hydroxyursolic acid, blockage of EGF-induced ERKs phosphorylation was observed which further directed inhibitory action on the activity of AP-1 and cell alteration in JB6 CL41 cells.[40]

10.3.2 ANTIDIABETIC

Polyphenols present in persimmon fruit acts as antidiabetic agent owing to their glucose-lowering and insulin-enhancing properties.[39] Betulinicacid has an inhibitory action on adipogenesis (by retarding the expression of adipogenic markers), and enhances osteogenesis (by upregulating the expression of osteogenic markers). Furthermore, an elevation in basal glucose uptake was also noticed in adipocytes (3T3-L1).[71] Diet enriched with persimmon leaf powder helped in improving glomerular hypertrophy by lowering the serum levels of urea nitrogen. Besides this, persimmon leaf reduced the concentration of lipid peroxide and H_2O_2 (hydrogen peroxide) in the kidney. In experimented mice (*db/db*) administrated with persimmon leaf, a significant elevation was observed in activities of antioxidative enzymes like glutathione peroxidase (GSSH), superoxide dismutase (SOD), and catalase along with an increase in expression of mRNA (respective genes) occurred in kidneys of respective mice.[18]

Likewise, according to the outcomes of another experimental trial conducted by Bae et al.,[5] it could be concluded that ingestion of persimmon leaves extract (aqueous) resulted in inhibition of α-glucosidase activity and elevated antioxidative conditions. A reduction in body weight gain, enhancement in glucose tolerance, augmentation of serum lipid profiling, maintenance in islet structure, and suppression of hepatic fat accumulation were observed in diabetic mice that were daily supplemented orally with persimmon leaf extracts (PLE). Dose-dependent reduction (IC_{50}=689.22 μg/mL) was noticed in fasting glucose level as two concentrations of persimmon extract were administrated to streptozotocin (STZ)-induced diabetic rats. The momentous decline in diet and drink intake (water) and increase in rat's body weight was also observed. These subjected extracts displayed hypolipidemic properties as an increase in serum HDL-c and reduction in LDL-c and triglyceride content was observed as compared to control group. Results from in vitro analysis of cell line (3T3-L1) revealed a noteworthy decline in differentiated cells in comparison with non-differentiated cells. Both outcomes of in vivo and in vitro analysis suggested that these extracts significantly acted as antidiabetic and hypocholesterolemic agents in experimented study trial.[54]

Administration of persimmon leaf powder (5% w/w)] to experimented mice (C57BL/KsJ-db/db) remarkably showed hypoglycemic potential due to a reduction in gluconeogenic enzymes, liver weight, and HOMA-IR. They also increased the mRNA content, glucokinase activity, and glycogen level in liver. Persimmon polyphenols also prompted a reduction in PPARγ (lipogenic transcriptional factor) along with a decrease in activity and expression of biocatalysts (enzymes) involved in lipogenesis.

Furthermore, lipid excretion via feces was also simultaneously increased due to the application of persimmon leaf powder. Reduction in levels of malondialdehyde (MDA) and enhancement in superoxide dismutase (SOD) was also observed in diabetic mice (induced diabetes via intraperitoneal administration of STZ) that were subjected to persimmon polyphenols.[60] Vomifoliol 9-O-α-arabino-furanosyl(1-6)-β-D-glucopyranoside a polyphenol isolated from leaves of persimmon are reported to have inhibitory action in a concentration-dependent manner on α-glucosidase activity and augmented peripheral glucose; therefore, could be helpful in mitigation of Type-2 diabetes.[98]

Persimmon fruit-based carotenoids meaningfully enhanced the accretion of β-cryptoxanthin (source of vitamin A) in the liver and lowered the activity of serum alanine aminotransferase (ALAT). These polyphenols also transformed gene expression profiling and elevated phosphorylation of insulin receptor β-tyrosine in animal (rats) model.[33] Additionally, bioactive compounds in persimmon fruit evidently exhibited inhibitory action on carbohydrate absorption, enhanced antioxidative defense system, and peripheral glucose utilization.[112] Persimmon proanthocyanidins impeded with the actions of digestive enzymes and advanced glycation end product (AGE) formation. Ameliorating effect of proanthocyanidins isolated from persimmon against diabetes has been revealed earlier due to a reduction in renal advanced glycation end-products (RAGE), serum glucose, urea nitrogen, and urinary and glycosylated proteins.[46] Conclusively, protecting effect of persimmon's proanthocyanins against diabetic rats revealed significant inhibitory action on the liberation of reactive oxygen species (ROS) and sterol regulatory element binding protein (SREBP)-1 and 2. Besides this, they up-regulated the expression of nuclear factor kappa-B (NF-kB) through activation of inhibitor kappa B alpha (Ik-Bα) protein. It also lessened the expression of COX-2 and inductive nitric oxide synthases (NOSs).[22]

10.3.3 OXIDATIVE STRESS

Administration of persimmon leaf extract (1000 mg/kg BW) to rats elevated the concentration of superoxide dismutase (SOD), xanthine dehydrogenase (XDH), GSH (hepatic glutathione content), and catalase activities along with decline in xanthine oxidase (XO) and MDA levels.

Similarly, reduction in serum aspartate transaminase (AST), ALP, alanine transaminase (ALT), GGT (γ-glutamyl-transferase), and bilirubin content were also observed due to ingestion polyphenols present in persimmon leaf extract.[81] Likewise, Bilal et al.[9] analyzed the defensive role of oleic and linoleic acid present in seeds of persimmon fruit against acetylcholinesterase (AChE) inhibition and oxidative stress. They were of the view that inhibitory effect of examined fatty acids were mainly due to their ability of scavenge superoxide anion radicals in a concentration-dependent manner.

Oral ingestion of persimmon polyphenols (1000 mg/kg/day) has also been reported to protect against cisplatin (CP)-induced oxidation in testicles of male rats. Furthermore, application of mentioned treatment inhibited the increase of thiobarbituric acid-reactive substances (TBARS) content and decreased the levels of glutathione peroxidase (GPx), SOD, GSH, and catalase. Bioactive compounds in persimmon further led to decrease of lipid peroxidation and improvement in serum testosterone content, total sperm count, and motility.[87] Some proanthocyanidins present berries (blackberries, grapes, and persimmon) having low polymerization are gaining importance in nutraceutical industry. Keeping this in view, Yokozawa et al.[103] conducted a trial comprising senescence-accelerated mouse prone/8 (SAMP-8) model. They administrated oligomers isolated from persimmon fruit and reported no stereotypical behavior in treated SAMP-8 rats. Moreover, significant elevation in density of axons in the hippocampal CA-1, phosphorylation of vascular endothelial growth factor receptor-2 (VEGFR-2) in the hippocampal CA-3, choroid plexus, and hypothalamus was noticed in oligomer-subjected rats as compared to control.

An experimental trial conducted by Sun et al.[93] comprising hydrogen peroxide stimulated murine cells (MC3T3-E1). In this efficacy trial, H_2O_2-induced oxidation was reduced due to therapeutic application of persimmon leaves flavonoids. Pretreatment of MC3T3-E1 cells with flavonoids extracted from persimmon leaves at a dose of 1.25, 2.50, and 5.00 μg/mL for 24 h before application of H_2O_2 (250 μM) for extra 6 h were adopted herein. Flavonoids from persimmon leaves pre-incubated with MC3T3-E1 cells did not show any cytotoxicity and enhanced the cell viability and $\Delta\Psi m$ dose dependently when treated with H_2O_2. Significant inhibition in iNOS, NO, and MDA content was noticed in proincubated flavonoids from persimmon leaves. In a nutshell, persimmon leaf flavonoids significantly suppressed the apoptotic conditions in hydrogen peroxide induced MC3T3-E1 cells as revealed from results of protein expression of NF-κB/p65 and mRNA expression of iNOS, COX-2, Bax, Bcl-2, and caspase-3. Anti-apoptotic mechanism of persimmon leaf-based flavonoids is linked to the inhibition of translocation of NF-κB/p65 toward the nucleus. Similarly, proanthocyanidin averts cellular injury induced by application of hydrogen peroxide due to enhancement in activities of anti-oxidative enzyme, reduction in mitochondrial membrane potential and, ratio of glutathione (GSH) to oxidized glulathione (GSSG).[51]

Proanthocyanidin extracted from persimmon peel have found to lower the expression of 8-hydroxy-2'-deoxyguanosine (8-OHdG) and elevated the nuclear sirtuin-1 (SIRT-1) expression in a dose-dependent manner.[22] Polyphenols (5 and 10 μg/mL) from persimmon peels considerably suppressed the cytotoxicity due to induction of glucose, reduced the intra-cellular levels of ROS in a concentration-dependent manner, and inhibited the contents of NO, superoxide, and peroxynitrite. Proanthocyanidins from persimmon reduced the overexpressions of iNOS and COX-2 proteins. In addition, they also repressed translocation of NF-kappaB.[3] Active ingredients present in persimmon amplified total antioxidative potential, reduced urinary oxidative stress, and urinary 8-isoprostane content.[74] Cells pretreated with flavonoids isolated from persimmon helped in lessening of hydrogen peroxide-induced cellular injuries and apoptotic conditions due to increase in the expression of Bcl-2. Furthermore, enhancement in redox imbalance was also noticed due to improvement in decline of endogenous antioxidants (catalase, GSH, and GPx) along with reduction in leakage of lactate dehydrogenase and decrease in amassing content of MDA.[34]

10.3.4 ANTIOBESITY

Persimmon fruit is known to be a significant antiobesity agent owing to its potential to lower the levels of triglycerides and accumulation of lipids in 3T3-L1 adipose cells of mouse.[53,54] Extracts collected from persimmon lead to significant reduction in the serum concentration of cholesterol (total and hepatic), thiobarbituric acid-reactive substances (TBARS) present in erythrocytes, levels of HMG-CoA reductase (regulating enzymes in cholesterol metabolism), and expression of sterol regulatory element binding protein-1c (SREBP-1c) gene.[43] Vinegar produced from persimmon fruit helps in reduction of body fat and weight, suppressed the secretions of interleukin-6 (IL-6), tumor necrosis factor-alpha (TNF-α), NF-k (nuclear factor kappa) of B-cells.[88] Similarly, Zou et al.[57] examined that tannins from persimmon remarkably inhibited the peroxisome proliferator-activated receptor-γ (PPAR-γ), intracellular fat accumulation, and expression of C/EBPα (CCAAT-[cytosine-cytosine-adenosine-adenosine-thymidine]/enhancer binding proteins-α). Moreover, they also inhibited the SREBP-1, enhanced the concentration of microRNA-27a&27b, and suppressed the adipocyte fatty acid binding protein-2 (A-FABP-2).[57] The protective effect of persimmon fruits in an efficacy trial consisting of C57BL/6 male mice momentously enhanced hepatic CYP7A1 (cholesterol 7 alpha-hydroxylase) expression.[67]

Persimmon have reported to momentously decrease the escalation of serum lipid biomarkers like total cholesterol (TC), low-density lipoprotein cholesterol (LDL-c), and triglycerides (TG) in male C57BL/6 mice. Results of real-time polymerase chain reaction (RT-PCR) conclusively showed that persimmon fruit-supplemented diets increased (three-folds) the cellular response of CYP7A1 (cholesterol 7 alpha-hydroxylase) gene in liver of experimented rats. CYP7A1 is known to be an important gene in maintaining the cholesterol homeostasis by up-regulating the synthesis of bile acids. Therefore, the elevated synthesis of bile acids from cholesterol could possibly be the reason for hypocholesterolemic potential of persimmon fruits.[58]

10.3.5 ANTI-INFLAMMATORY ROLE

Aqueous extract of persimmon fruit displayed anti-inflammatory role against the mast cell mediated-allergy by suppressing the release of β-hexosaminidase and histamine from allergic cells due to modulation of intracellular calcium (Ca) and cAMP levels. Persimmon polyphenols also reduced the expression of TNF-α and IL-1β.[44] Similarly, in an allergic BALB/c mice model, methanolic extract of persimmon caused meaningful inhibition in total amount of ovalbumin (OVA)-induced total inflammatory cells (lymphocytes, eosinophils, and macrophages) present in bronchoalveolar lavage fluids (BALFs). Furthermore, significant reduction was observed in IL-4, IL-5, and eotaxin contents due to administration of persimmon polyphenols.

In Alzheimer's disease [AD], a peptide Amyloid-β [Aβ] is responsible for initiating various pathological actions like inflammation, activation of microglial cells, and oxidative stress conditions, causing plaque formation and neuronal death. Flavonoids originated from persimmon have reported to minimize the production of amyloid-β, inhibition of neuroinflammation, enhancement in cerebrovascular functioning, and improvement of cognitive performance. These polyphenols inhibited the expression of BACE-1 (β-site amyloid precursor protein cleavage enzyme-1), lessened the burden of amyloid-β, and minimized the activation of microglial in senile plaques.[60,61] In recent times, it has also been explored that persimmon flavonoids effectively protected against atopic dermatitis (AD). This was mainly due to reduction in atopic dermatitis lesions on skin, serum immunoglobulin-E (Ig-E), serum histamine contents, and infiltration of cellular inflammation. Aqueous extract of persimmon have inhibitory effect on the TNF-α, IL-1β, and a chemokine through suppression of NF-kB and signal transducer and activator of transcription-1 in cells from immortal keratinocytes (HaCaT-cells).[62] Persimmon polyphenols are reported to have inhibitory action on production of TNF-α and IL-6, suppressed dermal infiltration, and decreased the physical scratching of lesions. They also reduced the dermal infiltration of mast cells, thickness of ear skin, production of serum IL-4, and IgE in mice having AD.[14]

Astragalin, an anti-inflammatory component abundantly present in leaves of persimmon significantly reduced emphysema tissue damage in alveoli of mice exposed to OVA. Supplementation of astragalin (20 mg/kg) to OVA-treated mice reduced the accumulation of mast cells in subepithelium and helped in reversing the airway constriction and thickening produced due to an episode of OVA.[48] Quercetin present in persimmon fruit have been reported to possess anti-inflammatory potential as it reduces the expression of COX-2, TNF-α, nitric oxide (NO), IL-1β [interleukin-one beta], IL-6 [interleukin-6], mitogen-activated protein kinase (MAPKs), and iNOS (inducible NO synthase iNOS) in mouse RAW 264.7 macrophages stimulated with LPS (lipopolysaccharide). Polyphenol quercetin also inhibits the proteins involved in MAPKs pathways such as JNK, p-p38, and ERK.[17] Astragalin also diminished the activity of MPO (myeloperoxidase) and the expression of TNF-α, IL-6, IL-1β, and infiltration of inflammatory cells in a concentration-dependent manner. Furthermore, astragalin effectively reduced the activity of NF-κB by suppressing the phosphorylation and degradation of IκBα and the nuclear translocation of p-65.[68]

Astragalin meaningfully decreased LPS-stimulated COX-2, cytokines/chemokines, iNOS, expression, and formation of NO in mouse [J774A.1] macrophages. Astragalin repressed LPS-stimulated NF-κBactivation due to suppression of deprivation of IκBα, translocation of NF-κB, and NF-κB dependent gene reporter assay.[58]

Likewise, betulinic and coussaric acid displayed anti-inflammatory potential through inhibition of NF-kB pathway. Moreover, they also significantly suppressed liberation of NO, IL-1β, TNF-α, IL-6, and prostaglandin E_2 (PGE_2) in lipopolysaccharide-stimulated macrophages (RAW 264.7). Additionally, both these compounds, that is, betulinic (BA) and coussaricacids (CA) reduced the expression of iNOS and COX-2. They further examined the potential of BA and CA on expression of heme oxygenase (HO)-1 in RAW 246.7 macrophages. Among both these compounds BA, showed expression of HO-1 protein in concentration-dependent manner as compared to CA. BA also suppressed LPS stimulated NF-kB activity along with production of cytokine and pro-inflammatory mediator. This is mainly due to reversal of SnPP, an inhibitor of heme oxygenase-1.[45]

It has also been proven from outcomes of various epidemiological studies that liver dysfunction and risk of insulin resistance are inversely interlinked with serum β-cryptoxanthin. Ingestion of β-cryptoxanthin may prevent from nonalcoholic steatohepatitis (NASH) that is proposed to be an issue developed due to oxidative stress conditions and insulin resistance from nonalcoholic fatty liver disease. In order to examine the potential of β-cryptoxanthin against diet-induced NASH, fat, and cholesterol-rich diet was administrated with or without β-cryptoxanthin (0.003%) to C56BL/6J mice for 84 days. Due to administration of this feeding pattern, β-cryptoxanthin ameliorated fat accumulation, activated fibrosis in CL diet-induced NASH, and stellate cells in experimented mice. Results of complete gene profiling revealed that β-cryptoxanthin minimized steatosis and was reported to be more effectual in suppressing expression of inflammatory gene change in NASH. β-cryptoxanthin proved to be less effective in case of gene expression in relation to lipid and cholesterol metabolism. β-cryptoxanthin inhibited the propagation of LPS-stimulated TNF-α-inducible gene in NASH. Elevated levels of biomarkers related to oxidative stress conditions like thiobarbituric acid reactive substances (TBARS) significantly decreased due to ingestion of β-cryptoxanthin in NASH. Hence, β-cryptoxanthin inhibited fibrosis and inflammation mainly due to inhibiting effect on increase and activation of immune cells and macrophages. Conclusively, they proposed that decrease in oxidative stress parameters could possibly be the anti-inflammatory mechanism in livers of mice with NASH.[50,70]

10.3.6 ANTIAGING

AD is known to be the most prevailing neurodegenerative ailment categorized by loss of neurons within the nervous system resulting in mental impairment. Administration of persimmon leaves ethyl acetate extracts (200 and 400 mg/kg) might significantly minimize the latency, markedly elevated the SOD and glutathione peroxidase (GSH-Px) contents, and reduced the MDA levels. Additionally, ethyl acetate extract of persimmon leaves (EAPL) reduced apoptotic conditions in neurons of hippocampus, inhibited the expression of caspase-3, p-JNK, and ratio of Bax to Bcl-2.[30]

Orally administrated oligomers isolated from persimmon fruits have reported to enhance spatial and object recognition injury of brain in senescence-accelerated mouse prone/8 (SAMP8) models. Persimmon fruit-based oligomers elevate the density of axons, therefore, increasing the expression of phosphorylated neurofilament-H in hippocampus subfield CA-1. Purposely, Yokozawa et al.[104] planned a study to authenticate the mechanism involved in use of oligomers in alleviating brain dysfunctionality due to aging factor. Ingestion of oligomers resulted in an enhancement of phosphorylation of VEGFR-2, validating the usage of persimmon-based oligomers as an effective neuroprotective agent. An increase in phosphorylated vascular endothelial growth factor receptor was more prominent in hypothalamus and choroid plexus as compared to other experimented regions of brain in SAMP8 mouse. Illuminating the association among memory loss due to aging and signaling of vascular endothelial growth factor receptor-2 could provide a new pathway in curtailment of memory loss due to aging. Furthermore, the lifespan of experimented animals were increased due to elevation in expression of SIRT-1; a vital factor known for extension of lifespan in brains. Earlier, Yokozawa et al.[105] also reported an improvement in behavioral changes and life span linked with the aging process. For this purpose, they selected the most used animal model (SAMP8) to investigate the effect of any specific compound on behavioral changes associated with aging. They were of the view that ingestion of persimmon oligomers did not stimulated rash stereotypical behaviors, for instance, jumping, rearing, and hanging from the lid of a cage. However, restriction in diet given to them amplified frequencies of these behaviors without a momentous alteration in motor functionality.

Experimented in vivo animal (rats) models subjected to persimmon flavonoids already having focal ischemia/reperfusion (I/R) injury caused by induction of middle cerebral artery occlusion (MCAO) revealed momentous reduction in lesions of brain hemisphere and elevated the neurodegenerative behavior of examined rats. These flavonoids also had protective effect on neurons in a concentration-dependent manner.[55] Persimmon-based oligomeric proanthocanidins augmented the defect in learning and memorizing due to aging process. Rats subjected to oligomers illuminated an increase in synaptophysin, neurofilament-H, and mircrotubule-linked proteins (MAP-2a, 2b).[59]

10.3.7 CARDIOVASCULAR ROLE

Persimmon fruit-based flavonoids have evidently been reported to minimize the serum contents of total cholesterol (TC), triacylglycerols (TAC), very-low-density lipoprotein cholesterol (VLDL-c), and low-density lipoprotein cholesterol (LDL-c). Consequently, they also elevate the serum concentration of high-density lipoprotein cholesterol (HDL-c) and apolipoproteins (A1 and B).[108] Persimmon fruit extract inhibits the effect of platelet-derived induced factor (PDGF) and expression of BB-induced MMP-1. Persimmon peel extract (PPE) suppressed the expression of PDGF–BB-induced MMP-1, invasion of HASMCs, inhibited rat aortic sprouting, and minimized the cell migration. Polyphenols present in PPE not only decreased the activity of Src-kinase but also ameliorated Akt signaling pathways and MAPK phosphorylation.[51,90] Application of persimmon peel extract on human umbilical vein endothelial cells revealed a significant elevation of NO metabolites, phosphorylation of Akt, and eNOS as well as reduction in secretion of ET-1.[79,80]

Serum concentration of VLDL-c, LDL-c, and total triglycerides were significantly decreased due to administration of persimmon-based flavonols, and flavonoids. Moreover, reduction in atherosclerotic conditions (lesions) was also noticed in aorta of rats subjected to flavonols and flavonoids.[89] Likewise, flavones extracted from persimmon fruit resulted in inhibition of ASK1 expression, which was elevated by TNF-α in aortic vascular smooth muscle cells (VSMCs) of rats.[75] Flavone from leaves of persimmon fruit exerts vascular protection due to inhibition

of VSMCs linked with hyperlipidemia.[76] Flavones suppress the in vitro apoptotic conditions in cultured neonatal rat cardiac myocytes which was achieved by induction of AGE products and hypoxia-reoxygenation.[77]

Hypocholesterolemic potential of persimmon fruit could also be authenticated from the results of an experimental trial comprising hyper-lipidemic hamsters. Accordingly, persimmon polyphenols increased serum high-density lipoproteins cholesterol (HDL-c) and reduced the elevated concentration of TC, LDL-c, and TG. Furthermore, wines made from red grapes and persimmon fruit enhanced vascular reactivity without any significant change in endothelial functions. Similarly, no noteworthy reduction was noticed between the application of persimmon or merlot wine groups, however, the procyanidin profiling of both administrated wines differed momentously while the dimer content were quantified as same.[92]

Consumption of higher amounts of fruits and vegetables in our daily diet are directly linked with reduction of total and LDL-c, therefore, reducing the onset of cardiovascular diseases in concentration-dependent manner.[101] Persimmon fruit as a whole is known to be a nutritious entity that is responsible for hypolipidemic effect. Main reason for this hypocho-lesterolemic property associated with persimmon fruit is the presence of phytochemicals that possess antioxidative characteristics.[26,49] Furthermore, the bioactivity of persimmon fruit could be attributed due to the presence of antioxidative minerals, vitamins, and polyphenols.[29] Other than this, water-soluble dietary fibers are also an important part of ingested fruit. It has been proven through various experimental trials that dietary fibers have modulating effect on various lipid-related metabolic syndromes, therefore, dietary fiber associated with persimmon fruit could be helpful in achieving the recommended dietary allowance of dietary fiber (30–45 g). Other than dietary fibers, various micronutrients are also abundantly present in persimmon fruit that could be helpful in preventing coronary heart diseases and its related complications.[29]

Various scientific researchers have been conducted in order to explore the lipid-lowering potential of whole persimmon fruit or its different parts (leaf, seed, and pulp) in hyperlipidemic rat models. Persimmon leaf powder somewhat augmented hepatic and plasma lipid contents by increasing excretion of lipids through feces.[35] These health-promoting benefits might be due to the presence of phenolic biomolecules (1.15 g/100 g) and high fiber contents (63.48 g/100 g).

Reports are present showing that people living in Asian countries use leaves of persimmon fruit in curing hypertensive conditions. Furthermore, the experimental trials reveal that angiotensin-converting enzyme (ACE) and renin–angiotensin system (RAS) are the most vital targeted enzymes responsible for managing hypertension. Likewise, Sa et al.[85] isolated polyphenolic anti-coagulants from leaves of persimmon fruit. They were of the view that isolated fractions not only delayed prothrombin time (PT) but also hindered thrombin time (TT) and activated partial thromboplastin (APTT). These effects might be due to presence of astragalin which is reported to inhibit the activity of ACE activity.[65] The significance of ACE inhibitors in treatment of metabolic ailments has been proven from various clinical trials.[46] Afterwards, Matsumoto et al.[69] also addressed the hypocholesterolemic potential and bile acid binding properties related to persimmon polyphenols. They concluded that persimmon-derived polyphenols elevated the bile acid excretion via feces. Reduction of lipid content in blood and liver is directly associated with excretion of bile acids. The mode of action regarding hypolipidemic properties was mainly ascribed due to upregulation of 7α-hydroxylase, low-density lipoprotein receptor, and sterol regulatory element-binding protein-2 (SREBP-2) gene. Incidence of stroke was also reduced due to administration of persimmon fruit owing to its inhibitory action on lipid peroxidation and free radical scavenging properties.[1]

10.3.8 HEPATOPROTECTIVE AND ANALGESIC ROLE

Protective role of vinegar produced from persimmon fruit against hepatic injury induced by ingestion of alcohol was examined in an experimental trial conducted by Lee et al.[52] To authenticate this hypothesis, they administrated persimmon vinegar (100 mg/kg BW) through oral gavage to rats having alcohol-induced hepatic injury. They were of the view that administration of persimmon vinegar reduced the expression in hepatic mRNA of IL-(1β, 12β), COX-2, and toll-like receptor (TLR)-4.[95] Persimmon polyphenols administration (100 mg/kg BW) has reported as antihyperalgesic agent due to inhibition of noxious due to inhibition of histamine and dextran induced rat paw edema.[20,82] Furthermore, taraxeren-3-one a bioactive compound found in various parts of persimmon fruit has been reported to reduce Carr-induced expression of iNOS and COX-2. These protective mechanisms of taraxeren-3-one could be linked with the reduction of MDA levels in edema paw through elevation in CAT, GSH, SOD, and GPx activities of liver.[12,94,99]

Persimmon fruit is a renowned commodity among Korean pharmaceutics due to its alleviating potential against diseases like hypertension, diabetes, cancer, and atherosclerosis. To evaluate the hepatoprotective effect of methanolic extract of persimmon leaf (PLME) and whole fruit (PFME), administrated these extracts to C57BL/6 hepatotoxic male mice in which toxicity was induced due to application of ethanol (5 g/kg BW). Post-oral administration of ethanol for 12 h, blood and liver samples were analyzed to assess the hepatotoxicity. Outcomes of this study revealed that PFME and PLME treatments reduced serum alanine aminotransferase (ALT) and aspartate aminotransferase (AST) activities as compared to ethanol treated groups. Both administrated extracts not only minimized the level of serum lactate dehydrogenase (LDH) but also increased the activities of alcohol dehydrogenase (ADH). Similarly, due to application of PFME and PLME, hepatic cholesterol and serum triglyceride levels were momentously reduced in respective groups. Concentration of antioxidative enzymes like catalase, GSH-Px , and GSH-reductase were elevated significantly due to administration of PLME and PFME. However, content of MDA also decreased significantly in PFME- and PLME-treated groups as compared to control. Persimmon-derived flavonoids have been reported to significantly restore the viability of HepG2 cells in a concentration dependent manner. These flavonoids also assuaged the intracellular oxidative stress, probably owing to effectual antioxidative properties within the cells.[72]

10.3.9 HYPERTENSION

Persimmon tannins had protective effect on pathological and physiological changes in stroke-prone spontaneously hypertensive rats and also retarded lipid peroxidation in brain homogenates of rats, in a dose-dependent manner.[36,95] The vasorelaxant and antihypertensive potentials of water-soluble proanthocyanidins, collected in persimmon leaf-based tea, were explored in instinctively hypertensive rats, human umbilical vein endothelial cells, and rat aortas. Oral ingestion of proanthocyanidins meaningfully reduced the systolic blood pressure (SBP) in rats after 4 h when compared with the control. It was observed after supplementation of proanthocyanidin rich diet to experimented rats that upstream kinase Akt (Ser-473) in umbilical cells and phosphorylation of endothelial NOS (Ser-1177) elevated as the study progressed. Outcomes of this particular study suggest that anti-hypertensive characteristics of proanthocyanidins present in tea prepared from persimmon leaf may be due to vaso-relaxation through an endothelium-dependent NO/cGMP pathway. Conclusively, persimmon leaf-based proanthocyanidins could be used as dietary intervention in lowering blood pressure.[37]

Likewise, effect of administration of unripen persimmon and soluble tannin isolated from persimmon on blood pressure was investigated in spontaneously hypertensive rats. Reduction in SBP was noticed after ingestion of both (PP and PST) treatments for 2–3 days and lately was again elevated after switching to basal intake. A significant increase was noticed in serum NO contents and expression of NOS3 gene in PP fed groups. This antihypertensive mechanism of PP might be due to the induction of NO formation. Furthermore, administration of isolated persimmon soluble tannins suggested that PST treatment mainly reduced the SBP due to angiotensin I-converting enzyme (ACE) inhibitory potential. Single dose of scopoletin, a bioactive component present in PP and PST, actually lowered SBP revealing its potential as antihypertensive agent.[106]

10.3.10 ANTIASTHAMA

Kaempferol-3-O-glucoside in BALB/c mice sensitized with OVA blocked the expression induction of the autophagy-related beclin-1 and light chains 3A/B [LC3A/B] within 4 h induced by H_2O_2. It also suppressed the autophagosome formation with inhibiting the induction of these proteins in OVA-challenged airway subepithelium. Induction of autophagy by spermidine influenced the epithelial induction of E-cadherin and vimentin that was blocked by treating astragalin.[86]

Oxidative stress conditions and inflammation of lungs are known as the primary cause of obstructive pulmonary diseases (OPD). In emphysema, the macrophages formation are the core reason for alveolar damage and pulmonary inflammation persimmon leaves and green tea seeds contain an anti-inflammatory flavonoid known as astragalin. This experimental trial explicated that astragalin had inhibitory action on infiltration of inflammatory cell induced by H_2O_2 (20 μM) and congested thickening of airway and alveolar emphysema that was induced in mice by OVA (20 μg). In bronchial airway, epithelial BEAS-2B cells hydrogen peroxide elevated the expression of MCP-1/ICAM-1/αv integrin. This induction was further inhibited via application of astragalin at a concentration of 10–20 mg/kg and 1–20 μM to OVA-treated mice and oxidant-stimulated cells, respectively. Orally administrated (20 mg/kg) astragalin minimized the initiation of F4/80/CD68/CD11b in airways of OVA-challenged mice. Moreover, damage of emphysema tissues was noticed in OVA-exposed alveoli. Supplementation of astragalin in experimented mice resulted in blockage of mast cell recruitment in the airway subepithelium. Oral administration of astragalin (20 mg/kg) decreased induction of α-SMA in inflammatory airways. Furthermore, reversal was observed in airway thickening and constriction, occurring due to induction of an episode of OVA. Depending upon the outcomes of abovementioned results, it could be concluded that astragalin, a flavonoid present in persimmon leaves, might act as a nutraceutical compound in curtailing asthma and OPD.[64] Amelioration of oxidative stress-related epithelial eosinophilia and apoptotic conditions due to disturbance in TLR4-PKCβ2-NADPH oxidase-responsive signaling. Consequently, astragalin can be a potent therapeutic compound antagonizing endotoxin-induced oxidative stress causing airway inflammation and dysfunction.[93,106]

Various pathophysiological events are directly interlinked to this complex ailment known as asthma. In an investigation conducted by Lee et al.,[107] it was concluded that methanolic extract of fruit *Diospyros* blancoi exerted protective properties against allergic asthma in a murine asthma model. For this purpose, BALB/c mice were treated with OVA to induce allergic airway inflammation in specific murine model. These rats were then subjected to 20–40 mg/kg *Diospyros* blancoi methanolic extract (DBE) for 72 h. Owing to application of DBE, significant suppression was noticed in the total concentration of inflammatory cells (eosinophils, lymphocytes, and macrophages) in BALF. Besides this, treatment of DBE resulted in momentous reduction in IL-4, IL-5, and eotaxincontents in BALF and serum (Ig)E and IgG1 levels. Histological examination of lung tissue showed noticeable alleviation of inflammation in allergen-induced lung eosinophilic and mucus-producing goblet cells in the airway. DBE-induced expression of heme oxygenase-1 and retarded the activity of matrix metalloproteinase-9. The present findings collectively suggest that DBE demonstrated anti-inflammatory properties against allergic bronchial asthema in airway inflammation mouse model.[56]

10.3.11 ANTICOAGULATION

Thrombus formation is primarily ascribed with alteration of blood state, damage of cardiovascular walls, and physiochemical properties. Normally, body maintains both coagulation and anticoagulation systems in equilibrium within any living being. Thrombosis could actually be prevented by inhibition of blood coagulation. Process of coagulation is actually a bit complex in human body and is formed by transformation of fibrinogen to fibrin that is catalyzed by thrombin.[42,66]

It has been stated that leaves of persimmon ameliorates the content of plasma thromboxane B2 and reduced the expression of arterial serotonin transporter mRNA without changing the concentration of soluble P-selectin or levels of serotonin. Supplementation of persimmon leaves and omega-3 fatty acids caused significant decline in concentration of leptin and C-reactive proteins. Furthermore, persimmon leaves also minimized the serum content of total and LDL-cholesterol.[63,84]

10.4 SUMMARY

Persimmon is a promising and significant source of bioactive moieties including proanthocyanidins, phenolic, tannins, flavonoids, carotenoids, and dietary fiber, and has been used to alleviate several maladies. These valued compounds have been reported to effectively protect from numerous disorders associated with oxidative stress. They play an imperative role against cancer insurgence, diabetes complications, inflammation, cardiovascular, aging, and so on. Furthermore, precisely research-based designed models should be planned in clinical studies to reinforce the data obtained from in vivo and ex vivo studies.

KEYWORDS

- **persimon**
- **nutritional profiling**
- **health perspectives**
- **plant pigment**
- **protective agent**
- **anthocyanins**
- **phenolic acids**

REFERENCES

1. Ahn, H.-S.; Jeon, T.-I.; Lee, J.-Y.; Hwang, S.-G.; Lim, Y.; Park, D.-K. Antioxidative Activity of Persimmon and Grape Seed Extract, In Vitro and In Vivo. *Nutri. Res.* **2002,** *22*, 1265–1273.
2. Akagi, T.; Ikegami, A.; Yonemori, K. DkMyb2 Wound-induced Transcription Factor of Persimmon (*Diospyros kaki* Thunb.), Contributes to Proanthocyanidin Regulation. *Planta* **2010,** *232*, 1045–1059.
3. Akter, M.-S.; Ahmed, M.; Eun, J.-B. Effect of Blanching and Drying Temperatures on the Physicochemical Characteristics, Dietary Fiber Composition and Antioxidant related Parameters of Dried Persimmons Peel Powder. *Int. J. Food Sci. Nutri.* **2010,** *61*, 702–712.
4. Ashry, O.-M.; Hussein, E.-M.; Abd El-Azime, A.-S. Restorative Role of Persimmon Leaf (*Diospyros kaki*) to Gamma Irradiation-induced Oxidative Stress and Tissue Injury in Rats. *Int. J. Rad. Biol.* **2017,** *93* (3), 324–329.
5. Bae, U.-J.; Park, S.-H.; Jung, S.-Y.; Park, B.-H.; Chae, S.-W. Hypoglycemic Effects of Aqueous Persimmon Leaf Extract in a Murine Model of Diabetes. *Mol. Med. Rep.* **2015,** *12* (2), 2547–2554.
6. Bei, W.; Peng, W.; Ma, Y.; Xu, A. Flavonoids from the Leaves of *Diospyros kaki* Reduce Hydrogen Peroxide-induced Injury of NG108-15 Cells. *Life Sci.* **2005,** *76* (17), 1975–1788.
7. Bei, W.; Zang, L.; Guo, J.; Peng, W.; Xu, A.; Good, D. A; Hu, Y.; Wu, W.; Hu, D.; Zhu, X.; Wei, M.; Li, C. Neuroprotective Effects of a Standardized Flavonoid Extract from *Diospyros kaki* Leaves. *J. Ethnopharmacol.* **2009,** *126* (1), 134–142.
8. Bibi, N.; Khattak, A.-B.; Mehmood, Z. Quality Improvement and Shelf Life Extension of Persimmon Fruit *Diospyros kaki. J. Food Eng.* **2007,** *79*, 1359–1363.
9. Bilal, S.; Khan, A.-L.; Waqas, M.; Shahzad, R.; Kim, I.-D.; Lee, I.-J.; Shin, D.-H. Biochemical Constituents and In Vitro Antioxidant and Anticholinesterase Potential of Seeds from Native Korean Persimmon Genotypes. *Molecules* **2016,** *21* (7). DOI: 10.3390/molecules21070893.

10. Bravo, L.; Saura-Calixto, F.; Goni, I. Effects of Dietary Fibre and Tannins from Apple Pulp on the Composition of Feces in Rats. *Bri. J. Nutri.* **1992,** *67*, 463–473.
11. Brusotti, G.; Montanari, R.; Capelli, D.; Cattaneo, G.; Laghezza, A.; Tortorella, P.; Loiodice, F.; Peiretti, F.; Bonardo, B.; Paiardini, A.; Calleri, E.; Pochetti, G. Betulinic Acid is a PPARγ Antagonist that Improves Glucose Uptake, Promotes Osteogenesis and Inhibits Adipogenesis. *Sci. Rep.* **2017,** *7* (1), 5777.
12. Chang, T.-N.; Huang, S.-S.; Chang, Y.-S.; Chang, C.-I.; Yang, H.-L.; Deng, J.-S.; Kuo, Y. H.; Huang, G.-J. Analgesic Effects and Mechanisms of Anti-inflammation of Taraxeren-3-One from Diospyrosmaritima in Mice. *J. Agri. Food Chem.* **2011,** *59* (17), 9112–9119.
13. Chinfang, L.; Shizue, K.; Jun, T.; Ruriko, I.; Tomoyoshi, S.; Akira, O.; Kiharu, I. Antihypertensive Effects of Unripe Persimmon (*Diospyros kaki* L. cv. Hiratanenashi) Fruit and Its Component in Spontaneously Hypertensive Rats. *Food Sci. Technol. Res.* **2012,** *18* (3), 391–398.
14. Cho, B.-O.; Che, D.-N.; Yin, H.-H.; Shin, J.-Y.; Jang, S.-I. *Diospyros lotus* Leaf and Grapefruit Stem Extract Synergistically Ameliorate Atopic Dermatitis-like Skin Lesion in Mice by Suppressing Infiltration of Mast Cells in Skin Lesions. *Biomed. Pharmaco.* **2017,** *89*, 819–826.
15. Cho, I.-H.; Gong, J.-H.; Kang, M.-K.; Lee, E.-J.; Park, J.-H.; Park, S.-J.; Kang, Y.-H. Astragalin Inhibits Airway Eotaxin-1 Induction and Epithelial Apoptosis Through Modulating Oxidative Stress-responsive MAPK Signaling. *Pul. Med.* **2014,** *14*, 122.
16. Cho, I.-H.; Choi, Y.-J.; Gong, J.-H.; Shin, D.; Kang, M.-K.; Kang, Y.-H. Astragalin Inhibits Autophagy-associated Airway Epithelial Fibrosis. *Res. Res.* **2015,** *16*, 51.
17. Cho, Y.-H.; Kim, N.-H.; Khan, I.; Yu, J.-M.; Jung, H.-G.; Kim, H.-H.; Jang, J.-Y.; Kim, H. J.; Kim, D.-I.; Kwak, J.-H.; Kang, S.-C.; An, B.-J. Anti-inflammatory Potential of Quercetin-3-O-β-D-("2"-galloyl)-Glucopyranoside and Quercetin Isolated from *Diospyros kaki* Calyx via Suppression of MAP Signaling Molecules in LPS-induced RAW 264.7 Macrophages. *J. Food Sci.* **2016,** *81* (10), C2447–C2456.
18. Choi, M.-S.; Jeong, M.-J.; Park, Y.-B.; Kim, S.-R.; Jung, U.-J. The Leaf of *Diospyros kaki* Thumb Ameliorates Renal Oxidative Damage in Mice with Type 2 Diabetes. *Prev. Nutri. Food Sci.* **2016,** *21* (4), 378–383.
19. Das, S.; Bhattacharya, S.; Pramanik, G.; Haldar, P. K. Antitumoureffect of Diospyros Cordifolia Bark on Ehrlich Ascites Carcinoma-bearing Swiss Albino Mice. *Nat. Prod. Res.* **2012,** *26* (17), 1631–1633.
20. Das, S.; Haldar, P.-K.; Pramanik, G.; Panda, S.-P.; Bera, S. Evaluation of Analgesic and Anti-inflammatory Activity of Diospyros Cordifolia Extract. *Afr. J. Trad. Comp. Alt. Med.* **2011,** *8* (1), 11–14.
21. Deng, H.; Wen, Q.; Luo, Y.; Huang, Y.; Huang, R. Influence of Different Extracts from Persimmon Leaves on the Antioxidant Activity in Diabetic Mice. *Zhong Nan Da XueXueBao Yi Xue Ban* **2012,** *37* (5), 469–473.
22. Dewanjee, S.; Gangopadhyay, M.; Das, A.-K. Multimodal Approaches of the Hydroalcoholic Extract of Diospyros Peregrina Fruits in Diabetic Therapy. *Nat. Prod. Res.* **2011,** *25* (8), 827–833.
23. Ding, Y.; Ren, K.; Dong, H.; Song, F.; Chen, J.; Guo, Y.; Liu, Y.; Tao, W.; Zhang, Y. Flavonoids from Persimmon (*Diospyros kaki* L.) Leaves Inhibit Proliferation and

Induce Apoptosis in PC-3 Cells by Activation of Oxidative Stress and Mitochondrial Apoptosis. *Chem. Biol. Int.* **2017,** *275*, 210–217.

24. Direito, R.; Lima, A.; Rocha, J.; Ferreira, R. B.; Mota, J.; Rebelo, P.; Fernandes, A.; Pinto, R.; Alves, P.; Bronze, R.; Sepodes, B.; Figueira, M. E. Dyospiros Kaki Phenolics Inhibit Colitis and Colon Cancer Cell Proliferation, but not Gelatinase Activities. *J. Nutr. Biochem.* **2017,** *46*, 100–108.
25. Dushimemaria, F.; Preez, C.-I.-D.; Mumbengegwi, D.-R. Randomized Anticancer and Cytotoxicity Activities of *Guibourtia coleosperma* and *Diospyros chamaethamnus*. *Afr. J. Trad. Comp. Alt. Med.* **2017,** *14* (4), 1–7.
26. Gorinstein, S.; Kulasek, G.-W.; Bartnikowska, E.; Leontowicz, M.; Zemser, M.; Morawiec, M. The Effects of Diets, Supplemented with Either Whole Persimmon or Phenol-free Persimmon, on Rats Fed Cholesterol. *Food Chem.* **2000,** *70*, 303–308.
27. Gorinstein, S.; Zachwieja, Z.; Folta, M.; Barton, H.; Piotrowicz, J.; Zemser, M. Comparative Content of Dietary Fiber Total Phenolics and Minerals in Persimmon and Apples. *J. Agric. Food Chem.* **2001,** *49*, 952–957.
28. Gorinstein, S.; Leontowicz, H.; Leontowicz, M.; Jesion, I.; Namiesnik, J.; Drzewiecki, J.; Park, Y.-S.; Ham, K.-S.; Giordani, E.; Trakhtenberg, S. Influence of Two Cultivars of Persimmon on Atherosclerosis Indices in Rats Fed Cholesterol-containing Diets. Investigation In Vitro and In Vivo. *Nutri.* **2011,** *27* (7–8), 838–846.
29. Hertog, M.-G.; Kromhout, D.; Aravanis, C.; Blackburn, H.; Buzina, R.; Fidanza, F. Flavonoid Intake and Long-term Risk of Coronary Heart Disease and Cancer in the Seven Countries Study. *Arch. Int. Med.* **1995,** *155*, 381–386.
30. Huang, S.-W.; Wang, W.; Zhang, M.-Y.; Liu, Q.-B.; Luo, S.-Y.; Peng, Y.; Sun, B.; Wu, D. L.; Song, S.-J. The Effect of Ethyl Acetate Extract from Persimmon Leaves on Alzheimer's Disease and Its Underlying Mechanism. *Phytomedicine* **2016,** *23* (7), 694–704.
31. Igual, M.; Castelló, M.-L.; Ortolá, M.-D.; Andrés, A. Influence of Vacuum Impregnation on Respiration Rate, Mechanical and Optical Properties of Cut Persimmon. *J. Food Eng.* **2008,** *86*, 315–323.
32. Innami, S.; Tabata, K.; Shimizu, J.; Kusunoki, K.; Ishida, H.; Matsuguma, M. Dried Green Leaf Powders of Jew's Mellow (Corchorus), Persimmon (*Diospyros kaki*) and Sweet Potato (*Ipomoea Batatas* Poir) Lower Hepatic Cholesterol Concentration and Increase Fecal Bile Acid Excretion in Rats Fed a Cholesterol-free Diet. *Plant Foods Hum. Nutri.* **1998,** *52*, 55–65.
33. Izuchi, R.; Nakai, Y.; Takahashi, H.; Ushiama, S.; Okada, S.; Misaka, T.; Abe, K. Hepatic Gene Expression of the Insulin Signaling Pathway Is Altered by Administration of Persimmon Peel Extract, a DNA Microarray Study Using Type 2 Diabetic Goto-Kakizaki rats. *J. Agric. Food Chem.* **2011,** *59* (7), 3320–3329.
34. Jung, S.-T.; Park, Y.-S.; Zachwiej, Z.; Folta, M.; Barton, H.; Piotrowicz, J.; Katrich, E.; Trakhtenberg, S.; Gorinstein, S. Some Essential Phytochemicals and the Antioxidant Potential in Fresh and Dried Persimmon. *Int. J. Food Sci. Nutri.* **2005,** *56* (2), 105–113.
35. Jung, U.-J.; Park, Y.-B.; Kim, S.-R.; Choi, M.-S. Supplementation of Persimmon Leaf Ameliorates Hyperglycemia, Dyslipidemia and Hepatic Fat Accumulation in Type 2 Diabetic Mice. *PLoS One* **2012,** *7*, e49030.

36. Kameda, K.; Takaku, T.; Okuda, H.; Kimura, Y.; Okuda, T.; Hatano, T.; Agata, I.; Arichi, S. Inhibitory Effects of Various Flavonoids Isolated from Leaves of Persimmon on Angiotensin-converting Enzyme Activity. *J. Nat. Prod.* **1987,** *50* (4), 680–683.
37. Kawakami, K.; Aketa, S.; Sakai, H.; Watanabe, Y.; Nishida, H.; Hirayama, M. Antihypertensive and Vasorelaxant Effects of Water-soluble Proanthocyanidins from Persimmon Leaf Tea in Spontaneously Hypertensive Rats. *Biosci. Biotechnol. Biochem.* **2011,** *75* (8), 1435–1439.
38. Kawakami, K.; Nishida, H.; Tatewaki, N.; Nakajima, Y.; Konishi, T.; Hirayama, M. Persimmon Leaf Extract Inhibits the ATM Activity During DNA Damage Response Induced by Doxorubicin in A549 Lung Adenocarcinoma Cells. *Biosci., Biotechnol., Biochem.* **2011,** *75* (4), 650–655.
39. Khan, M.-M., Tran, B.-Q.; Jang, Y.-J.; Park, S.-H.; Fondrie, W.-E.; Chowdhury, K.; Yoon, S.-H.; Goodlett, D.-R.; Chae, S.-W.; Chae, H.-J.; Seo, S.-Y.; Goo, Y.-A. Assessment of the Therapeutic Potential of Persimmon Leaf Extract on Prediabetic Subjects. *Mol. Cells* **2017,** *40* (7), 466–475.
40. Khanal, P.; Oh, W.-K.; Thuong, P.-T.; Cho, S.-D.; Choi, H.-S. 24-hydroxyursolic Acid from the Leaves of the *Diospyros kaki* (Persimmon) Induces Apoptosis by Activation of AMP-activated Protein Kinase. *Planta Med.* **2010,** *76* (7), 689–693.
41. Kim, A.-H.; Kim, H.-J.; Ryu, R.; Han, H.-J.; Han, Y.-J.; Lee, M.-K.; Choi, M.-S.; Park, Y.-B.-A Mixture of Ethanol Extracts of Persimmon Leaf and *Citrus junos* Sieb Improves Blood Coagulation Parameters and Ameliorates Lipid Metabolism Disturbances Caused by Diet-Induced Obesity in C57BL/6J Mice. *J. Microbiol. Biotechnol.* **2016,** *26* (2), 295–308.
42. Kim, D.-C.; Ku, S.-K.; Bae, J.-S. Anticoagulant Activities of Curcumin and Its Derivative. *Rep. Biochem. Mol. Biol.* **2012,** *45*, 221e6.
43. Kim, G.-N.; Shin, M.-R.; Shin, S.-H.; Lee, A.-R.; Lee, J.-Y.; Seo, B.-I.; Kim, M.-Y.; Kim, T.-H.; Noh, J.-S.; Rhee, M.-H.; Roh, S.-S. Study of Antiobesity Effect through Inhibition of Pancreatic Lipase Activity of *Diospyros kaki* Fruit and Citrus Unshiu Peel. *Biomed. Res. Int.* **2016,** 1723042.
44. Kim, H.-H.; Kim, D.-S.; Kim, S.-W.; Lim, S.-H.; Kim, D.-K.; Shin, T.-Y.; Kim, S.-H. Inhibitory Effects of *Diospyros kaki* in a Model of Allergic Inflammation, Role of cAMP, Calcium and Nuclear Factor-κB. *Int. J. Mol. Med.* **2013,** *32* (4), 945–951.
45. Kim, K.-S.; Lee, D.-S.; Kim, D.-C.; Yoon, C.-S.; Ko, W.; Oh, H.; Kim, Y.-C. Anti-inflammatory Effects and Mechanisms of Action of Coussaric and Betulinic Acids Isolated from *Diospyros kaki* in Lipopolysaccharide-stimulated RAW 264.7 Macrophages. *Molecules* **2016,** *21* (9). DOI: 10.3390/molecules21091206.
46. Kim, M.-S.; Kim, S.-H. Inhibitory Effect of Astragalin on Expression of Lipopolysaccharide-induced Inflammatory Mediators Through NF-κB in Macrophages. *Arch. Pharmacal. Res.* **2011,** *34* (12), 2101–2107.
47. Kim, S.-H.; Cho, S.-S.; Simkhada, J.-R.; Park, S.-J.; Lee, H.-J.; Kim, T.-S.; Yoo, J.-C. Effects and Action Mechanism of *Diospyros kaki* on the Differentiation of Human Leukemia HL-60 Cells. *Oncol. Rep.* **2010,** *23* (1), 89–95.
48. Kim, Y.-H.; Choi, Y.-J.; Kang, M.-K.; Park, S.-H.; Antika, L.-D.; Lee, E.-J.; Kim, D.-Y.; Kang, Y.-H. Astragalin Inhibits Allergic Inflammation and Airway Thickening in Ovalbumin-challenged Mice. *J. Agric. Food Chem.* **2017,** *65* (4), 836–845.

49. Kim, Y.-J.; Yokozawa, T. Modulation of Oxidative Stress and Melanogenesis by Proanthocyanidins. *Biol. Pharma. Bull.* **2009,** *32* (7), 1155–1159.
50. Kobori, M.; Ni, Y.; Takahashi, Y.; Watanabe, N.; Sugiura, M.; Ogawa, K.; Nagashimada, M.; Kaneko, S.; Naito, S.; Ota, T. β-cryptoxanthin Alleviates Diet-induced Nonalcoholic Steatohepatitis by Suppressing Inflammatory Gene Expression in Mice. *PLoS One* **2014,** *9* (5), e98294.
51. Konić-Ristić, A.; Srdić-Rajić, T.; Kardum, N.; Aleksić-Veličković, V.; Kroon, P.-A.; Hollands, W. J.; Needs, P. W.; Boyko, N.; Hayran, O.; Jorjadze, M.; Glibetić, M. Effects of Bioactive-rich Extracts of Pomegranate, Persimmon, Nettle, Dill, Kale and Sideritis and Isolated Bioactives on Arachidonic Acid Induced Markers of Platelet Activation and Aggregation. *J. Sci. Food Agric.* **2013,** *93* (14), 3581–3587.
52. Lee, H.-M.; Park, M.-Y.; Kim, J.; Shin, J.-H.; Park, K.-S.; Kwon, O. Persimmon Vinegar and Its Fractions Protect Against Alcohol-induced Hepatic Injury in Rats Through the Suppression of CYP2E1 Expression. *Pharma. Biol.* **2016,** *54* (11), 2437–2442.
53. Lee, J.-H.; Lee, Y.-B.; Seo, W.-D.; Kang, S.-T.; Lim, J.-W.; Cho, K.-M. Comparative Studies of Antioxidant Activities and Nutritional Constituents of Persimmon Juice (*Diospyros kaki* L. cv. Gapjubaekmok). *Prev. Nutri. Food Sci.* **2012,** *17*, 141–151.
54. Lee, J.-S.; Lee, M.-K.; Ha, T.-Y.; Bok, S.-H.; Park, H.-M.; Jeong, K.-S.; Woo, M.-N.; Do, G.-M.; Yeo, J.-Y.; Choi, M.-S. Supplementation of Whole Persimmon Leaf Improves Lipid Profiles and Suppresses Body Weight Gain in Rats Fed High-fat Diet. *Food, Chem. Toxicol.* **2006,** *44* (11), 1875–1883.
55. Lee, K.-Y.; Jung, J.-Y.; Lee, M.-Y.; Jung, D.; Cho, E.-S.; Son, H.-Y. Diospyros Blancoi Attenuates Asthmatic Effects in a Mouse Model of Airway Inflammation. *Inflammation* **2012,** *35* (2), 623–632.
56. Lee, N.-K. Oleaginous Yeast Rhodosporidiumtoruloides as a Tool for Rapid Evaluation of Anti-Obesity Candidates, Inhibitory Effect of Persimmon Leaf Fermentate on Lipid Accumulation. *J. Microbiol. Biotechnol.* **2017,** *27* (10), 1773–1777.
57. Lee, Y.-A.; Cho, E.-J.; Yokozawa, T. Protective Effect of Persimmon (*Diospyros kaki*) Peel Proanthocyanidin Against Oxidative Damage Under H_2O_2-induced Cellular Senescence. *Biol. Pharma. Bull.* **2008,** *31* (6), 1265–1269.
58. Lee, Y.-A., Cho, E.-J.; Tanaka, T.; Yokozawa, T. Inhibitory Activities of Proanthocyanidins from Persimmon against Oxidative Stress and Digestive Enzymes Related to Diabetes. *J. Nutri. Sci. Vitaminol.* 2007, *53* (3), 287–292.
59. Lee, Y.-A.; Cho, E.-J.; Yokozawa, T. Oligomeric Proanthocyanidins Improve Memory and Enhance Phosphorylation of Vascular Endothelial Growth Factor Receptor-2 in Senescence-accelerated Mouse Prone/8. *Brit. J. Nutri.* **2010,** *103* (4), 479–489.
60. Leontowicz, H.; Gorinstein, S.; Lojek, A. Comparative Content of Some Bioactive Compounds in Apples, Peaches and Pears and Their Influence on Lipids and Antioxidant Capacity in Rats. *J. Nutri. Biochem.* **2002,** *13* (10), 603–610.
61. Li, F.; Liang, D.; Yang, Z.; Wang, T.; Wang, W.; Song, X.; Guo, M.; Zhou, E.; Li, D.; Cao, Y.; Zhang, N. Astragalin Suppresses Inflammatory Responses via Down-regulation of NF-κB Signaling Pathway in Lipopolysaccharide-induced Mastitis in a Murine Model. *Int. Immunopharmacol.* **2013,** *17* (2), 478–482.
62. Li, C.-M.; Leverence, R.; Trombley, J.-D.; Hagerman, A.-E. High Molecular Weight Persimmon (*Diospyros kaki* L.) Proanthocyanidin, A Highly Galloylated, A-Linked

Tannin with an Unusual Flavonol Terminal Unit, Myricetin. *J. Agric. Food Chem.* **2010,** *58* (16), 9033–9042.

63. Lu, X.; Mo, X.; Guo, H.; Zhang, Y. Sulfation Modification and Anticoagulant Activity of the Polysaccharides Obtained from Persimmon (*Diospyros kaki* L.) Fruits. *Int. J. Biol. Macromol.* **2012,** *51* (5), 1189–1195.
64. Ma, Jie; Liu, X.-Y.; Noh, K.-H., Kim, M.-J., Song, Y.-S. Protective Effects of Persimmon Leaf and Fruit Extracts Against Acute Ethanol-Induced Hepatotoxicity. *Prev. Nutri. Food Sci.* **2007,** *12* (4), 202–208.
65. Manach, C.; Scalbert, A.; Morand, C.; Rémésy, C.; Jiménez, L. Polyphenols, Food Sources and Bioavailability. *Am. J. Clin. Nutri.* **2004,** *79,* 727–747.
66. Mathers, C.-D.; Loncar, D. Projections of Global Mortality and Burden of Disease from 2002 to 2030. *PLoS Med.* **2006,** *3,* e442.
67. Matsumoto, K.; Watanabe, Y.; Ohya, M.-A.; Yokoyama, S. Young Persimmon Fruits Prevent the Rise in Plasma Lipids in a Diet-induced Murine Obesity Model. *Biol. Pharma. Bull.* **2006,** *29* (12), 2532–2535.
68. Matsumura, Y.; Kitabatake, M.; Ouji-Sageshima, N.; Yasui, S.; Mochida, N.; Nakano, R.; Kasahara, K.; Tomoda, K.; Yano, H.; Kayano, S.-I.; Ito, T. Persimmon-derived Tannin Has Bacteriostatic and Anti-inflammatory Activity in a Murine Model of Mycobacterium Avium Complex (MAC) Disease. *PLoS One* **2017,** *12* (8), e0183489.
69. Matsumoto, K.; Yokoyama, S.; Gato, N. Bile Acid-binding Activity of Young Persimmon (*Diospyros kaki*) Fruit and Its Hypolipidemic Effect in Mice. *Phytother. Res.* **2010,** *24,* 205–210.
70. Menendez, J.-A.; Vellon, L.; Colomer, R.; Lupu, R. Oleic Acid, the Main Monounsaturated Fatty Acid of Olive Oil, Suppresses Her-2/neu (erbB-2) Expression and Synergistically Enhances the Growth Inhibitory Effects of Trastuzumab (Herceptin) in Breast Cancer Cells with Her-2/neu Oncogene Amplification. *Ann. Oncol.* **2005,** *16* (3), 359–371.
71. Miller, P.-E.; Snyder, D.-C. Phytochemicals and Cancer Risk, a Review of the Epidemiological Evidence. *Nutri. Clin. Prac.* **2012,** *27*, 599–612.
72. Lee, M.-S.; Lee, I.-I.; Young, Y.-K.; Kim, J.; Heo, H.-J.; Kim, D.-O. Inhibitory Effect of the Ethyl Acetate Fraction from Astringent Persimmon on H2O2-induced Oxidative Stress in HepG2 Cells. *Food Sci. Biotechnol.* 2014, *23* (4), 1247–1252.
73. El Hadi, M.-A.-M.; Zhang, F.-J.; Wu, F.-F.; Zhou, C.-H.; Tao, J. Advances in Fruit Aroma Volatile Research. *Molecules* **2013,** *18* (7), 8200–8229.
74. Mure, K.; Takeshita, T.; Morioka, I.; Arita, M. Effects of Kakisu (Persimmon Vinegar) on Plasma Antioxidant Power and Urinary 8-isoprostane Level. *Nihon Eiseigaku-Zasshi* **2007,** *62* (1), 32–38.
75. Ouyancg, P.; Zhang, B.; Bei, W.-J.; Lai, W.-Y.; Peng, W. L.; Xu, D. L. Effects of Flavone from Leaves of *Diospyros kaki* on Expression of Apoptosis Signal-regulating Kinase 1 and Rat Vascular Smooth Muscle Cells Proliferation by Tumor Necrosis Factor Alpha In Vitro. *Zhong Yao Cai* **2007,** *30* (7), 819–822.
76. Ouyang, P.; Bei, W.; Lai, W.; Peng, W. Effects of Flavone from Leaves of *Diospyros kaki* on Rat Vascular Smooth Muscle Cells Proliferation Stimulated by Native Low-density Lipoprotein In Vitro. *Zhong Yao Cai* **2004,** *27* (8), 600–602.
77. Ouyang, P.; Bei, W.-J.; Lai, W.-Y.; Xu, D.-L.; Peng, W.-L. In Vitro Effects of Flavone from Leaves of *Diospyros kaki* on Rat Cardiac Myocyte Apoptosis Induced by

Hypoxia-reoxygenation and Advanced Glycation End Products. *Di Yi Jun Yi Da XueXueBao* **2003,** *23* (7), 680–682.

78. Palafox-Carlos, H.; Ayala-Zavala, J.-F.; González-Aguilar, G.-A. The Role of Dietary Fiber in the Bioaccessibility and Bioavailability of Fruit and Vegetable Antioxidants. *J. Food Sci.* **2011,** *76* (1), R6–R15.
79. Park, S.-B.; Park, G.-H.; Song, H.-M.; Son, H.-J.; Um, Y.; Kim, H.-S.; Jeong, J.-B. Anticancer Activity of Calyx of *Diospyros kaki* Thunb. through Downregulation of Cyclin D1 via Inducing Proteasomal Degradation and Transcriptional Inhibition in Human Colorectal Cancer Cells. *BMC Compl. Alt. Med.* **2017,** *17* (1), 445.
80. Plaza, L.; Colina, C.; Ancos, B.-D; Sanchez-Moreno, C.; Pilar Cano, M. Influence of Ripening and Astringency on Carotenoid Content of High-pressure Treated Persimmon Fruit (*Diospyros kaki* L.). *Food Chem.* **2012,** *130* (3), 591–597.
81. Rahman, M.-A.; Islam, A.-K.-M.-S.; Khair, A.; Bala, B.-K. Effect of Non-polar Gases on the Storage of Persimmon Fruits at Different Temperatures. *Pak. J. Biol. Sci.* **2002,** *5*, 84–87.
82. Rauf, A.; Uddin, G.; Khan, H.; Siddiqui, B.-S.; Arfan, M. Anti-hyperalgesic Activity of Crude Extract and 7-methyljuglone of *Diospyros lotus* Roots. *Nat. Prod. Res.* **2015,** *29* (23), 2226–2229.
83. Rice-Evans, C.-A.; Miller, N.-J.; Paganga, J. Antioxidant Properties of Phenolic Compounds. *Trend. Plant Sci.* **1997,** *2*, 152–159.
84. Ryu, R.; Kim, H.-J.; Moon, B.; Jung, U.-J.; Lee, M.-K.; Lee, D.-G.; Ryoo, Z.; Park, Y.-B.; Choi, M.-S. Ethanol Extract of Persimmon Tree Leaves Improves Blood Circulation and Lipid Metabolism in Rats Fed a High-Fat Diet. *J. Med. Food* **2015,** *18* (7), 715–723.
85. Sa, Y.-S.; Kim, S.-J.; Choi, H.-S. The Anticoagulant Fraction from the Leaves of *Diospyros kaki* L. Has an Antithrombotic Activity. *Arch. Pharmacal. Res.* **2005,** *28* (6), 667–674.
86. Sakanaka, S.; Tachibana, Y.; Okada, Y. Preparation and Antioxidant Properties of Extracts of Japanese Persimmon Leaf Tea (Kakinoha-cha). *Food Chem.* **2005,** *89*, 569–575.
87. Saral, S.; Ozcelik, E.; Cetin, A.; Saral, O.; Basak, N.; Aydın, M.; Ciftci, O. Protective Role of *Diospyros lotus* on Cisplatin-induced Changes in Sperm Characteristics, Testicular Damage and Oxidative Stress in Rats. *Andrologia* **2016,** *48* (3), 308–317.
88. Seo, H.; Jeon, B.-D.; Ryu, S. Persimmon Vinegar Ripening with the Mountain-cultivated Ginseng Ingestion Reduces Blood Lipids and Lowers Inflammatory Cytokines in Obese Adolescents. *J. Exer. Nutri. Biochem.* **2015,** *19* (1), 1–10.
89. Singh, S.; Joshi, H. *Diospyros kaki* (Ebenaceae), A Review. *Asian J. Res. Pharma. Sci.* **2011,** *1* (3), 55–58.
90. Son, J.-E.; Hwang, M.-K.; Lee, E.; Seo, S.-G.; Kim, J.-E.; Jung, S.-K.; Kim, J.-R.; Ahn, G.-H.; Lee, K.-W.; Lee, H.-J. Persimmon Peel Extract Attenuates PDGF-BB-induced Human Aortic Smooth Muscle Cell Migration and Invasion Through Inhibition of c-Src Activity. *Food Chem.* **2013,** *141* (4), 3309–3316.
91. Suh, J.-H.; Virsolvy, A.; Goux, A.; Cassan, C.; Richard, S.; Cristol, J.-P.; Teissèdre, P.-L.; Rouanet, J. M. Polyphenols Prevent Lipid Abnormalities and Arterial Dysfunction in Hamsters on a High-fat Diet, A Comparative Study of Red Grape and White Persimmon Wines. *Food Func.* **2011,** *2* (9), 555–561.

92. Suh, J.-H.; Virsolvy, A.; Goux, A.; Cassan, C.; Richard, S.; Cristol, J.-P.; Teissèdre, P. L.; Rouanet, J.-M. Polyphenols Prevent Lipid Abnormalities and Arterial Dysfunction in Hamsters on a High-fat Diet, A Comparative Study of Red Grape and White Persimmon Wines. *Food Func.* **2011,** *2* (9), 555–561.
93. Sun, L.; Zhang, J.; Fang, K.; Ding, Y.; Zhang, L.; Zhang, Y. Flavonoids from Persimmon (*Diospyros kaki*) Leaves (FPL) Attenuate H_2O_2-induced Apoptosis in MC3T3-E1 Cells via the NF-κB Path Way. *Food Func.* **2014,** *5* (3), 471–479.
94. Trongsakul, S.; Panthong, A.; Kanjanapothi, D.; Taesotikul, T. The Analgesic, Antipyretic and Anti-inflammatory Activity of *Diospyros Variegatakruz. J. Ethnopharmacol.* **2003,** *85* (2–3), 221–225.
95. Uchida, S.; Ohta, H.; Niwa, M.; Mori, A.; Nonaka, G.; Nishioka, I.; Ozaki, M. Prolongation of Life Span of Stroke-prone Spontaneously Hypertensive Rats (SHRSP) Ingesting Persimmon Tannin. *Chem. Pharmaceut. Bull.* **1990,** *38* (4), 1049–1052.
96. Uchida, S.; Ozaki, M.; Akashi, T.; Yamashita, K.; Niwa, M.; Taniyama, K. Effects of (-)-Epigallocatechin-3-O-gallate (Green Tea Tannin) on the Life Span of Stroke-prone Spontaneously Hypertensive Rats. *Clin. Exp. Pharmacol. Phys. Suppl.* **1995,** *22* (1), S302–S303.
97. Uddin, G.; Rauf, A.; Siddiqui, B.-S.; Muhammad, N.; Khan, A.; Shah, S.-U. Antinociceptive, Anti-inflammatory and Sedative Activities of the Extracts and Chemical Constituents *of Diospyros lotus* L. *Phytomedicine* **2014,** *21* (7), 954–959.
98. Wang, L.; Xu, M.-L.; Rasmussen, S.-K.; Wang, M.-H. Vomifoliol 9-O-α-arabinofuranosyl (1→6)-β-D-glucopyranoside from the Leaves of *Diospyros kaki* Stimulates the Glucose Uptake in HepG2 and 3T3-L1 Cells. *Carbohydr. Res.* **2011,** *346* (10), 1212–1216.
99. Willett, W.-C. The Role of Dietary n-6 Fatty Acids in the Prevention of Cardiovascular Disease. *J. Cardiovasc. Med.* **2007,** *8* (1), S42–S45.
100. Woodcock, M.-E.; Hollands, W.-J.; Konic-Ristic, A.; Glibetic, M.; Boyko, N.; Koçaoglu, B.; Kroon, P. -A. Bioactive-rich Extracts of Persimmon, but not Nettle, Sideritis, Dill or Kale, Increase eNOS Activation and NO Bioavailability and Decrease Endothelin-1 Secretion by Human Vascular Endothelial Cells. *J. Sci. Food Agric.* **2013,** *93* (14), 3574–3580.
101. Woodside, J.-V.; Young, I. -S.; McKinley, M.-C. Fruits and Vegetables, Measuring Intake and Encouraging Increased Consumption. *Proc. Nutri. Soc.* **2013,** *72* (2), 236–245.
102. Yokozawa, T.; Kim, Y.-A.; Kim, H.-Y.; Lee, Y.-A.; Nonaka, G. Protective Effect of Persimmon Peels Polyphenol Against High Glucose-induced Oxidative Stress in LLCPK1 Cells. *Food Chem. Toxicol.* **2007,** *45*, 1979–1987.
103. Yokozawa, T.; Park, C.-H.; Noh, J.-S.; Roh, S.-S. Role of Oligomeric Proanthocyanidins Derived from an Extract of Persimmon Fruits in the Oxidative Stress-related Aging Process. *Molecules*, **2014,** *19* (5), 6707–6726.
104. Yokozawa, T.; Lee, Y.-A.; Cho, E.-J.; Matsumoto, K.; Park, C.-H.; Shibahara, N. Anti-aging Effects of Oligomeric Proanthocyanidins Isolated from Persimmon Fruits. *Drug Disc. Therapeut.* **2011,** *5* (3), 109–118.
105. Yokozawa, T.; Lee, Y.-A.; Zhao, Q.; Matsumoto, K.; Cho, E.-J. Persimmon Oligomeric Proanthocyanidins Extend Life Span of Senescence-accelerated Mice. *J. Med. Food* **2009,** *12* (6), 1199–1205.

106. Yu, J.-H.; Jin, M.; Choi, Y.-A.; Jeong, N.-H.; Park, J.-S.; Shin, T.-Y.; Kim, S.-H. Suppressive Effect of an Aqueous Extract of *Diospyros kaki* Calyx on Dust Mite Extract/2,4-Dinitrochlorobenzene-induced Atopic Dermatitis-like Skin Lesions. *Int. J. Mol. Med.* **2017,** *40* (2), 505–511.
107. Yuan, B.; Xu, H.-L.; Leng, S.-P. Content and Chemical Composition of Carotenoids in Persimmon Fruit. *Chin. Agric. Sci. Bull.* **2006,** *22*, 277–280.
108. Zhang, K.; Zhang, Y.; Zhang, M.; Gu, L.; Liu, Z.; Jia, J.; Chen, X. Effects of Phospholipid Complexes of Total Flavonoids from Persimmon (*Diospyros kaki* L.) Leaves on Experimental Atherosclerosis Rats. *J. Ethnopharmacol.* **2016,** *191*, 245–253.
109. Zhao, D.; Zhou, C.; Tao, J. Carotenoid Accumulation and Carotenogenic Genes Expression During Two Types of Persimmon Fruit (*Diospyros kaki* L.) Development. *Plant Mol. Biol. Rep.* **2011,** *29* (3), 646–654.
110. Zheng, Q.-L.; Nakatsuka, A.; Itamura, H. Involvement of Negative Feedback Regulation in Wound-induced Ethylene Synthesis in 'Saijo' Persimmon. *J. Agric. Food Chem.* **2006,** *54*, 5875–5879.
111. Zhou, C.; Zhao, D.; Sheng, Y.; Liang, G.; Tao, J. Molecular Cloning and Expression of Squalene Synthase and 2,3-oxidosqualene Cyclase Genes in Persimmon (*Diospyros kaki* L.) Fruits. *Mol. Biol. Rep.* **2012,** *39* (2), 1125–1132.
112. Zou, B.; Ge, Z.; Zhu, W.; Xu, Z.; Li, C. Persimmon Tannin Represses 3T3-L1 Preadipocyte Differentiation via Up-regulating Expression of miR-27 and Down-regulating Expression of Peroxisome Proliferator-activated Receptor-γ in the Early Phase of Adipogenesis. *Eur. J. Nutri.* **2015,** *54* (8), 1333–1343.

INDEX

B

C

M

N

O

P

S